Global Issues in Water Policy

Volume 22

More information about this series at http://www.springer.com/series/8877

François Molle • Carles Sanchis-Ibor
Llorenç Avellà-Reus
Editors

Irrigation in the Mediterranean

Technologies, Institutions and Policies

Springer

Editors
François Molle
UMR-Geau, Institute of Research for Development (IRD)
University of Montpellier
Montpellier, France

Carles Sanchis-Ibor
Valencian Centre for Irrigation Studies (CVER)
Universitat Politècnica de València
València, Spain

Llorenç Avellà-Reus
Valencian Centre for Irrigation Studies (CVER)
Universitat Politècnica de València
València, Spain

ISSN 2211-0631 ISSN 2211-0658 (electronic)
Global Issues in Water Policy
ISBN 978-3-030-03696-6 ISBN 978-3-030-03698-0 (eBook)
https://doi.org/10.1007/978-3-030-03698-0

Library of Congress Control Number: 2018966878

This Springer imprint is published by the registered company Springer Nature Switzerland AG.
The registered company address is: Gewerbestrasse 11, 6330 Cham, Switzerland

Foreword

Irrigation was initiated about 6000 BC in ancient Egypt and Mesopotamia, when farmers started diverting water from the flooding Nile or Tigris/Euphrates rivers, conveying them through networks of small channels to irrigate their fields and secure production in an otherwise arid environment.

This gave rise to great civilizations that flourished thanks to their ability to control water supply. Since then, around the Mediterranean, water became an element of strategic importance, and sophisticated sets of institutions, laws, rules, and regulations have been progressively put in place to ensure efficient allocation, distribution, and use of water for agriculture. Nowhere else can we find water institutions that are so intimately associated with people, cultures, and societies.

Over the last century, in the same region, history has accelerated, and many struggle to understand the role and place of irrigation in today's societies. The heritage from the past, the close links between water, agriculture, and society as a whole, and the social and economic roles of irrigation are still felt strongly, to the point that today's decisions on irrigation are often based on yesterday's situation. But the Mediterranean region is undergoing very rapid changes: how is irrigation adapting to the new situation?

After the Second World War and until the end of the twentieth century, large-scale public irrigation schemes have dominated the scene in the Mediterranean countries, mobilizing all available water, leading to increased competition among users and environmental impacts. Modern institutions and rules have been put in place to manage these schemes, but have often struggled to establish their legitimacy among water users, in contrast with traditional schemes. Frequent changes in approaches have not helped, evolving in a relatively short time from top-down command and control to participatory irrigation management, later followed by irrigation management transfer and public-private partnerships.

Technological changes have also transformed irrigation and the way it impacts water resources. The groundwater revolution that spread rapidly among the countries of the region provided farmers with the security and flexibility in water supply that large-scale surface irrigation schemes were not able to guarantee. As soon as the technology became affordable, nothing could stop farmers from drilling wells and pumping water. Without established rules, this race to the well quickly translated into general overexploitation of groundwater, depletion of aquifers, and, in coastal areas, seawater intrusion and salinization.

At the same time, the external factors that influence agriculture and irrigation have also rapidly evolved. Agriculture has progressively lost its central role in the Mediterranean societies (albeit with big differences between countries), and irrigation is increasingly perceived as a thirsty sector offering low return for an ever-scarcer resource. Demographic pressure has continued to increase in several countries, leading to growing competition for water between agriculture, cities, and industries. Climate change is already hitting the region hard, with changes in patterns of rainfall distribution and alarming forecasts on reduction of an already fully committed water resource base.

Conflicting visions of the role, importance, and impact of irrigation in the economy (national or local) of the countries of the Mediterranean region are feeding endless debates about the policies, strategies, investment, and incentives needed to ensure a sustainable future to irrigation and water. On the one side, irrigation is seen as a purely economic enterprise that needs to respond to market logics, whereby allocation of water resources is determined by productivity and opportunity costs. Under this view, irrigation should compete with other uses, through pricing mechanisms that reflect the true value of water. On the other side, in many places, irrigation is considered as an important driver of agricultural and social development in rural areas which needs to be protected and further developed. The equation is further complicated by the complexity of the water cycle, by the impacts of policies and technologies (e.g., drip irrigation technology) on water availability and the use and the associated political agendas and sectoral interests.

Irrigation needs to adapt to this fast-changing context. On the ground, such adaptation takes place every day, as farmers seek to find solutions to concrete water problems. But whether the policies and institutions that govern irrigation are adapting as rapidly remains an open question. Finding water governance models that are perceived as equitable and acceptable by users is the new challenge.

In this context, the detailed scientific review of the irrigation technologies, institutions, and policies in the countries of the Mediterranean region presented in this book is timely and welcome. Through country case studies, the authors progressively untangle the socioecological complexity of irrigation; explore its relation with water, agriculture, people, politics, and societies; and help us understand the present in order to be better prepared for the future.

FAO, Rome Jean-Marc Faurès

Preface

This book sets out to describe, analyze, and discuss recent changes and future trends in irrigation policies, institutions, and technologies in the Mediterranean region. Our purpose is to scrutinize the current policy responses to the economic, environmental, and social changes taking place in this widely irrigated region, cradle of a millenary water culture, and highly strategic in geopolitical terms.

Mediterranean irrigation is diverse due to, among other factors, the relative importance of water in the economy of each country; varied levels of aridity; heterogeneous economic, social and technological development; and differences in political and social organization. However, most of the Mediterranean countries face similar problems to meet their water demands because of the scarcity and variability of renewable resources, growing water requirements from nonagricultural sectors, increasing environmental concerns related to water quality and environmental degradation, a social demand for larger public participation, and important technological changes. This book, focused on eight selected countries (Tunisia, Morocco, Spain, France, Italy, Turkey, Israel, and Egypt), provides a comparative perspective that both thoroughly explores national particularities and identifies the common challenges faced by the irrigation sector in these countries.

This book has been written at a critical moment, when the continued application of a supply-side water management model is revealing its unsustainable nature in numerous places, when significant technological changes are taking place in the irrigation sector, when new forms of management and governance are widely held as badly needed, and when climate change is compounding many of the difficulties that have characterized irrigation policies and practices in the past decades. Today, in the Mediterranean region, irrigation uncovers various political dilemmas on water management, mobilizes diverse social groups, triggers territorial and land conflicts, and stimulates new technological development. It is time, then, to analyze the particular trajectory of various Mediterranean countries in these uncertain transformations and to identify the best strategies to avert or overcome future risks.

The different chapters of this book present irrigation policies, institutions, and technologies in the form of key syntheses by country, in order to facilitate their use as authoritative references. All the country-level chapters have a similar structure to ensure comparability and provide readers with an in-depth understanding of the irrigation sector in these different national contexts. A final chapter compares findings and outlines a critical transversal view of the most relevant dynamics observed.

The book has three editors, but, as the readers will note, only two of them sign the introduction and conclusion chapters. Llorenç Avellà, who enthusiastically initiated the project, had to prematurely abandon its coordination for health reasons. These words aim at acknowledging his experienced view and his permanent support to the book, even during his long convalescence periods.

The three editors acknowledge the contribution of Mohamed Al-Dbiyat and Awadis Arslan, whose chapter on the Syrian irrigation policies had to be canceled due to the dramatic conflict that has devastated this country. We thoroughly thank all the (anonymous) reviewers of the chapters of this book and appreciate the valuable recommendations of Ariel Dinar and Rathinasamy Maria Saleth, which helped to shape up the final configuration of the book, in terms of structure and contents. Moreover, the confidence of José Albiac (*gracias Pepe!*) was determinant in starting the project for this book and extremely helpful in solving some critical obstacles. We also thank Joseph Daniel and Fritz Schmuhl for their interest in this project and their patience throughout the long production process.

Montpellier, France François Molle
València, Spain Carles Sanchis-Ibor

Contents

Chapter 1
Introduction

Carles Sanchis-Ibor and François Molle

Abstract This chapter presents an introductory description of political, technological and institutional change in the field of irrigated agriculture in the Mediterranean area. After a brief description of the main characteristics of irrigation in this world region, the chapter introduces the eight national chapters: the cases of Tunisia and Morocco in the Maghreb; Spain, France and Italy in the European Union; and Turkey, Israel and Egypt on the oriental shores of the Mediterranean Sea.

Keywords Irrigation institutions · Irrigation technologies · Water policies · Water management

1.1 Mediterranean Irrigation

Irrigation is often seen by academia and decision-makers as pertaining to the realm of engineering. It is seemingly a matter of skill in evenly spreading an adequate quantity of water on a cultivated plot. But if irrigation is conceptualized as *the process by which capital, technology, energy, labor, land, and water (considered as a collective resource) are mobilized to grow crops for home consumption and markets, while generating positive and negative socio-economic and environmental externalities,* then its complexity and multiple linkages are manifest. Irrigation combines the worlds of agriculture and water management, two major activities that connect the farm scale to the global level of world markets, and cut across intricate

C. Sanchis-Ibor (✉)
Valencian Centre for Irrigation Studies (CVER), Universitat Politècnica de València, València, Spain
e-mail: csanchis@hma.upv.es

F. Molle
UMR-Geau, Institute of Research for Development (IRD), University of Montpellier, Montpellier, France
e-mail: francois.molle@ird.fr

F. Molle et al. (eds.), *Irrigation in the Mediterranean*, Global Issues in Water Policy 22, https://doi.org/10.1007/978-3-030-03698-0_1

social, economic, environmental, and political realities. The sophisticated, multifunctional, and multiscalar dimensions of irrigation are inscribed in the mosaic of Mediterranean lands.

Irrigation is a particularly salient feature of Mediterranean countries owing to shared climatic characteristics. Precipitations largely fall outside of the spring and summer periods that are favorable to agricultural production, and water resources are predominantly produced in little cultivated mountain ranges (the Pyrenees and Alps, the Rif and the Atlas, Ethiopian highlands (for the Nile), Mount Lebanon, and western Anatolian mountains (for Turkey). By reducing the uncertainty caused by rainfall scarcity and stochasticity, irrigation has been the central historical response to these adverse natural conditions.

Unsurprisingly, irrigation connects us to ancient civilizations (e.g. Egypt and Mesopotamia), the Greek and Etruscan/Roman eras and, more importantly, to the spread of techniques associated with the Arab conquest (Churchill Semple 1929; Glick 1970; Watson 1974). Beyond the magnificent remains and some historical structures still in use, there is also a valuable institutional legacy: collective management institutions that have enhanced social capital and preserved and improved irrigation systems over several centuries, and whose valuable role has often been overlooked by the modern states.

After World War II, most Mediterranean countries developed large-scale public irrigation to a level that dwarfed the communal/traditional schemes that had been developed during earlier centuries. Engineering took control of the sector, devising expansive plans and guiding the construction of large waterworks, in the name of nation-building or development. In their 1972 review of irrigation in Mediterranean OECD member countries, Béraud and Gabriel (1972) focused on the issues of growth, expansion, cost, and productivity. At a time of unprecedented and unquestioned growth, the main problems these authors identified were those of meeting the needs of supply and demand through investments.

Today, river basins have closed in much of the Mediterranean and most major aquifers are being depleted at alarming rates. Irrigation must vie for resources that are increasingly diverted to urban and industrial, and sometimes environmental, needs. Water quality issues have also become prominent. The irrigation sector, as the recipient of the largest share of total water withdrawals[1] (Fig. 1.1), is asked to adjust, reform, save water, and to increase its productivity.

Consequently, Mediterranean states are currently providing public financial support to a massive transformation of irrigation technology in order to raise productivity, develop private agriculture, and allegedly to reduce 'losses' and improve efficiency. From the pioneer effort of Israel to the recent drip fever in Morocco, pressurized networks have been expanding in all Mediterranean countries, albeit with large differences (Table 1.1). This technological change and the expected water savings are reviving the critique of the old water efficiency paradigm (Seckler 1996; Ward and Pulido 2008; Perry and Steduto 2017), as we will see later.

[1] 70% of withdrawals worldwide, but 92% of the consumption of water resources (Allan et al. 2015).

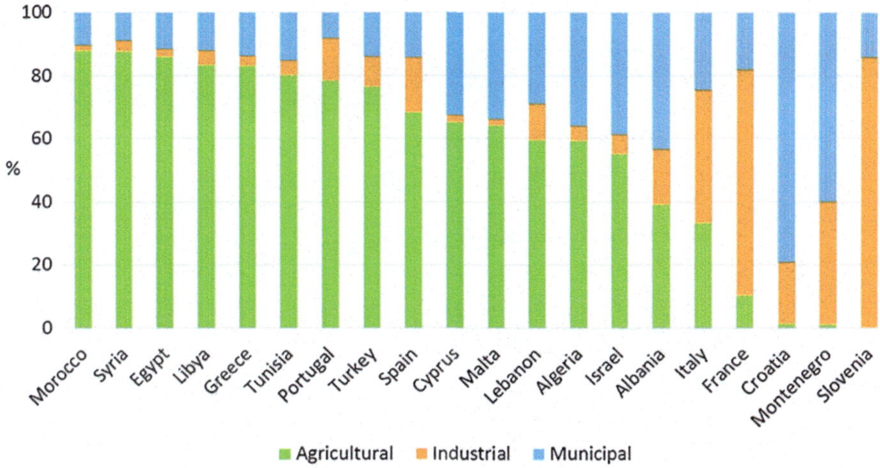

Fig. 1.1 Water use per sector in different Mediterranean countries. (Source: AQUASTAT 2014)

Powerful socio-economic interests and (geo)political strategies still support the mobilization of water resources in the region. In the past 30 years, increasingly cheap pumping devices have paved the way for another wave of expansion driven by individual initiative, mostly through the use of groundwater (Table 1.1). Groundwater is a major source for irrigation in France, Tunisia, Algeria, and Lebanon. As we will see, this development is associated with numerous examples of the Tragedy of the Commons (Hardin 1968) and results in environmental damage to wetlands and in seawater intrusion in coastal areas.

Because of aquifer overexploitation and basin closure processes, non-conventional water resources – wastewater, with various levels of treatment, and desalination – are emerging as new ways of ensuring the expansion or consolidation of irrigation. However, with some exceptions (and very crude statistics available), their use is still relatively limited.

Irrigation creates complex socio-ecological systems that connect, in particular, land, water, energy, food, the environment, and climate. Engineering objectives, economic plans, (geo)political ambitions, social concerns, and environmental principles are frequently drawn together and often clash. In order to untangle this conundrum, this book endeavors to present an in-depth analysis of political, technological, and institutional change in the field of irrigated agriculture in eight Mediterranean countries with a longstanding tradition in the development and practice of irrigation. The chapters adopt a historical perspective – with an emphasis on recent decades – to review the evolution of irrigation technologies and parallel changes in institutions and policies. They investigate in particular the institutional reforms that have taken place and assess their outcomes. The suitability of current technologies, institutions, and policies to achieve efficient and sustainable water management is analyzed for each country. We provide a comparative perspective and a multidisciplinary approach, encompassing technical, agronomic, economic, social, and political

Table 1.1 Main characteristics of irrigation in Mediterranean countries

	Total irrigable area (Mha)	Total irrigated area (Mha)	Surface irrigation area (ha)	Sprinkler irrigation (ha)	Micro-irrigation (ha)	Total pressurized (ha)	% pressurized area	Groundwater irrigation (% tot area)	Volume diverted (Bm^3)	Volume diverted %tot	Source	Year
Turkey		6.09	4,080,000	1,157,000	852,600	1,933,757	33,0	7	30	74	*DSI*	*2017*
Spain	6.751[¥]	3.780	958,000	885,000	1,793,000	3,636,000	73,7	22			*ICID*	*2015*
Egypt		3.610	2,797,200*	397,100	505,400	3,699,700	25,0		61	76	*MWRI*	*2015*
France	2.31[¥]	1.57[¥]	69,195	1,371,310	129,495	1500,805[μ]	95,5		##2.8	9	*Census*	*2010*
Italy	4.004[¥]	2.419	1,036,340	958,535	422,534	2,417,409	57,1	22	16	47		*2010*
Morocco		1.458	1,044,075	125,763	288,162	1,458,000	28,4	30+			*FAOSTAT*	*2011*
Greece	1.521[¥]	1.430	657,800	757,900	14,300	1,430,000	54,0				*EASAC*	*2010*
Algeria		1.176	686,000	270,000	220,000	1176	41,7				*FAOSTAT*	*2012*
Tunisia	0.469	0.425	102,000	131,175	191,250	424,425	76,0	44#	2.14	80	*FAOSTAT*	*2015*
Israel		0.231	1000	60,000	170,000	231,000	99,6				*FAO/ICID*	*2000*
Lebanon		0.130	57,000	28,100	28,000	113,100	49,6				*FAO* 2012	*2010*
Total		22.30	11,488,610	6,141,883	4,614,741	10,756,624	48,4					

*Using the distribution given by 76%, 11%, and 13% given by http://www.fao.org/nr/water/aquastat/countries_regions/EGY
[¥]EUROSTAT; [μ] the distribution by type of installed technology has been applied to the total actual irrigated area
#for total use; ## http://www.eaufrance.fr/IMG/pdf/prelevements2013_201701.pdf (50% of withdrawals for cooling of power plants)

aspects in an attempt to unravel the intertwining of technical, environmental, and institutional transformations.

The first two chapters are geographically focused on the Maghreb and analyze the cases of Tunisia and Morocco. The book then crosses the Strait of Gibraltar to review the irrigation sector in three member countries of the European Union: Spain, France, and Italy. Finally, three chapters examine irrigation on the oriental shores of the Mediterranean Sea, discussing the cases of Turkey, Israel, and Egypt. A concluding chapter written by the editors closes the book and provides a critical and transversal analysis of irrigation policies, institutions, and technologies in the Mediterranean.

1.2 The Maghreb

The southwestern shore of the Mediterranean, the land called Tamazgha in Berber culture, is known as the Maghreb in Arabic, meaning “land of the sunset.” The region encompasses a large portion of the outer margin of the Sahara Desert, traced by the Atlas mountain range. The summits of the Atlas divide this region, leaving a sub-humid or semi-arid corridor beside the sea, and arid inner plains leaning towards the Sahara.

Irrigation has been a determinant strategy of the adaptation of Maghrebi societies to these conditions of aridity. Traditional schemes still include magnificent examples of water mines or *qanat*, called *khettaras* in Morocco or *foggaras* in Algeria and Tunisia, valuable small-scale mountain irrigation systems, old earthen channels (*seguias*), and ancient diversion weirs in permanent and ephemeral rivers. Throughout the twentieth century, colonial regimes and national liberation movements firmly banked on the development of large irrigation systems, based on modern engineering and associated with a significant degree of state control.

The “not one drop lost to the sea” philosophy has imbued irrigation policies and technological development in these countries, and it is still present in national plans and projects, more or less intertwined with efficiency and green rhetoric. This reflects the critical role of irrigation in the battered economy of the Maghreb. Agriculture generates 10% to 20% of GDP and exports, both in Tunisia and Morocco (depending on the annual rainfall), and 19% and 33% of their respective workforce is engaged in this sector. Irrigation is therefore essential to sustain the rural population of Tunisia (3.7 million) and Morocco (14 million), and consequently is a political priority above many other considerations. Under such conditions, the shift to demand-side management is hampered by significant inertia and productivist views.

Chapter 2, by Abddelkader Hamdane, explores irrigation development and management in Tunisia. The author first recalls the historical roots of Tunisian irrigation, influenced by the knowledge and technologies assimilated with the Arab and Andalusi acculturation. The chapter then describes how Tunisia started to develop large-scale irrigation schemes under the French Protectorate, with technologies adapted to each region. Dam building was promoted in the northern regions, where surface water was abundant, whereas groundwater abstraction was developed in the southern desert regions. Food self-sufficiency was, Hamdane remarks, the justification for continuing this development model after independence, basing the management of the irrigation system on firm state control inspired by socialist principles.

The chapter adopts a critical approach to stress the weaknesses of the management of irrigation in Tunisia. Hamdane describes an obsolete institutional framework, still based on the old Water Act of 1975 and excessively dependent on the initiative of the state, which manages the majority of irrigation schemes. Water user associations lack community spirit, have suffered from political meddling, have very limited finances, and are on average performing poorly despite several reforms. The deficit in participatory groundwater governance notably affects aquifer conservation. The deficiencies of this social and institutional framework, and the persistence of a management model focused on the mobilization of water resources, means that the sustainability of irrigation systems is approaching a critical threshold.

Hamdane also points out how the government has been promoting drip irrigation through public funding. This technological change was the main action defined in the National Irrigation Water-saving Program. In contrast, despite early experience with the reuse of treated wastewater, this option now faces significant obstacles.

The chapter then explores the environmental limitations and the technical and institutional challenges of the irrigation sector. The salinity of water for irrigation is one of the major problems faced by farmers, particularly worrisome in the many areas where saline water tables are rising. Soil conservation is also a critical problem in Tunisia, with pollution caused by over-fertigation. The author concludes by demanding institutional reform and a turn towards more effective demand-side management in response to these problems and the threats posed by climate change and groundwater overexploitation.

The case of Morocco, which has similarities with Tunisia, is analyzed in Chap. 3 by François Molle, Oumaima Tanouti, and Nicolas Faysse. These authors give a retrospective view of Moroccan irrigation policies, examining the role of the state in the mobilization of water resources for agriculture over the last century. They show how the technocrats of both the French Protectorate and the early independent state, aware of the country's agricultural potential, viewed California as a model to be emulated. This dream guided a strategic bet on dam building in order to develop new irrigated lands and consolidate existing systems. The creation of various regional administrations, the *Office Régional de Mise en Valeur Agricole*, allowed the state to control the new, large-scale irrigation schemes.

The Water Act of 1995 has been perceived as a determinant milestone of Moroccan water policy. It signaled a shift to demand-side management and ushered in some remarkable institutional reforms. However, the authors explain that this legislative change did not significantly affect the irrigation sector. Indeed, since the turn of the century, the irrigation sector has sustained its expansion thanks to state support. This chapter shows how groundwater exploitation has been subsidized without any limiting legislation, despite provisions in the legal framework. The failed governance that undergirds the groundwater rush reflects broader politics that favors capital-intensive irrigation while being fearful, in a volatile context, of curtailing access to a resource that is, for many smallholders, a livelihood sine qua non.

At the same time, micro-irrigation technology has been widely introduced in the modern and traditional irrigated areas. Fostered by an incontestable discourse of "modernization" and "efficiency," drip irrigation has been intensively promoted, but the authors demonstrate that this technological change, far from achieving the promised water-saving goals, is favoring irrigation expansion and intensification, eventually causing an increase in water *consumption* in most areas.

The chapter emphasizes the discrepancy between institutions, legislation, and discourses, on the one hand, and reality on the ground, on the other. The last section illustrates this by pointing to the environmental limits of a prolonged mobilization of water resources, made visible in numerous pollution, degradation, and sustainability problems often caused by the lack of an integrated vision of hydrological systems. The limited attention paid up to now by Moroccan authorities to the challenges of climate change, despite the fragile position of the national system of water resources, is also underlined.

1.3 Southern European Countries

In 1886, William Hammond Hall published a report on the irrigation practices, customs, and laws of "the three countries of the world from whose experience we may hope to learn something" to improve Californian irrigation (Hall 1886). This pioneer comparative analysis of irrigation institutions was focused on the same three European countries selected for this book: France, Italy, and Spain.

The countries have similar historical, technical, and political contexts, and now share the common European regulations. They also share a particular dichotomy: a marked contrast between their northwestern lands (Atlantic in the cases of Spain and France) and their south/southeastern regions, more purely Mediterranean. This division is rooted in agro-climatic conditions as well as in the institutional and technological nature and performance of irrigation. The largest parts of these regions, the *Mezzogiorno*, the *España Seca,* and the historical *Occitanie* have semiarid conditions that turn them into what several early experts, such as Jaubert de Passà (1823) and Jean Brunhes (1902), called "the classic land of irrigation." But, beyond the aridity indicators, this perception is sustained in France, Italy, and Spain by the strength of collective irrigation institutions that govern water conveyance and distribution, in many cases since the medieval period. The Spanish *Comunidades de Regantes,* the French *Associations Syndicales Autorisées* and the Italian *Consorzi di Bonifica e Irrigazione* are the cornerstones of the irrigation sector of these countries, representing the majority of irrigators and having a significant role in decision-making regarding technology and local water management.

Throughout the twentieth century, public administrations were firmly committed to the mobilization of water resources for irrigation in these countries, although with differing intensities and institutional and economic instruments. Large-scale infrastructure and irrigation facilities were built by government agencies, such as the French *Sociétés d'Aménagement Regionales*, the Italian *Casa per il Mezzogiorno,* and the Spanish *IRYDA* and *Instituto de Colonización*, contributing to the significant expansion of the irrigated areas and (initially) increasing the reliability of the water supply. As in other Mediterranean countries, the prolonged public promotion of irrigation and the weakness of control mechanisms led to groundwater overexploitation, increased scarcity, regional and inter-sectoral conflicts, and economic setbacks related to the process of basin closure.

In the three cases, by the turn of the century, the rapid dissemination of micro-irrigation and sprinkler technologies and the reform of the national legal water frameworks changed the orientation of irrigation policies, slowly moving to demand-side management goals. This was the spirit behind the Spanish 1985 Water Act and its subsequent reforms, the Italian 1994 Galli Act, and the French 1992 Water Act (and its 2006 actualization fully incorporating the WFD). The Water Framework Directive (WFD) of the European Union is a critical milestone in this new direction, moving beyond mere demand-side management and proposing an ambitious conceptual shift in which water is considered as an eco-social asset.

By giving priority to the environmental function of water, the WFD put the European Mediterranean countries in a difficult position by challenging the long-standing preeminence the irrigation sector had enjoyed during the twentieth century. The WFD introduced highly demanding but necessary goals where a tension arises between the agricultural sector, which perceives water as a critical input for production, and those who defend environmental objectives and criticize the impact of intensive irrigated agriculture.

In Chap. 4, García-Mollá et al. analyze the case of Spain, where irrigation not only plays a major role in the agricultural sector but has also achieved a pivotal position in the regional and national political debates of recent decades. These originated in southeastern regions, where the economic weight and potential of the irrigation sector fueled a sustained expansion of irrigated lands, despite a context of scarcity and recurrent droughts, leading to river basin overbuilding and groundwater overexploitation. The two main political parties each proposed a means to provide an additional billion cubic meters of water to these coastal regions: a large transfer from the Ebro River versus the construction of numerous desalination plants. Both supply-side oriented projects failed for different reasons.

The chapter describes how, to overcome this hydro-political crisis, the administration and the farmers' associations found alternative ways to meet the increasing demands of the irrigation sector – in these regions and in others not involved in the Cold War for water. The authors describe three non-conflicting strategies promoted by the public authorities: introducing water markets to introduce flexibility into allocations for irrigation, stimulating treated wastewater reuse, and modernizing irrigation systems, particularly through the massive transformation of gravity schemes into drip irrigation. This water-saving technology has been equally attractive to administrations seeking to reduce water consumption and to farmers' organizations that aspire to consolidate or expand their irrigated land, while allowing growth in productivity. The chapter also analyzes some weaknesses of this irrigation policy, which is currently under review in the country.

The authors express a critical view of groundwater development for irrigation, based on the contrasting situations of the Eastern and Western Mancha, where different management strategies have been applied with distinct results. The pressure on groundwater, and on the whole water resource system that feeds irrigation, is expected to increase as a result of climate change. This raises questions and challenges that are considered in the last section of the chapter.

In Chap. 5, Loubier, Ruf, and Garin outline the recent developments, present situation, and future challenges of irrigation in France. They explore the historical evolution of irrigation to highlight the critical role of the institutions of collective management. Until World War II, these organizations (ASA) were the pillar on which irrigation policy rested, but the Fourth Republic set up public regional companies to develop hydraulic infrastructure and new irrigated areas under public management, consolidating a dual model in Mediterranean areas.

However, in the last 30 years, individual irrigation based on groundwater abstraction has spread markedly. The authors explain how this rapid expansion and the lack of a collective management tradition have led to problematic water management, in that withdrawal authorizations (together with illegal pumping) have frequently exceeded the available resources. The chapter analyzes the institutional architecture of water management and emphasizes how the administration is now promoting the creation of collectives (*Organisme Unique de Gestion Collective, OUGC*) responsible for managing a quota and implementing restrictions in times of water shortages. The OUGCs face significant difficulties in effectively controling individual groundwater pumping – a problem that Loubier and colleagues illustrate with local case studies.

The chapter provides the main indicators of the irrigation sector and recent technological change in French irrigation systems. It explains how, in recent years, sprinkler irrigation associated with groundwater use and grain crops (mainly corn), has been widely introduced in the farming systems of Atlantic France, while drip has displaced gravity irrigation in the Mediterranean area. Traditional surface irrigation has only been preserved in a limited number of historical systems. This technological change, witnessed in many other Mediterranean countries, took place with interesting nuances in France. Here, the water-saving discourse has been somehow absent, and the technological shift occurred because gravity irrigation was perceived as constraining and labor/time-consuming, while pressurized systems also helped to meet the standards demanded by the market. However, Loubier and colleagues also point to the government's recent concern for the protection of traditional gravity systems because of their heritage and environmental value.

The authors underline the impact of the recent environmental constraints in the allocation of water for irrigation, reducing agricultural quotas. These force farmers to invest in storage infrastructure, develop water-saving strategies, or negotiate with other sectors. Loubier et al. conclude that greater institutional and policy coordination is needed, not only among water users and the different levels of the administration and policy-making (including the European level and its Common Agricultural Policy) but also with the energy sector, to secure irrigation water and anticipate climate change impacts.

In Chap. 6, Gabriele Dono, Simone Severini, Davide Dell'Unto, and Raffaele Cortignani describe irrigation in Italy – a nation with differing developments, techniques, and institutional contexts in the northern and southern regions. This dichotomy permeates the whole chapter and is reflected in the cultural and agrarian trajectories of each region. While in the northern region, large irrigation infrastructure and collective institutions were developed since the medieval period, in the Mezzogiorno, the predominance of large landholdings hindered the creation of collective irrigation structures and only a few small-scale collective systems were created, some of them through historical Arab influence, such as the singular *orti sarraceni* of Tricarico in the Basilicata region (Graziadei 2014).

After a brief review of the historical origins of irrigation in Italy, the chapter analyzes the role of the state in the regulation and promotion of irrigation in modern Italy since unification. Particular attention is paid to the reclamation policy and *Bonifica integrale* principle that sought the promotion of drainage and irrigation for new agricultural development up to WWII, with controversial results and effects limited to the northern regions.

The authors describe how, after the war, the state concentrated its efforts on the southern regions, through the construction of an inter-regional system of water transfers meant to correct "hydrological imbalances" and the promotion of irrigation through the creation of collective schemes. The current implications of this irrigation policy are then illustrated with the case of the Capitanata scheme, in the Basilicata region. This irrigation district faces problems and limitations common to many other public irrigation developments, where the total needs of the projected systems exceed actual resource availability, causing management and maintenance problems, generating groundwater overexploitation, and, finally, stimulating a shift to drip irrigation to increase on-farm efficiency.

Dono and colleagues describe the collective irrigation institutions, the *Consorzi di Bonifica e Irrigazione*, most of which (414 out of 489) are located in northern Italy, including their financial and economic functioning. They expand their analysis of the institutional framework for irrigation and the role of the public sector, and express concern at the lack of an authority capable of mediating conflict between the regions involved in water transfers.

The chapter also explores Italy's shift to demand-side management, through an analysis of legislative changes. The authors describe the transitions to sprinkler and drip irrigation in the northern and southern regions, respectively; the limitation imposed by ecological flows; and the turn to the more integrated management of water resources. These changes have taken place in a context of significantly decreased irrigated land, which the authors relate to water shortages and recent changes in the European Union's Common Agricultural Policy.

The last section of the chapter identifies the main current and future challenges. The incapacity of present institutions to control groundwater overexploitation, increasing salinization, and the predicted impact of climatic change constitute a triad of interconnected factors that spell trouble for the sector. Finally, the authors discourage the use of volumetric pricing as a solution, and recommend the promotion of treated wastewater and the development of collective governance to face these problems.

1.4 The Orient

The last group of study cases includes three oriental Mediterranean countries –Turkey, Israel, and Egypt– which have interesting trajectories but few elements in common. Their water resource systems, social contexts, and historical experiences are specific and contrasting, and have produced different irrigation policies and technological developments.

Probably the only – or at least the most significant – element common to these countries is the intense involvement of the state in irrigation development, considered as a strategic sector for the consolidation of their respective national projects. The Republic founded by Atatürk, the Zionist mission, and Nasser's regime understood and utilized the political and territorial value of irrigation, launching and conducting large-scale internal colonization projects.

In the Egyptian and Israeli national projects, food self-sufficiency was an important incentive to expand irrigation to desert areas, but such motivations—what today we would more likely call food sovereignty—have vanished, especially for Israel. Neoliberal policies have reduced national agricultural budgets, opened frontiers, and stimulated capital-intensive and export-oriented agriculture. Traditional cultivation patterns have been altered, giving rise to new crops and irrigation techniques, as well as new land tenure and labor relationships in the case of Egypt.

In Chap. 7, Sevilay Topcu, Aysegul Kibaroglu, and Zeynep Kadirbeyoglu analyze the irrigation policies, institutions, and technologies developed in Turkey during the twentieth century, under the strict control of the state. They describe how the investigation and exploitation of water resources was one of the pillars of the construction of modern Turkey, since the foundation of the Republic. The chapter identifies two milestones of this hydraulic mission. First, the creation of the DSI—modeled after the US Bureau of Reclamation—and the General Directory for Rural Services, both designed to promote and govern the expansion of irrigation at different scales. Second, the development of the GAP project, the flagship of the Turkish hydraulic bureaucracy, which massively mobilized water resources for irrigation and energy production, and served as a model for subsequent regional projects throughout Anatolia.

Topcu and colleagues describe Turkish irrigation systems and the technologies used for irrigation, emphasizing the lasting predominance of surface gravity systems, despite the generous financial stimulus given by the state to promote pressurized irrigation after the most recent droughts. They also highlight the need to renovate most of the country's conveyance systems due to their deterioration and obsolescence.

Particular attention is paid to the irrigation management transfer process developed since 1993, following the Mexican experience. This shift from state control to collective management, despite some occasional failures, has resulted in the more efficient co-management of irrigation and a substantial increase in the fee collection rate.

The authors adopt a critical view of the environmental problems caused by the prolonged application of the Turkish supply-driven state model of development. The impact of agrochemicals used in irrigation, the discharge of wastewater into the drainage system, and excessive irrigation causing waterlogging and soil salinization are some of the bad practices affecting natural and agricultural systems. Topcu and colleagues denounce the persistence of a water policy still oriented towards the construction of new infrastructure to expand irrigation, despite the fact that the overall cropping intensity is as low as 65%. They criticize the lack of a sound, integrated river basin management model, despite evidence of an environmental crisis that is expected to be compounded by climate change.

In Chap. 8, Doron Lavee, Eran Feitelson, and Hadas Joseph-Ezra focus on irrigation policies and practices in Israel. This chapter shows a nation that has succeeded in advancing irrigated agriculture on a wide scale on arid and semiarid lands, with the intensive use of technology and capital, and a firm, state-led irrigation policy.

After briefly describing the agro-climatic conditions and the water resources system, Lavee and colleagues periodize the evolution of the irrigation sector since the British Mandate, after Feitelson (2013). They divide the implementation of irrigation policies into four eras. The first, the *hydraulic imperative* period, focused on the promotion of irrigation and the mobilization of water resources at a regional scale as part of the nation-building and agricultural development efforts; the second, after the enactment of the 1959 Water Law, included a turn from a discourse of plenty to one of scarcity, which materialized in institutional reforms and the construction of a nation-wide water conveyance system. During the third era, irrigation was considered as a buffer sector and water tariffs were used to reduce the agricultural demand for water, stimulating the use of treated wastewater. Finally, in 2005, Israel entered the desalination era, which has reshuffled the distribution of water among the different productive sectors.

The chapter explores the institutional framework of the irrigation sector, from the local community level to the state structure, and its modification in 2005, when the Water Commissioner, attached to the Ministry of Agriculture, was replaced by a Water Authority placed under the Ministry of Water and Energy. This important change is shown to have been connected to the redefinition of inter-sectoral water allocation priorities and of water tariffs and levies. More broadly, it reflected the change from a statist-socialist policy to a neo-liberal, market-oriented ideology, and from an agricultural sector devoted to food self-sufficiency to one oriented towards export.

The authors also underscore the importance of technology in Israeli irrigation. The country has had a pioneering role in the development and dissemination of micro-irrigation systems, and more recently of monitoring systems and technologies adapted to desalinization or wastewater use for irrigation. The massive use of treated wastewater in agriculture is described in detail, analyzing quality problems and the institutional and legal frameworks that regulate the use of this resource in order to limit environmental and health hazards.

Finally, the chapter identifies the major challenges that must be met in the coming years, combining water quantity and water quality problems, and the impact of climate change, possibly compounded by tensions with neighboring countries.

In Chap. 9, François Molle undertakes a critical analysis of Egyptian irrigation. Egypt is, perhaps, the most unique waterscape of the Mediterranean basin, and also the nation where irrigation, agriculture, and water management are most markedly identified. For millennia, the agriculture of this one-river nation strictly followed the rhythm of the recurrent flood season, until the High Aswan Dam (HAD) was built to regulate the river flow and allow the introduction of double-cropping and the expansion of irrigated lands.

This chapter shows the hydrological complexity of the Nile Delta "irrigation machine," and the fragile balance between salinity, surface water, irrigation, and drainage in this area, in a context of the conjunctive use of canals, drains, and groundwater. The functioning of this huge hydraulic network has been complicated by policies to expand irrigation to adjacent desert areas.

Land expansion strategies have been a constant in Egyptian irrigation policies. Land reclamation and colonization plans, justified by a Malthusian logic as well as by political calculus, have significantly enlarged the irrigated area of the country. But, as Molle shows, this hydraulic mission has faced severe technical and social difficulties, and is also rife with stories of corruption and land-grabbing. Despite the mixed record of past projects, notably in the oases of the western desert, and the uncertainty with regard to the available resource and logistic capacity, the chapter shows how the rhetoric of land reclamation is still alive in the political and social imagination, as plans to conquer and "make the desert bloom" are in progress.

Molle then analyzes the ambitious Irrigation Improvement Projects that have guided irrigation modernization in Egypt in recent decades, describing how their implementation has been bedeviled by numerous lingering problems that undermine farmers' confidence and identifying the reasons why the expected water savings have not materialized. The chapter also reveals the complexity and misunderstandings around the water balance of the Delta – an example of the fact that plot-level efficiency gains cannot be extrapolated to the entire hydrologic system (Seckler 1996).

The complexity of the Nile irrigation system is paralleled by similarly intricate institutional architecture. The chapter reviews a number of institutional innovations and reforms that sought to develop participatory management but concludes that coordination between farmers and managers is still weak and that bureaucratic and political support for the reforms is insufficient. Molle describes in particular several attempts to establish different levels of WUAs since 1970 that faced difficulties achieving significant levels of participatory governance and overcoming the top-down, hyper-centralized ministerial structure.

Finally, the chapter questions whether the new National Water Resources Plan 2037 departs from earlier plans, possibly instilling a new sense of limits and realism in the face of the threats posed by growing pressure over water in the Upper Nile (Ethiopia and Sudan).

In the concluding chapter, we tease out the main on-going dynamics, as well as the challenges, faced by irrigation in the Mediterranean. We question the status of irrigation in the overall sectoral allocation of water resources; the growing recognition of the ecological functions of water; the respective merits and potential of supply- and demand-based management; and the achievements and reforms of large-scale public irrigation. We explore whether and how irrigation practices have been affected by technological change, institutional reform, and the implementation of economic tools. We highlight the limited available knowledge on issues related to data collection and analysis, reform assessment, and policy processes. Irrigation trends, such as the development of groundwater use or of a corporate sector, are frequently environmentally unsustainable and socially inequitable, and these vulnerabilities are likely to be heightened, in the future, by climate change and market volatility. Increasingly important and equally fragile, irrigation in the Mediterranean stands at a crossroads.

References

Allan, T., Keulertz, M., & Woertz, E. (2015). The water–food–energy nexus: An introduction to nexus concepts and some conceptual and operational problems. *International Journal of Water Resources Development, 31*(3), 301–311.

AQUASTAT. (2014). http://www.fao.org/nr/water/aquastat/water_use/index.stm, Last Accessed 21 Nov 2018.

Béraud, D., & Gabriel, N. (1972). *Politiques de l'irrigation dans les pays méditerranéens membres de l'OCDE. L'eau* (Options Méditerranéennes 14, pp. 24–29). Paris: CIHEAM. http://om.ciheam.org/om/pdf/r14/CI010476.pdf

Brunhes, J. (1902). *L'irrigation, ses conditions géographiques, ses modes, organisation dans la péninsule ibérique et dans l'Afrique du Nord.* Paris: Naud.

Churchill Semple, E. (1929). Irrigation and reclamation in the ancient Mediterranean region. *Annals of the Association of American Geographers, 19*(3), 111–148.

FAO. (2012). *Liban: Recensement Agricole 2010.*

Feitelson, E. (2013). The four eras of Israeli water policies. In N. Becker (Ed.), *Water policy in Israel: Context, issues and options* (pp. 15–32). Dordrecht: Springer.

Glick, T. F. (1970). *Irrigation and society in medieval Valencia.* Cambridge, MA: Harvard University Press.

Graziadei, A. (2014). *Traditional irrigation in the shaping of cultural landscapes. The case study of Tricarico, southern Italy.*In C. Sanchis-Ibor, G. Palau-Salvador, L.P. Martínez, & I. Mangue (Eds.), Irrigation, society and landscape. Tribute to Tom F. Glick (pp. 869–893). València: Universitat Politècnica de València.

Hall, W. H. (1886). *Irrigation development. History, customs, laws, and administrative systems relating to irrigations, water-courses, and waters in France, Italy, and Spain.* Sacramento: State Office.

Hardin, G. (1968). The tragedy of the commons. *Science, 162*(3859), 1243–1248.

Jaubert de Passà, F. (1823). *Voyage en Espagne, dans les années 1816, 1817, 1818, 1819 ou, recherches sur les arrosages, sur les lois et coutumes qui les régissent, sur les lois domaniales et municipales, considérés comme un puissant moyen de perfectionner l'agriculture française: précédé du rapport fait à la société royale et centrale d'agriculture.* Paris: Madame Huzard.

Perry, C. J., & Steduto (2017). Does Improved Irrigation Technology Save Water? A review of the evidence. *Discusion paper on irrigation and sustainable water resources management in the Near East and North Africa.* Rome, Food and Agriculture Organization. https://doi.org/10.13140/RG.2.2.35540.81280

Seckler, D. (1996). *The new era of water resources management.* Research Report 1. Colombo: International Irrigation Management Institute (IIMI).

Ward, F., & Pulido, M. (2008). Water conservation in irrigation can increase water use. *Proceedings of the National Academy of Sciences, 105*(47), 18215–18220.

Watson, A. M. (1974). The Arab Agricultural Revolution and its diffusion, 700-1100. *The Journal of Economic History, 34*(1), 8–35.

Chapter 2
Tunisia

Abdelkader Hamdane

Abstract Tunisia is structurally affected by water scarcity and irrigation practices are profoundly marked by the culture and history of the Mediterranean and desert areas. Since 1956, the new Tunisian state has mobilized the financial resources necessary for the development of major public water projects. It also encouraged farmers to develop individual irrigation projects. Despite the relatively limited extension of irrigated areas, irrigation is considered as a strategic sector (food security, economic and social role).

In Tunisia, total water withdrawals are close to the potential of the resource. A "weak form of water demand management" prevails, characterized by the implementation of technical, economic, and regulatory instruments primarily aimed at reducing water losses and inadequate water use and encouraging the use of nonconventional water resources. Weaknesses in agricultural development are found in the majority of large-scale irrigated schemes, where agricultural intensification and the productivity of water remain to be enhanced.

Keywords Water demand management · Water saving · Agricultural intensification · Irrigation policy · Tunisia

2.1 A Short History of Hydro-Agricultural Development

2.1.1 History at a Glance

In Tunisia, irrigation water management practices are profoundly marked by the culture and history of the Mediterranean and, in desert areas, those of the Sahara. The extreme characteristics of the Mediterranean climate mean that this management has been a crucial element of civilization and development since ancient times (Fig. 2.1).

A. Hamdane (✉)
University of Carthage, National Agronomic Institute of Tunis (INAT), Tunis, Tunisia
e-mail: abdelkader.hamdane69@uniagro.fr

F. Molle et al. (eds.), *Irrigation in the Mediterranean*, Global Issues in Water Policy 22, https://doi.org/10.1007/978-3-030-03698-0_2

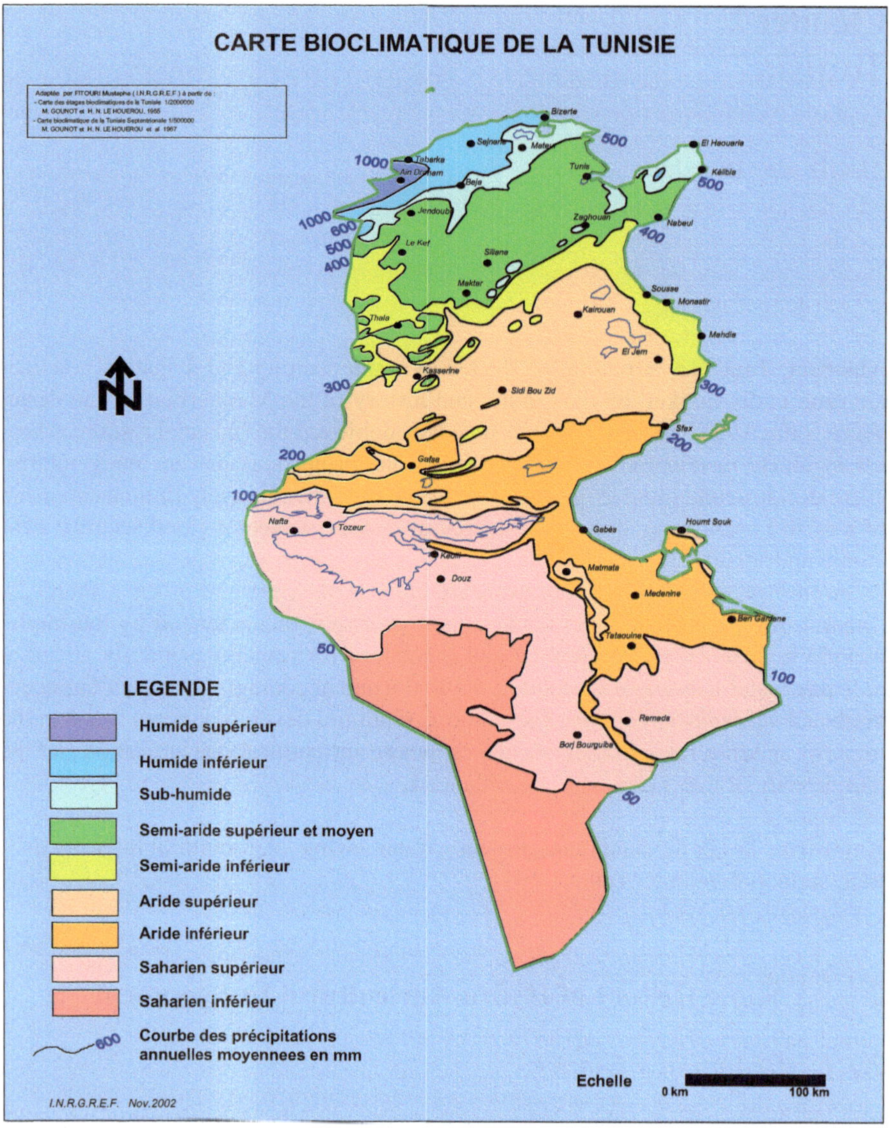

Fig. 2.1 Bioclimatological map of Tunisia (INRGREF 2002)

Hence today we find a range of infrastructures, in almost every region, which serve one or several purposes (e.g., storage, transfer, spate irrigation, distribution to urban areas, and agriculture) and which together form an impressive accumulation of works built over the centuries. Then the hydraulic infrastructure was reworked, developed, and modernized, entailing a series of innovations but also some patent failures, with continual readjustments as requirements evolved. Throughout the ages, irrigation has consequently accounted for a very significant share of the volumes of the water thus mobilized.

Hydraulic development traditions and irrigation practices were first disseminated in Tunisia by the Berber people, well before Phoenician times, but it was during the Punic Era that abstraction and rainwater harvesting techniques were imported from Oriental and Mediterranean civilizations. These techniques were subsequently consolidated and spread to every region in the country from the Roman period onward (see the treaty on agronomy by Mago in the fifth century BC). Although some significant water transfer structures were built to supply towns with water (e.g., the 132 km-long Zaghouan-Carthage aqueduct, built in the middle of the second century AD under Hadrian), Roman hydraulic works in Tunisia mainly consisted of small-scale local developments designed to make use of local resources and the rainwater collected and stored (Gaukler 1897).

From the seventh century onwards, Arab and Islamic influences gave way to a new legal and institutional approach to water management, with the rehabilitation of ancient structures and the extension of existing hydraulic techniques to other parts of the country, as well as new contributions, transfers, and exchanges with other countries in Islamic and Oriental civilizations, bringing Tunisia into contact with world hydraulic engineering heritage.

In the southern deserts, hydraulic societies developed community-based networks in the oases. From the thirteenth century, a strict water distribution system in the Tozeur oasis, promoted by the jurist Ibn Chabbat, provided a model for water supply according to a fixed schedule of rotation between users. Despite the scattered nature of agricultural activities in rural areas and the predominance of pastoralism in the country's hinterland, flood or spate irrigation led to the development of several flourishing hydraulic centers in the high steppe areas and the Kairouan plain in Central Tunisia. In the Sahel (the coastal area), runoff water was collected in reservoirs, supporting extensive olive groves over the centuries (El Amami 1984).

Likewise, innovative technology was introduced in Tunisia via successive waves of Andalusi immigration. The *noria*, an improved version of the Persian wheel, was the first attempt to mechanize water lifting, the basis for all irrigated agriculture in the north of the country and in the coastal zones, while water abstraction via underground galleries (*foggaras*) made it possible to supply some towns with drinking water and to irrigate the oases. In 1622, the Andalusis initiated the construction of a dam with a bridge and locks on the Medjerda, Tunisia's largest river, at El Bathane. This structure supplied two main pipe systems built to irrigate the Tebourba olive grove over a 2500 ha area (Kress 1977). It was in fact the first major hydro-agricultural development in Tunisia's modern history (Besbes et al. 2014).

With the Andalusi influence, Tunisia also felt the consequences of an "agricultural revolution" that swept the entire Muslim world and spurred its demographic expansion. This agricultural revolution led to the extension and improvement of irrigation systems and the introduction of a variety of plants largely requiring irrigation (rice, sorghum, durum wheat, sugarcane, cotton, aubergines, water melons, spinach, artichokes, citrus, bananas, etc.), plus others used as medicine, condiments, and dyes (henna and indigo), most of which were imported from India (Watson 1984).

2.1.2 The Modern Era and Hydro-Agricultural Developments

With the dawn of the French Protectorate in 1881 came a publicly funded policy to drill relatively deep wells to tap aquifers. This was rolled out across the south and center of the country to help farmers expand their olive groves, other tree crops, and fodder crops. At the same time, there was an extension of surface areas irrigated by shallow wells dug at the initiative of farmers to access groundwater, mainly in the regions of Tunis and Cap Bon, specialized in vegetable crops and citrus. In the 1950s, intensive groundwater abstraction led to the overexploitation of aquifers, witnessed for the first time in Tunisia and consequently raising the question of its conservation, especially in the coastal area of the Cap Bon and the Soukra region near Tunis.

The policy of developing publicly funded collective groundwater-based irrigation systems to supply several tens of hectares continued throughout the twentieth century in most of the country. As a result, Tunisia now looks like a succession of separate irrigated islets.

After the Second World War and given the then very harsh economic and social conditions, the policy to develop large storage dams was favorably received in the Medjerda basin, home to the country's largest river and where large swathes of fertile land are found. Here, only irrigation could offset erratic rainfall in the main rain-fed areas. These factors led to the emergence of the first regional hydro-agricultural development plans.

(a) *In the North*: the program to build large dams with a total usable capacity of 400 Mm^3/year and the Lower Medjerda valley irrigation scheme (40,000 ha irrigable from the Neber dam, the Laroussia reservoir and the old Taullerville canal) were the largest hydro-agricultural projects based on collective networks in the period from 1950 to 1980. This project was part of the "Tunisian modernization and equipment plan" (Chevalier 1950) but was not completed by the colonial administration (Fig. 2.2)[1] (Poncet 1956). It was continued by the Tunisian state with the coming of independence and provided a framework for the development of the Medjerda and its tributaries for irrigation purposes (Besbes et al. 2014).

(b) *In the Center*: the oueds in this region are very erratic and do not permit the dam-type solutions adopted in the North. However, groundwater was extracted for use in numerous public irrigation schemes totaling 11,000 ha. The public boreholes enabled the development of small- and medium-sized areas from 50 ha to around 200 ha. Spate irrigation has been used in the area for thousands of years and is still used on the Kairouan plain: modernization work was carried out to better control aggradation and floodwater distribution mechanisms.

[1] The funding of the project's initial phases came from the Marshall Plan to rebuild Europe after the Second World War.

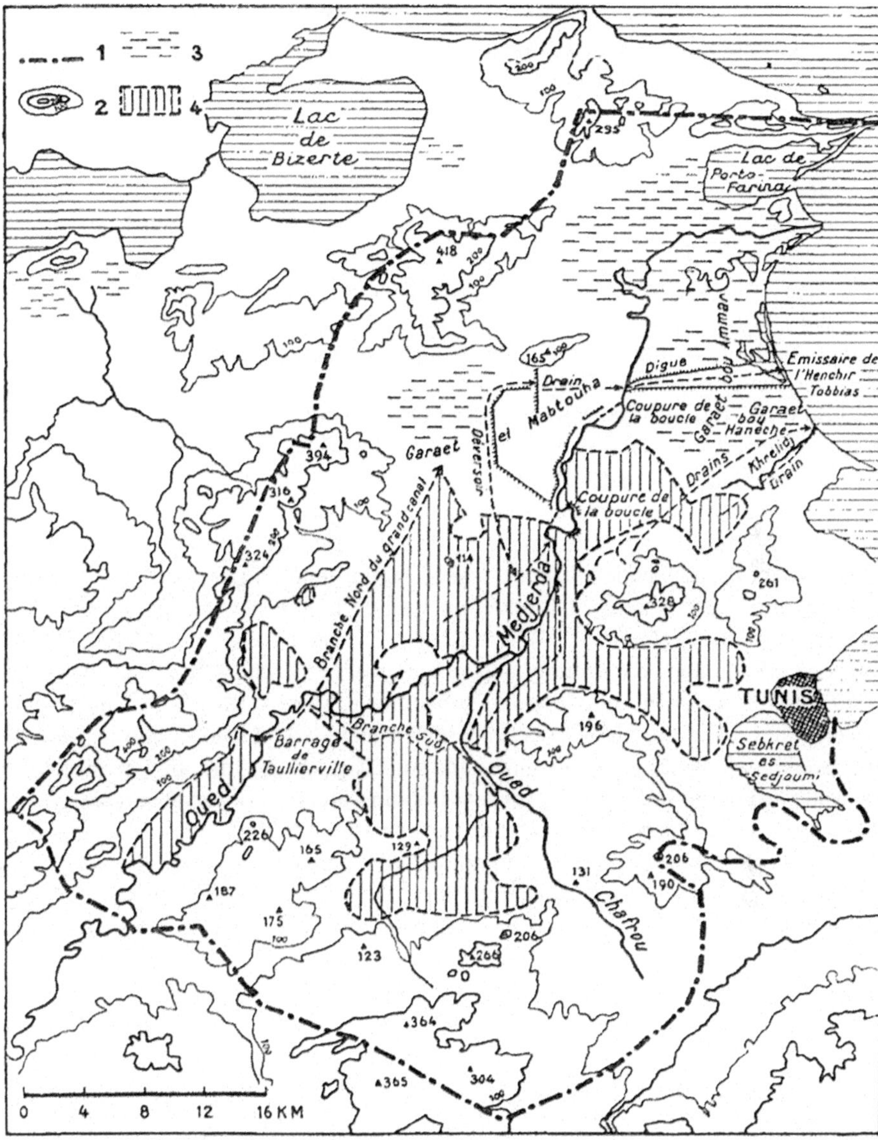

Fig. 2.2 Main works planned and irrigable areas for development of the Lower Medjerda Valley (OMVVM 1958)

(c) *In the South*: the main projects are the works to safeguard the older oases where natural resources have dried up and the creation of new structures in Djerid and Nefzaoua. New, nonrenewable water resources became available with the creation of very deep boreholes, some of which were artesian and provided hot water (70 °C in the Continental Intercalaire Aquifer). In the Jeffara-Gabes basin, there has been work to spread and retain floodwaters from the Dahar wadis.

2.2 The Postindependence Water Policy and the Dominance of Supply-Side Management

2.2.1 Water Policy and Planning

After independence (1956) the new Tunisian state mobilized the necessary national and international resources to complement the major projects initiated during the colonial period. The *Office de Mise en Valeur de la Vallée de la Medjerda* (OMVVM – Medjerda Valley Development Agency) was set up in 1958 to operate hydraulic structures, conduct technical studies for new projects, advise and assist farmers in the agricultural development of the area, and implement a land reform. The importance given to the Medjerda hydraulic development project was such that the OMVVM initially reported directly to the presidential administration. It was then transferred to the Ministry of Agriculture which went on to play a key role in the implementation of regional development policies.

The "Decennial development outlook for 1962–1971" (Secretary of State for Planning and the Treasury 1961) defined the initial precepts for social and economic development in independent Tunisia, based on the founding principles of "decolonization and development of production means, human advancement, structural reform and self-development." When it came to water, these principles contained some specific goals: "the scarcity and irregularity of rainfall forced Tunisia to use every possible means to maximize its capacities to store surface water and exploit groundwater." The measures to be undertaken included (i) studies and research to improve knowledge of water resources and soils; (ii) urban hydraulic infrastructures for water supply to the cities, to prevent flooding and ensure urban sanitation; and (iii) agricultural hydraulic works with the construction wells and boreholes, four large dams (Kasseb, Bou Hertma, Joumine and Nebhana), as well as smaller dams and structures, to irrigate 60,000 ha.

In the early 1970s, the key notion underpinning agricultural policies was planning with a view to ensuring a higher level of food self-sufficiency. In a dry climate, intensified farming through irrigation was intended to boost agricultural output and help develop rural areas across the country. Tunisia introduced large-scale hydro-agricultural planning to fulfill the political goal of food self-sufficiency. In this respect, the hydraulic schemes introduced in the colonial era had paved the way for large-scale development schemes.

While maintaining an extensive publicly managed collective water system, the State also encouraged the private sector (the operators actually exploiting the land developed by the State) through public technical services and financial incentives and support to develop modern, productive irrigated agriculture. Large-scale hydraulic development was identified as the priority for improvement of the northern region, and the preference for major structures impacted the technical choices made in this part of Tunisia. The development of groundwater resources to establish public irrigated areas and privately run schemes was an approach particularly favored in the coastal regions, the Center, and the South.

The originality of the Tunisian approach, dictated by the scarcity of water resources and a deficit in the national food balance, is no doubt reflected in this early awareness of the major challenges raised by water in terms of regional development and spatial distribution of resources and needs: the "hydraulic option" was upheld and bolstered by a policy of large-scale transfers to deficit areas, establishing an interconnection at the national scale designed to overcome inter-regional disparities in access to water (Besbes et al. 2014). Apart from drinking water supply, water transfers were meant to meet the needs of agriculture, a major consumer of water. This policy led to the design and development of vast hydraulic infrastructure program based on three water master plans elaborated in the 1970s, respectively, for the North (with the PDEN) (MA 1970), the Center (PDEC), and the South (PDES).

2.2.2 The Institutional and Legal Framework

2.2.2.1 The Water Code

In Tunisia, the primary legal instrument governing water management is the Water Code enacted by law no. 75-16 of 31 March 1975. Taking inspiration from the principles of Islamic law and common law, as well as the reform of water rights that came with the French colonization, this Code also introduced the following fundamental principles with direct impacts on the water sector as a whole and on irrigation in particular:

The public ownership of water resources: The Code restricts the protection of individual rights and freedoms in favor of the needs and imperatives of the general interest and public utility.

The central role of the public administration is in the planning, mobilization, control, and monitoring of water use from both quantitative and qualitative points of view.

The principle of obtaining maximum efficiency from every cubic meter of water was adopted nationally and for all sectors of use.

The protection of the water environment, with the prohibition of any action likely to cause direct or indirect pollution of water habitats.

2.2.2.2 Administrative, Technical, and Professional Organizations

Since 1975, management of the public hydraulic domain (PHD) has been the remit of the Ministry of Agriculture, which is in charge of water. It is assisted by two commissions: (i) the *Comité national de l'eau* (National Water Board), known as the *Conseil National de l'Eau* from 2001 onward, whose role is to guide water policy involving a broad spectrum of stakeholders and operators in the sector, and (ii) the Public Hydraulic Domain (PHD) Commission.

Although a large portion of the PHD is still supervised by this Ministry, the Ministry of Environment has been involved since the 1990s in environmental protection matters (National Environmental Protection Agency: ANPE) and in urban sanitation and wastewater treatment (National Sanitation Agency: ONAS). The Ministry of Public Works also oversees certain flood control operations in urban areas. This is a fairly long-standing arrangement and reflects the priority given to agriculture, which is subject to very unfavorable climatic variations and which therefore monopolizes a large share of the available water resources and infrastructures. Furthermore, the Ministry of Agriculture is still the only technical administration with local offices covering the entire country and able to ensure control over the resource.

Within the Ministry of Agriculture, there are several technical divisions and associated public establishments involved in water prospection, hydraulic works, rural engineering, management of the public hydraulic domain, etc. By the nature of its core tasks, the Ministry of Agriculture is responsible for the general supervision of the agricultural sector via several specialized bodies working in the irrigated agriculture sector: autonomous public bodies with an important specific role in agricultural planning in the irrigated areas; inter-branch groups tasked with the integration of various stakeholders in the value chains, market regulation, and the promotion of product quality; agronomic research and rural engineering institutes; and the technical centers in charge of experimentation and technological adaptation, information and technical support to producers, training of technicians, and so on.

The *Commissariat Régional au Développement Agricole* (CRDA, or Regional Agricultural Development Commission) is the body representing the Ministry of Agriculture in each region. Each of the 24 governorates has its own CRDA bringing together the main services provided by the Ministry of Agriculture. This is a public administrative establishment (PAE) with its own legal identity and financial autonomy but officially reporting to the Ministry of Agriculture. It also heads the implementation of national agricultural and hydraulic policies within the governorate, including the development of irrigated schemes and agricultural water management.

2.2.3 *Changes in the Institutional Framework and Management Structures*

2.2.3.1 Irrigation Development Boards: State Planning, Development, and Management

Implementation of the hydro-agricultural policy was supported by the institutional setup established over the first few decades after independence. The irrigated area development boards at the regional level (OMV-PPI) played a key role in applying the regional masterplans for agricultural development and irrigation management. Thirteen of these boards were established after the restructuring of the OMVVM (Medjerda valley development board) in the 1970s. These public industrial and commercial organizations spearheaded irrigation policy and were involved in various tasks similar to those of the OMVVM, some of which concerned agricultural development and were subsequently extended to private irrigated areas. The principal missions of the OMV-PPI were the integration of all activities related to irrigation development at the regional level: construction of irrigation schemes, management of collective hydraulic networks (operation and maintenance, fees collection, etc.), participation in the implementation of the agrarian reform, agricultural and irrigation extension, agricultural financing, marketing, etc. The OMV-PPIs were absorbed by the CRDAs in 1989.

The impact of these boards was relatively positive in terms of local development and food security. Despite undeniable technical success, the beneficiaries had little involvement in the collective management of water systems and the maintenance cost, and successive rehabilitation projects were mainly borne by the State, since the fees charged remained very modest. Continued implicit subsidies from the State for public irrigation became increasingly heavy as irrigated areas were extended.

2.2.3.2 Collective Water Management

In Tunisia, irrigation is based on traditional know-how and has long been practiced using common natural water resources in the oases or private shallow groundwater wells.

The Water Code, amended by law no. 87-35 of 6 July 1987, endorsed the long-standing model of Collective Interest Associations (CIA) established in 1936 and in particular defined the roles of these associations in water management, including operation, maintenance, and use of networks affecting the public domain that the CIAs have the right to access, irrigation or land reclamation through drainage, the management of domestic water systems in rural areas, etc.

According to law no. 2004-24 of 15 March 2004, amending and completing the previous law, the different water users' associations are now called *Groupement de Développement dans le Secteur de l'Agriculture et de la Pêche* (GDA or agriculture and fishing sector development group), and their role is extended to protect national resources (water, soils, forest, etc.) and to rationalize their use and conservation. They are tasked with equipping the areas under their remit with basic agricultural and rural infrastructures, participating in the supervision of their members and advising them on the most appropriate agricultural techniques, carrying out income-generating activities to provide services upstream and downstream of agricultural production, etc.

2.2.4 Land Reform in Irrigated Areas

Land reform in irrigated areas began in 1958 covering the large tracts of public land under the responsibility of the OMVVM. It was extended in 1963 to include all public areas in the country. The main goals of the reform were the distribution of plots of irrigable land taken from large state-owned estates and the limitation of irrigated property, the reorganization of private land ownership, and the obligation to exploit developed land. In 1977, a dedicated agency – the *Agence de la Réforme Agraire dans les Périmètres Publics Irrigués* (Agency for Land Reform in public irrigated areas), which became the *Agence Foncière Agricole* (Agricultural Land Agency) in 1999, was set up to implement the land reform policy in irrigated areas where collective hydraulic facilities were provided by the State.

If the land reform did succeed in reorganizing ownership in public irrigated areas, its results with regard to social objectives and the promotion of intensive agriculture were very mixed. The political project of a land reform that went back to the 1950s–1960s seemed to have become less consistent over time and less relevant in terms of its social aims, giving way to a new "liberal" policy designed to lease large irrigated areas owned by the State to private agricultural interests [agribusiness companies (SMVDA), farmers]. The land reform thus became "obsolete" but should now be adapted to the new context and updated to better reflect the social and economic changes occurring in irrigated areas.

2.3 Demand-Side Management and Recent Reforms

2.3.1 The Need for a Shift Toward Demand-Side Management

In Tunisia, total water withdrawals are on average close to the potential of the resource, and the search for a more secure water supply naturally raises the issue of the reallocation of the resource among the various users and sectors. In reality, if the available resource is considered as constant, the reallocation of water to satisfy the priority demand for domestic uses and ensure supply to productive economic sectors (industry and tourism) can only occur to the detriment of the agricultural sector. The prospects of increasingly scarce good quality water for irrigation as well as the competition between sectors already observed in several coastal regions very much underpinned the general framework of the «*Etude du Secteur de l'Eau* » (water sector study), a strategic study conducted by the Ministry of Agriculture in the 1990s.

This diagnosis of the water sector clearly indicated that the water policy implemented over the past several decades, primarily based on supply-side management, needed to be replaced with demand management, an important component of Integrated Water Resources Management (IWRM) which in particular aims to (i) reinforce measures to monitor, control, and protect resources; (ii) promote efficient use by controlling demand and getting maximum value from all usages; (ii) promote decentralization by strengthening the roles of users and of private operators; (iv) give economic, social, and environmental value to water; and (v) control risks. Within the framework of these general principles, this study put forward possible solutions for the medium and long term, all with significant impact on the irrigation sector (DGRE 1999).

Taking the above-mentioned challenges into consideration, Tunisia has been gradually shifting toward a still timid demand management policy (Cf. the law 2001- 116 reforming the Water Law on water conservation, participation of the private sector in nonconventional water resource management, self management of water resources, etc.). At the same time, it continues to develop additional resources, sometimes at a very high cost, especially in areas where there is a water deficit. A "weak form of water demand management" prevails, however, characterized by the implementation of technical, economic, and regulatory instruments primarily aimed at reducing water losses and inadequate water use in irrigation systems, stabilizing consumption in certain areas and encouraging the use of nonconventional water resources (Hamdane 2006). Against the backdrop of these early stages of demand-side management, the main political reforms and instruments applied are described in what follows.

2.3.2 *Decentralization and Associative Management of Irrigation Infrastructures*

2.3.2.1 Widespread Collective Management

As the institutional framework, with the Office de Mise en Valeur/CRDA, began to evolve (see Sects. 2.3.1 and 2.3.2), a vacuum appeared with relation to the management of irrigation between users. Over time, the need emerged for GDAs to serve as intermediaries to rebuild the link between water users and the public authorities. Furthermore, the support to irrigated agriculture had faded away, and new areas, with no tradition or structure for support, were pretty much neglected.

With regard to hydraulic management, the operational difficulties faced by the CRDA were the main reason for the large-scale development of users associations in all public irrigated areas after 1998. In small- and medium-sized systems, the GDAs were responsible for the operation and maintenance of all existing collective infrastructures. In larger schemes, the GDAs were entrusted with secondary and tertiary infrastructures (network subdivisions), while the main infrastructure was kept under the responsibility of the CRDA's departments concerned. The CRDA sold bulk water to the GDA.

All of this means that Tunisia has a relatively advanced legal and institutional framework. However, that framework is marked by (i) legislation, that is, to a certain degree, obsolete with regard to some aspects of water management and despite the various amendments to the Water Code (autonomy of the associations, clarification of the role of the various stakeholders, etc.), and (ii) during the second half of the twentieth century, institutional instability with successive cycles of organization/reorganization of management structures (the various Offices/CRDA, AIC/GDA), a policy that significantly reduced the likelihood of institutional progress in the irrigation sector.

2.3.2.2 The Strategy to Ensure Long-Term Sustainability of the GDAs

The number of GDAs in charge of public irrigation infrastructures stood at 200 in 1995. This number rose to 1253 GDAs by 2014 with these groups responsible to a varying extent for a total surface area of 213,000 ha of public irrigated areas. At present, 88% of these GDAs are considered to be active, while 12% are subject to temporary interruption of their activities for different technical and social reasons.

According to the GDA evaluation system (DGGREE[2] 2012), the technical and financial management of irrigation networks by GDAs is very inconsistent, and the performance of these groups in terms of service quality is still relatively disparate: 24% of the GDAs are seen as performing well, 56% of them show average performance, and 20% of them are poorly performing or have ceased operations for various technical, social, and organizational reasons.

Their main weaknesses lie in a diminished community spirit for some GDAs, in the interventionism and interference of the public authorities and local or regional authorities (especially during the pre-revolutionary period), and in the lack of GDA capacity for financial and administrative management or for the necessary maintenance and servicing of water systems. This latter shortcoming is often linked to the limited financial capacities of the GDAs and their difficulty in paying for outsourced maintenance services that are also often not available in remote rural areas, where the market is still marginal for this kind of service (CIRAD 2011). Several assistance, support, and research projects initiated by international cooperation organizations (WB, AFD, BAD, JIBIC, KfW, etc.) are currently being set up to reinforce the GDAs' capacities and ensure the sustainability and durability of the services they provide.

The strategy to ensure the sustainability of the GDAs adopted by the government has enabled the development of five cross-cutting concepts, namely, (i) the necessary institutional support for GDAs at national and regional level and an update of the GDA regulatory framework to take into account the specific nature of water management, (ii) the improvement of GDA financial control procedures, (iii) the professionalization of GDAs to gradually bring an end to the voluntary basis of current GDA management, (iv) involvement of the local private sector in water system operation and maintenance activities, and (v) responsibility for the protection, conservation, and efficient use of water resources in addition to the operation and maintenance of water systems (DGGREE/KfW 2013).

For the longer term, there are plans to study the conditions required to adapt other public irrigated area management models to the various irrigation contexts found in Tunisia: regional boards, public operating company (the SECADENORD[3] model), and public-private partnership (PPP).

[2] DGGRE: Direction General of Rural Engineering and Water Management – Ministry of Agriculture.

[3] Public company charged of management of the Northern transfer canalisations for drinking water and irrigation.

2.3.2.3 Participatory Management of Groundwater

Participatory management of groundwater (35% of the country's irrigable area) is not a common practice in Tunisia, despite the expansion of individual irrigation based on deeper aquifers in several parts of the country and the overexploitation of groundwater resources in some coastal regions and in the Center. Various strategies are currently being discussed at national level and often consider collective approaches involving the individual groundwater users (see "institutional support to public policies of water resource management for rural and agricultural development," known as PAPS-Eau: National strategy to conserve groundwater 2010). The participatory management experience concerning the Bsissi-Oued El Akarit aquifer in the Governorate of Gabes is interesting in terms of its success in reducing the number of wells operated, thanks to local community ("tribal") spirit, to effective awareness raising among members organized by an association of groundwater users (Laghrissi 2013; Lavenus et al. 2016), and to the administrative and technical support of the CRDA.

2.3.3 *Infrastructure Rehabilitation and Modernization*

Public irrigated areas are equipped with various types of distribution network which were designed according to the technology that prevailed at the time of their construction:

- Networks dating from before the 1970s are usually made up of reinforced concrete or prestressed concrete open channels, with a suitable hydraulic control system and rotation as the preferred distribution method, at least during peak periods (Oasis, Central Tunisia, and Lower Medjerda Valley). Lining of the channels was widely applied.
- The networks built over the last four decades are usually either low-pressure pipe systems (one bar) with rotational distribution, used for gravity-fed irrigation (coastal areas in the east), or medium-pressure systems (three bars) suited to sprinkler irrigation via mobile sprinkler lines (large areas in the North and Kairouan plains).

The fairly rapid shift toward modern "Californian" gravity-fed type of irrigation, manual or automatic sprinkler, and micro-irrigation in individual plots made it necessary to adapt the older systems designed for conventional gravity-fed irrigation. These adaptations involved pressurization equipment and individual on-farm storage and are partly funded by farmers in addition to the irrigation equipment specific to the plot. Farmers appreciate these technological developments for their quality of service in terms of flow rate, pressure, and continual supply but also labor savings.

With these new technologies and because of the deferred maintenance and aging of the collective facilities in public irrigated schemes, substantial investment in rehabilitation, replacement, and modernization is increasingly needed in many areas, if greater efficiency is to be achieved while lowering operating and production costs.

The modernizing of irrigation systems and substantially modifying one or several components in the development began in the 2000s and was witnessed in three main systems that are representative of the regional contexts: (i) the water-saving project in the small- and medium-sized hydraulic schemes in Central Tunisia (modernization of 11,000 ha: conversion to pressurized systems, empowering GDAs, etc.); (ii) the improvement project for irrigated areas in the oases in the South (APIOS-23000 ha: lining earth canals, subsurface drainage, etc.); and (iii) the modernization project of the former large-scale scheme in the Lower Medjerda Valley (first phase of 4000 ha: conversion to pressurized systems, strengthening of GDAs, water pricing reform, etc.). The modernization projects were financed by KfW and JICA, and there is no real contestation or debate regarding the relevance of these programs.

It should be noted that in the future, rehabilitation and modernization projects will increase in number and cover larger surface areas, affecting most regions in the country, since irrigation infrastructure is aging and has been amortized (Hamdane and Bachta 2015).

2.3.4 The National Water-Saving and Efficiency Improvement Program

The national irrigation water-saving program (PNEE) was adopted in 1995 with "the overall goals to rationalize irrigation water use on plots, maximize the economic value of water, and maintain irrigation water demand at a level compatible with available water resources."

The program gathered significant momentum with the modernization of irrigation water systems at plot level, further to the political decision to increase the level of subsidies for irrigation equipment (60%, 50%, and 40%, respectively, for small, medium, and large agricultural holdings). Between 1995 and 2014, total investments made through this program were estimated at $715 million, of which $240 million were financial incentives from the State. The high level of public aid is, in fact, one of the basic critiques against the PNEE expressed by some international financial agencies which are not involved in the program.

However, over a 20-year period and thanks to the PNEE, the conditions for irrigated agriculture have considerably changed from both a technical and economic viewpoint. New stakeholders have emerged on the institutional scene. The private sector, most especially the agri-food business and other downstream sectors, has taken on an unquantified but substantial role in irrigation development, especially for large farms. After evaluation of the program, we can note the following impacts (MARH 2016):

Table 2.1 Changes in irrigation technology (Country as a whole)

Year	Improved surface irrigation	Sprinkler irrigation	Micro-irrigation
1995	45%	47%	8%
2014	24%	31%	45%

Source: DGGREE (2015)

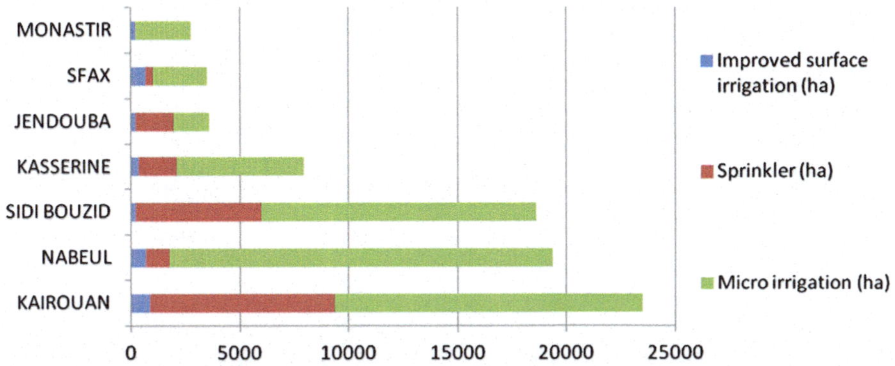

Fig. 2.3 Distribution of areas with water-saving irrigation techniques in some significant areas of the North and Center (Source: DGGREE 2015)

- In 20 years, the total surface area equipped with irrigation water-saving devices has almost tripled, and micro-irrigation has gradually replaced other types of equipment so that it now accounts for the largest share of the irrigated surface area (Table 2.1 and Fig. 2.3).
- On average, half of all irrigated systems have seen their water abstraction stabilize, while a third has seen it fall. There has, however, been an increase in water withdrawals for a small minority. The results are mixed in terms of a reduction in water use but should be compared with the high increase in crop yields in 70% of the farms.
- The value obtained from irrigation water has at least doubled for all crops. The PNEE has led to a considerable increase in water productivity, which has more than doubled in 20 years for fodder crops, vegetables, and tree crops, especially sprouted barley, chili, vines, and apple trees (see Table 2.2).

There is still some way to go when it comes to proficiency in the use of modern equipment. There are several other options that target more efficient use of irrigation water, and they should be explored and developed for the future (management tools, deficit irrigation, choice of crops that optimize water use, etc.), which means significant research and development efforts. Improved water management at farm level is still in its infancy, and the possibilities for reducing consumption and getting maximum value from the scarce resources available are real challenges for the future of the irrigation sector in Tunisia.

Table 2.2 Evolution of the economic performance of crops with and without the PNEE

	Production value per unit of water applied ($ /m^3)	Water cost/total expenses	Yield per unit of water applied (kg/m^3)
Cereal crops			
With PNEE	0.155	19.8%	1.756
Without PNEE	0.89	25%	1.073
Fodder crops			
With PNEE	0.78	13.9%	10.841
Without PNEE	0.48	16.6%	5.155
Vegetable crops			
With PNEE	0.43	14.1%	5.622
Without PNEE	0.29	13.5%	3.439
Tree crops			
With PNEE	0.462	14.7%	1.950
Without PNEE	0.165	25.4%	0.880

Source: DGGREE (2015)

2.3.5 *Pricing Policy and Recovery of the Cost of Water*

2.3.5.1 Water Pricing in Public Irrigated Schemes

Between 1990 and 2000, action was undertaken to strengthen irrigation water pricing in public schemes, including the small areas supplied by deep aquifers, working on three aspects: transparency over cost prices, flexibility (region-based pricing, variation according to the irrigated crops and the cost of supplying water, etc.), and related national objectives (preferential pricing for strategic crops and recycling of treated wastewater). From 1990 to 2000, there was a steady increase in water prices, amounting to a rate of 9% a year in constant value. Alongside these measures, considerable efforts have been made to test tiered pricing systems in large areas in the North (Abbas et al. 2005) and to generalize metering at the farm level.

The overall increase in prices between 1990 and 2002 served to cover a considerable portion of the rising water system operation and maintenance costs. The cost-recovery rate thus went up from 57% to 90% over the same period, and then prices were frozen as of 2002. The policy of a continued price increase was not well received, and there was sometimes strong resistance among irrigation users. Compensation measures, such as the application of preferential rates for cereal and fodder corps with low added value (rebate of 50% on standard rates, constant rate of 0.01 $/m^3 for treated wastewater, etc.), the deregulation of the price of irrigated production, and awareness raising among irrigation users about water savings on their plots, all gradually eased this resistance. However, the political decision to freeze irrigation water rates as of 2002 had a considerable impact on the financial balance of the GDAs, weakening their ability to fund servicing and maintenance work. "Political interventions" regarding pricing, bypassing regulatory bodies, viewing minimizing water charges as a way of improving farmers' revenues, has often had a negative impact on the sector.

Nonetheless, discussions on the development and application of new water pricing methods are underway, giving special consideration to the diversity in irrigation at local and regional level. Work has been carried out in this respect since 2015 (DGGREE/KfW).

The average price for irrigation water is estimated at 0.048 $/m^3$; the actual rate varies between 0.012 $/m^3$ (oases in the South) and 0.075 $/m^3$ (large areas in the North). Overall, the share of water fees in the total cost of growing crops remains significant at an average 14–15%, except for cereals where it reaches 20% despite the preferential rates for these crops.

2.3.5.2 State Fees in Private Areas

Tunisian regulations on the public hydraulic domain stipulate that authorizations and concessions concerning the abstraction of natural water come with a fee, to be paid to the State, based on the volume of water that may be extracted and applied regardless of the actual use of the water (Water Law: articles 63 and 53). This fee is currently set at a symbolic rate of 0.002 $/m^3$ with a minimum of $6.5 for agricultural use and 0.022 $/m^3$ with a minimum of $54 for other uses. This rate has no relation to the opportunity cost of water in regions where it is a rare commodity. In addition, there is a very low collection rate for these fees from users of individual or collective wells, boreholes, and abstraction points. The income generated by the State from these fees is very low, standing at between 5% and 10% of the theoretical amounts for groundwater. In addition, the authorities are not very responsive, which is both a consequence and an indicator of the major difficulties the State is faced with in managing the public hydraulic domain. Although financially marginal, the non-collection of these fees is a powerful symbol and constitutes an implicit subsidy benefiting water users in general and groundwater users in particular.

At present, there are no plans to introduce systematic volumetric metering at private water abstraction points because of the large number of devices required, their high cost (who would pay?), and burdensome maintenance constraints. Indirect measurement (through the area cultivated or the capacity of the well) appears to be the best advisable method in the country's current context.

2.3.6 *Reuse of Treated Wastewater in Agriculture*

Tunisia's experience with the reuse of treated wastewater for agricultural purposes goes back to the early 1960s (citrus fruit protection perimeter, Soukra-Tunis). In 2014, the total volume of treated wastewater produced by the ONAS was 240 Mm^3 and the volume reused 57 Mm^3, i.e., a recycling rate of 24%. The share for irrigation is just 26 Mm^3, a volume used to water around 9000 ha of crops, golf courses, and green areas. Tunisia has a quality standard for treated wastewater for irrigation and a set of specifications defining the role and responsibility of the various stakeholders involved in its reuse. The crops irrigated are defined in an order issued by the Ministry of Agriculture on 21 June 1991 and concern fodder crops and fruit trees (citrus, olives, pomegranates, peaches, etc.) for 55%, with the remainder being cereal and industrial crops. There is virtually no uncontrolled use of wastewater around towns because all urbanized areas have secondary-level water treatment plants.

Several studies and evaluations have been conducted on the reuse of treated wastewater in Tunisia (AHT 2009) and describe the emergence of technical, economic, and institutional constraints after the extension of this form of recycling. There are in fact an increasing number of difficulties with the process: irregular quality of the wastewater treated at secondary level only and supplied to irrigation, restricted crop choices for irrigation users, and so on. Contractualization between different stakeholders in the sector (ONAS, CRDA/GDA, and irrigation users) is considered pivotal in overcoming these difficulties.

2.4 Status of the Irrigated Sector

2.4.1 Mobilization of Water and Development of the Irrigable Potential

2.4.1.1 Context and Water Resources

Total rainwater resources for the country are estimated at 36 Bm^3/year, with 13 Bm^3/year of "green water" used by 5 million hectares of rain-fed agriculture. This potential reaches 19 Bm^3/year if we include evapotranspiration from vegetation on pasture land (5 million hectares) (Chahed et al. 2008).

Potential blue water resources are estimated at 4.86 Bm^3/year (see Table 2.3), divided between surface water (2.7 Bm^3/year or 56% of usable resources) and

Table 2.3 Development of conventional water resources in Mm^3

Nature of water resources	Potential resources (1)	Exploitable resources (2)	Mobilized resources (3)				
			1990	2000	2005	2010	2015
(A) Surface water	2700 (56%)	2500	1180	1876	2200	2400	2500
Major dams		2170	1170	1688	1927	2080	2170
Small dams		195	5	125	160	190	195
Lakes/reservoirs		135	5	63	113	130	135
(B) Groundwater	2155 (44%)	2155	1550	1860	1955	2015	2100
Shallow aquifers		745	700	780	805	810	815
Deep aquifers		1410	850	1080	1150	1205	1285
Total resources (A+B)	4855 (100%)	4655	2730	3736	4155	4415	4600
Mobilization rate (3/2)	–	–	(59%)	(80%)	(89%)	(95%)	(98%)

DGRE/DGBGTH

groundwater (2.17 Bm^3 or 44% of usable resources). In fact, these average values conceal wide regional disparities, and some groundwater resources have already been subject to intensive exploitation or are quasi-fossil water (MEDD/ANPE 2008).

Nonconventional resources are mainly limited to treated wastewater, estimated at around 240 Mm^3 in 2014 (ONAS 2015). Availabilities in 2030 are estimated at 450 Mm^3 and will only account for 9% of all water resources available at that time.

2.4.1.2 Irrigable Land Potential

In Tunisia, water resources – both groundwater and surface water – are limited, so the potential irrigable area with full controlled or intensive irrigation (requiring permanent water supply) stood at 410,000 ha in 2010 (425,000 ha in 2015), i.e., 8% of the actual agricultural area (UAA). Semi-intensive or supplementary irrigation and spate irrigation (with irregular water resources) have an estimated potential of around 150,000 ha. The irrigated areas are scattered across the country with concentrations in the Medjerda valley in the North, the Kairouan plain in the Center, and the oases in the South (see Fig. 2.4).

In terms of dynamics, the area with fully controlled irrigation was approximately 60,000 ha at the end of the 1960s and now reaches 410,000 ha. It is subdivided into (i) "public irrigated areas" (PIA) which are developed by direct investment of the State in collective infrastructures: 216,000 ha, 90% of which is managed by 1253 GDAs and 10% directly by the CRDAs, and (ii) the "private areas"[4] made up of individual systems generally supplied by shallow or deep wells and developed on the initiative of the farmers themselves (194,000 ha), with significant financial incentives from the State covering up to 25% of all private investments (Cf. Fig. 2.5).

2.4.2 Programs and Investment Budgets

In the 1970s, Tunisia opted for a water resource planning approach and developed water master plans for each of the three main natural regions of the country (North, Center, and South), backed by subregional strategies for water mobilization after 1990. Public investment for water has reached an average of 35% of total investment in the agricultural sector over the past few decades (Cf. Fig. 2.6).

From the above, we can conclude that huge efforts have been made to mobilize water: in 2012 hydraulic infrastructure included 33 large- and medium-sized dams (with a total capacity of 2.27 Bm^3), 253 small dams, 837 lakes/reservoirs, 5512

[4] The terms "public" and "private" mainly refer to the abstraction and conveyance infrastructures. Irrigated farms, including internal irrigation installations, are for the majority classified as "private."

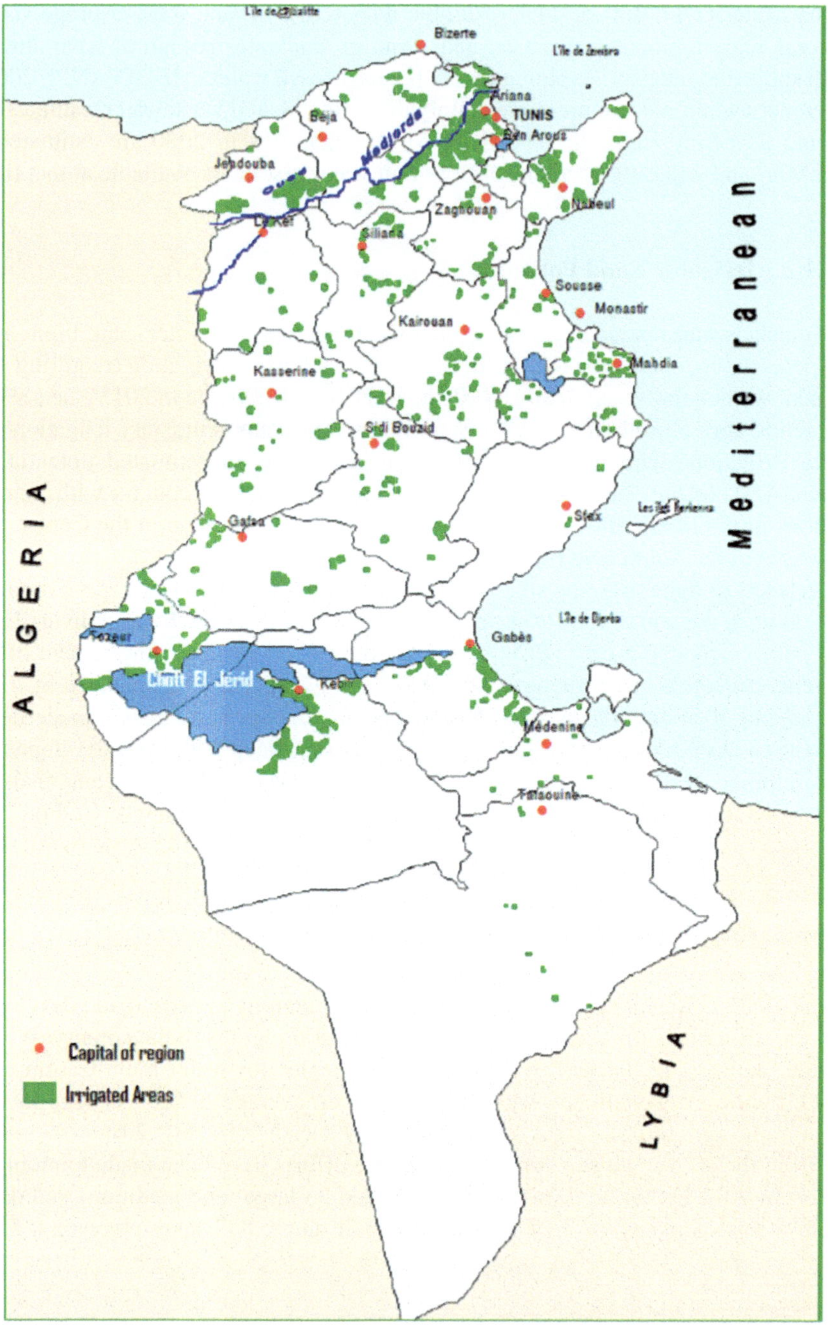

Fig. 2.4 Map of main irrigated areas in Tunisia (Source: DGGREE)

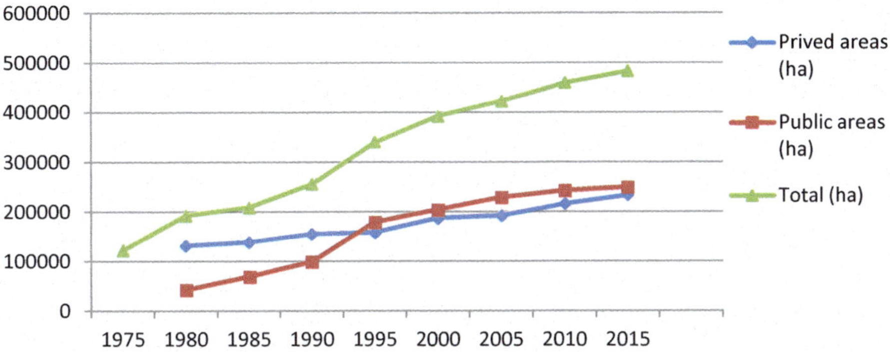

Fig. 2.5 Evolution of the total irrigable surface area (intensive and semi intensive) in (ha). (Source: DGEDA [General Management of Agricultural Studies and Development at the Ministry of Agriculture] 2012)

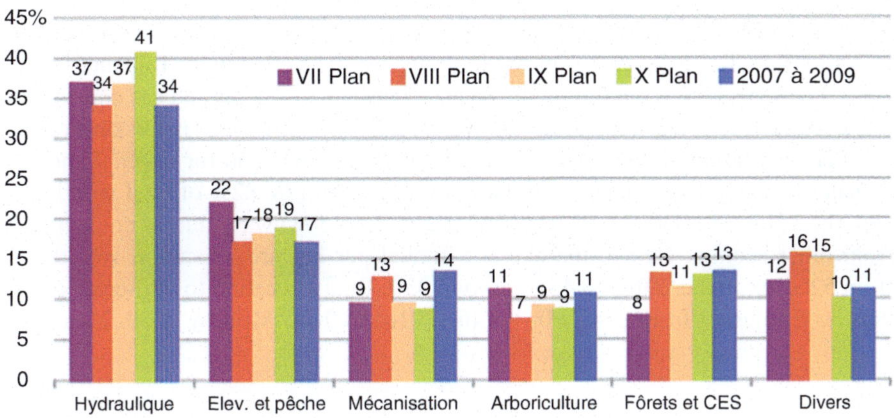

Fig. 2.6 Proportion of hydraulic investment in total investment in the agricultural sector, per Economic and Social Development Plan (Source: DGEDA, FAO-AFD)

boreholes, and 130,000 shallow wells.[5] The overall water resource mobilization rate[6] now exceeds 90%, and the country is thus approaching a relatively critical threshold for the mobilization of its resources. The negative impacts of climate change on resources have not yet been grasped at this level.

[5] Numbers are largely underestimated because of the difficulty to update such information after the 2011 Revolution.

[6] In reality, "mobilization of water resources" concerns resources made available in a sustainable manner. The overexploitation of groundwater resources only increases available resources for a limited time.

2.4.3 Agricultural Water Demand and Abstraction

Despite the lack of comprehensive data, in 2010, the total abstraction of water resources was estimated at 2688 Mm^3, 548 Mm^3 of which was to satisfy the demand for drinking water and 2140 Mm^3 the irrigation demand. The latter was met by groundwater (75%), surface water (23%), and treated wastewater (2%) (see Fig. 2.7).

The average share of irrigation in total withdrawals is only two thirds of what it was planned to be for reasons that are explained in another section. But average numbers conceal the fact that the main problem in Mediterranean environments is the very high yearly fluctuation of rainfall and runoff, which poses a challenge to dam management. The recent dry years 2015–2016 and 2016–2017, for example, illustrate this point.

The largest share of agricultural water demand comes from the coastal areas in the east, where population density is high, and in the oases where water consumption is very high. The coastal areas use more water than available, which means that water has to be brought from other better endowed regions in the north: in fact, the whole country is marked by long water transfers from the west and the north toward the east (164 Mm^3 in 2015 for irrigation and 400 Mm^3 in total on average).

Various studies have analyzed water strategies and prospects in Tunisia (Treyer 2002; ITES 2011). For the future, the balance between resources and use considered in the forward-looking study entitled "Water XXI" (MARH 1995) banks on a rather moderate evolution in domestic, industrial, and tourist demand for water. Although demand from the irrigation sector continues to prevail (78–82% of total withdrawals), the volumes allocated to irrigation are revised downward due to competition from other sectors. Hence, the average annual per hectare allocation would fall from 5200 m^3 in the 2010s to 4350 m^3 by 2030. Total allocation to the sector could thus be readjusted from 2140 Mm^3 in the 2010s to 2035 Mm^3 by 2030. Measures

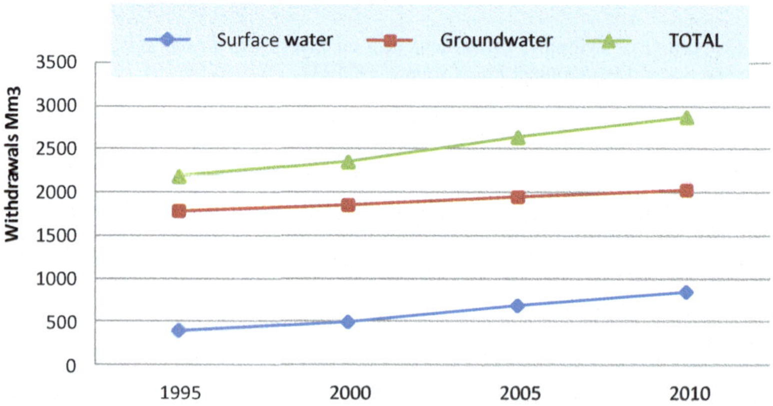

Fig. 2.7 Trend in water abstraction rates per type of water resource (Source: DGBGTH – DGRE)

encouraging farmers to use water-saving techniques and adopt crops that require less water could partially offset these imbalances which, otherwise, will likely be sourced from groundwater. Another alternative for the irrigated sector would be to make more use of treated wastewater: a volume of 220 Mm^3 of adequate quality should be made available for irrigation by 2030.

2.4.4 *Irrigated Agriculture and Intensification of Irrigated Areas*

2.4.4.1 Cultivated Surface Areas

Results from the latest study on irrigated areas, covering the 2011–2012 agricultural campaign, show that the total irrigable surface area subject to intensive and semi-intensive irrigation stands at 469,000 ha, of which 248,000 ha are found in public irrigated areas (i.e., 53% of the total surface area) and 221,000 ha in private areas (47% of the total surface area).

During the same campaign, the area of crops actually irrigated was assessed at 416,000 ha (taking into account "double cropping" practices and not including inter-cropping). Average cropping intensity[7] was 89%. This rate stood at 104% for the 2010–2011 campaign and 96% for 2009–2010. A more detailed analysis of the available data shows that cropping intensity is significantly higher in private irrigated areas, where the rate comes to 98% for 2011–2012, versus 80% in public irrigated areas.

It should be pointed out that the area actually irrigated is likely to increase in the future under the influence of two main factors: (i) one is related to the increase of the irrigable area allowed by the mobilization of still residual and local groundwater or surface water resources (DGGREE 2009), and (ii) on the other hand, the improvement of cropping intensity in existing perimeters, still lower than expectations because of significant constraints concerning the access to land, the financing of irrigated agriculture, the adequate water management of networks in order to secure water supplies, the lack of lead export markets, etc.

2.4.4.2 Irrigated Crop Type

Land use in irrigated areas is dominated by tree crops (olives, citrus, dates, pome fruit, and stone fruit), for which there has been the most significant increases. Over the 2011–2012 campaign, tree crops accounted for almost 40% of cultivated land, followed by vegetables (tomatoes, potatoes, chili, etc.) at 30%. Tree crops dominate because they better adapt to various climate and soil conditions. Changes in land use in the period from 2000 to 2012 are shown in Table 2.4.

[7] Ratio of the total surface area of irrigated crops to the total irrigable surface area

Table 2.4 Changes in land use (ha) in the period from 2000 to 2012

Crop type	2000	2005	2010	2012
Tree crops	155,000 (39%)	166,000	193,000	204,000 (41%)
Vegetables	123,000 (31%)	140,000	147,000	148,000 (29%)
Cereals and fodder	116,000 (29%)	120,000	170,000	148,000 (29%)
Industrial crops	4000 (1%)	4000	6000	4000 (1%)
Total	**398,000**	**430,000**	**516,000**	**504,000**

Source: PI Surveys by DGEDA

Analysis of the available data, most notably the results of studies of irrigated areas, shows that the same crops have been grown for several decades, with no new crop introduced on a large scale.

Greenhouse cultivation developed after 1970 and now accounts for 13,000 ha, with 10,000 ha in the Center, largely concentrated in the governorates of Monastir and Sfax. Crops grown under plastic tunnels cover an estimated 3000 ha, mainly located in the governorate of Sfax. Geothermal water in certain parts of the South (Tozeur, Kebili, and Gabes) has enabled vegetable growing under greenhouses on 150 ha.

The overall estimates available put yields at between 3.6 and 4 t/ha for cereals, 35 t/ha for fodder crops, 40–70 t/ha for field-grown tomatoes, 15–20 t/ha for potatoes, 30 t/ha for citrus, 15–20 t/ha for pome fruits, and 3–7 t/ha for date palms. However, a sustained growth in yields for certain tree crops and vegetables can now be observed, following the widespread introduction of micro-irrigation in the North and Center.

Innovations in cropping practices have not been particularly fast. However, the adoption of micro-irrigation by the majority of tree crop and vegetable growers, within the framework of the PNEE over the last 20 years, appears to have been a positive factor in the intensification of the irrigated production system.

2.4.4.3 The Problem of Low Agricultural Intensity

The general conclusion from the above analysis is that the current intensity and the value gained from irrigated agriculture remains quite low overall, when measured against the efforts made to develop land and water availability in certain regions (BM 2006). We can identify a paradox with newly irrigated farms in semi-arid areas. Generally speaking, irrigation allows for a threefold increase in gross output, but all too often, this only eventuates gradually. Even after several years, we frequently observe the presence of "residual" areas of dry crops in the middle of irrigated areas (Boussard 1972).

Analysis of the areas effectively irrigated shows that average horizontal intensity varies according to the year's rainfall but has remained relatively stable over the past few decades. Average cropping intensity is 90%, but in reality it can vary from 40% to 120%, depending on the situation in the area and the socioeconomic context.

Design studies put this potential rate at 120–130% for the majority of irrigated areas, which would give a potential irrigated area of more than 500,000 ha, compared to 400,000 ha currently subject to intensive irrigation with the same water allocation.

This underutilization of resources, especially in certain large-scale schemes of the North, means that there are substantial volumes of water, especially in reservoirs, which had been allocated to irrigation but which have not been utilized as planned. For example, the two largest irrigated areas in the country, the Lower Medjerda valley (Ariana-Manouba) and the Upper Medjerda valley (Jendouba), only divert on average half of the volumes targeted at the planning stage, despite more than 30 years of "maturity."

As for the yields attained, it is estimated that the levels reached only account for 40–60% of the potential. There is considerable scope for improvement in "vertical intensity," which could be exploited in the future to achieve better productivity from irrigation water and higher incomes for farms, especially smallholders.

All in all, we estimate that the overall cropping intensity in irrigated areas amounts to approximately 60% of the potential. This unsatisfactory achievement is due to several constraints that include water shortages and soil salinisation due to a lack of drainage facilities, but also reluctance by some farmers to engage in capital-intensive agriculture when input and output markets are too uncertain, who prefer to stick to traditional agricultural practices (Hamdane and Bachta 2015).

2.4.5 *Irrigated Farming Systems and Structures*

Public policy on irrigation development and land reform, along with the dynamics of private agriculture, has resulted in a landscape where small- and medium-sized farms dominate (see Table 2.5). Large-scale irrigated agriculture mainly emerged when state-owned estates established with colonial land confiscated in the 1960s were leased out to private companies (SMVDA). Although they are relatively

Table 2.5 Irrigated surface area according to the size of farms

	1994–1995 survey			2004–2005 survey		
Farm size	Area (1000 ha)	%	Irrigated/ cultivated (%)	Area (1000 ha)	%	Irrigated/ cultivated (%)
Less than 5 ha	71.9	24.4	17.1	82.6	25.0	16.0
5–10 ha	52.3	17.8	9.7	65.6	19.8	9.8
10–50 ha	99.7	34.0	5.9	108.8	32.9	6.3
50–100 ha	19.0	6.5	4.2	20.9	6.4	4.5
Over 100 ha	50.9	17.3	5.7	52.6	15.9	6.4
Total	293.8	100	7.5	330.6	100.0	7.8

Survey DGEDA

efficient in technical terms, these business-oriented farms have not had the expected impact on development, and their influence within irrigated areas remains very limited.

On the other hand, the development of irrigation has led to the emergence of quite different landscapes and farming systems from one region to another, through a combination of soil and anthropogenic factors (Elloumi 2016):

- In the Medjerda Valley, multi-cropping systems have developed with cereal and fodder in the upper valley and vegetables and tree crops in the mid and lower valley.
- The north-eastern coastal areas are dominated by an intensive system of citrus cultivation and vegetable production (tomatoes, strawberries, etc.).
- Central Tunisia has vegetables crops (chili, melons, watermelons, etc.) and high-added value tree crops (table grapes, peaches, apples, etc.).
- The Sahel is home to greenhouse crops and early crops.
- The oasis system in the South is typified by date palms, tree crops (e.g., pomegranates), and fodder (alfalfa).

2.4.6 Overall Performance of the Irrigated Sector

Despite the limited size of irrigated areas, which only account for 8% of the country's cultivated area, the irrigated sector is strategic in terms of its impact on food security and its economic and social role:

Economic weight: the irrigated sector has diversified its production and is one of the pillars of the agricultural economy, which remains largely rain-fed. The irrigated sector accounts for 35–40% of all agricultural output, 95% of vegetable crops, 70% of tree crops, and 30% of all dairy output. From a socioeconomic viewpoint, the irrigated sector contributes to 20% of agricultural exports (citrus, dates, early crops, etc.). Apart from its role in the economy, irrigation helps farmers remain on their farms by providing a relatively regular income, which is on average three times higher than in rain-fed agriculture. Intensity could definitely be stepped up in irrigated areas, while low water productivity means that resources are not exploited to their full potential. Nonetheless, the technical advances made in irrigated agriculture over the last 20 years are undeniable. Nearly one in four farmers in Tunisia is now involved in irrigation, and this proportion is even higher for small farms.

Social impact: the irrigated sector accounts for approximately 20% of the labor force, confirming the sector's role in employment. Irrigation has an impact on employment in all upstream and downstream sectors (supplies, agri-food business, services, etc.) with a positive influence on rural development in regions with irrigated agricultural activity.

Political weight: Although the weight of agricultural GDP is generally falling, irrigated agriculture still forms the core of small-scale agriculture, which has maintained its position despite the major technological and economic changes that have transformed the sector over the past few decades. It is unfortunate that the emergence of irrigated agricultural poles, which drain considerable investment to the sector, is still not fully integrated with small-scale farming. Moreover, GDAs are emerging as farmer representation groups and management structures in a sector that generally suffers from a lack of professional representation.

2.5 Sustainability Challenges for Resources and Uses and Institutional Reform

2.5.1 Salinity and Irrigated Land Use

In Tunisia, conventional water resources are naturally characterized by their varying degree of salinity. The water available often exceeds international health or agricultural standards on salinity. This affects crop yields and, in the long term, is likely to damage the structure and fertility of irrigated soils, sometimes irreversibly (CRUESI 1970). Presently, approximately 60,000 ha of irrigated land are sensitive to salinization and 75,000 ha subject to a rise of saline water tables in certain areas of the North and South. It is estimated that around 60% of the soils in public irrigated areas in Tunisia are moderately or highly sensitive to secondary salinization after irrigation; this rate reaches 86% in private areas (DGACTA 2005). Salinization linked to hydromorphy concerns 26% of public irrigated areas to a varying degree. Over the long term, these constraints are likely to cause sustainable damage to the country's irrigation potential, leading to a reduction in the intensity of irrigated cropping or loss of soil fertility in certain areas developed at great expense.

Currently, attempts to use increasingly saline water on newly developed land come with several risks, and management of previously developed land sometimes causes serious issues with regard to sustainability: unsuitability of cropping practices in a context of high salinity, insufficient soil leaching, inadequate maintenance of drainage systems, and so on. These problems are heightened in private irrigated areas on coastal aquifers, in public irrigated schemes in the Medjerda Valley, and in the oases.

2.5.2 Fertility Loss and Pollution of Irrigated Land

A survey carried out in 2007 (DGACTA-Directorate of soils), as part of a study to assess types of soil degradation in irrigated areas, revealed the following: (i) lower fertility in irrigated soils is a dominant type of degradation linked to the modification of certain physical properties in 50% of the public irrigated areas surveyed and 63% of private areas; (ii) chemical pollution is mainly caused by the excessive or inadequate use of mineral fertilizers and crop protection products. In the long term, it is a threat to 28% of public areas where there are very localized, low to medium pollution risks; (iii) biological pollution is apparent where soils are affected by disease (nematodes, fusariosis, etc.). This is primarily observed in vegetable-growing areas subject to intense cropping (Ras Djebel, Chott Mariem, etc.).

2.5.3 Uncontrolled Urban Expansion

Tunisia is experiencing increasingly rapid urbanization with considerable impacts on regional development.

Irrigated areas form islets of economic growth that often require relatively high urban concentration to provide the usual range of social and economic services, along with the specific services required in these areas. However, unplanned urban expansion, combined with the development of industrial zones, is encroaching on agricultural areas and, in some cases, on private irrigable land or even public schemes (e.g., *Grand Tunis*, despite regulations to protect this type of land). Where land developed for irrigation is converted, agricultural production disappears completely from certain urban fringes, and there is disruption of agricultural activities in suburban fringes. These phenomena affect almost 28% of the large public irrigated areas in one way or another.

2.5.4 Intensive Use of Nonrenewable Resources in the South

Southern Tunisia has a significant share of the country's total groundwater resources (38%), especially deep aquifer resources (58%). The volume exploited from stocks with few or no recycling possibilities comes to 800 Mm^3/year, mainly used for agriculture, compared to just 650 Mm^3 in the 1980s.

This situation is mainly found in the oases and irrigated areas in the South (MEDD 2010). Despite planning and rigorous monitoring of resources, signs of overexploitation and loss of artesianism can now be seen with worrying consequences for the future of some areas (especially with the wide, unauthorized extensions of the oases in Kebili). This is likely to seriously affect the quality of the resource and increase the energy burden of its exploitation after a rapid fall in the piezometric level of the water table. These are very serious constraints for the future of the South and a challenge to the sustainability of oasis agriculture, the basis of the region's economy (MEDD/ANPE 2008).

Working with the three countries concerned by groundwater in the Sahara (Tunisia, Algeria, and Libya), the Sahara and Sahel Observatory (OSS) introduced the SASS program in the 1990s (i) to develop awareness within the basin through an update of the status of water resources and the long-term development scenarios and (ii) to prepare the setup of a consultative body involving all three countries to lead discussions and joint initiatives of mutual benefit (Besbes et al. 2003).

2.5.5 The Overexploitation of Groundwater Resources

In Tunisia, the use of groundwater resources is subject to fairly unrestricted access, with relative checks and ex post verifications by the authorities. Groundwater pumping has developed over the past few decades to satisfy the demand for irrigation water, mainly in private areas and especially in the coastal areas and the Center, leading to overexploitation of groundwater in some of these private areas, a drop in piezometric levels, salinization of water resources (seawater intrusion in some coastal areas in the governorates of Cap Bon and Bizerte) and irrigated land, and an increase in pumping costs, borne by farmers. Groundwater resources are currently used in more than 60% of irrigated areas and have a considerable effect on the country's food security. The productivist agricultural policy implemented over the past few decades, with financial incentives from the State, drove the intensive exploitation of readily available groundwater resources (Lavenus et al. 2016).

Several measures have been adopted to stem this trend, some of which are of a legal nature (declaration of perimeters of protection/prohibition), while others concern supply-side management (groundwater recharge or water transfers from the North to ensure conjugated management of groundwater and surface water in certain protection perimeters) and demand-side management (expanding water-saving systems at plot level). However, the results are quite mixed, although there are some positive trends in certain regions: stabilization of abstraction levels (Monastir, Sidi Bouzid) or, indeed, a fall (Nabeul, Kasserine). In other areas (Kairouan), the phenomenon is worsening (MARH 2016) (Fig. 2.8).

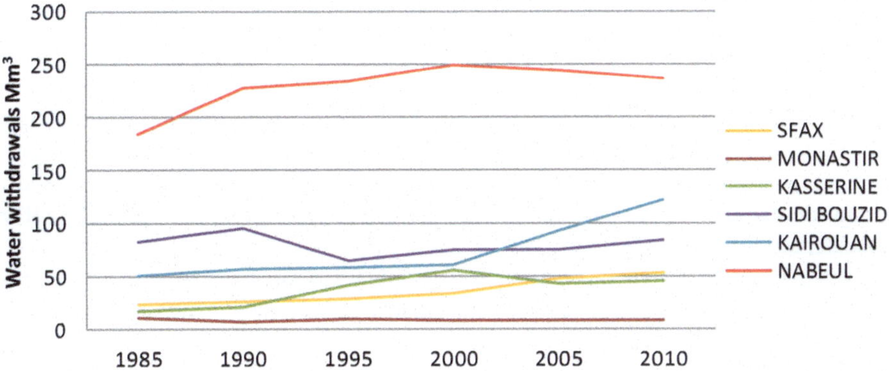

Fig. 2.8 Trends in groundwater abstraction in certain governorates (Source: Groundwater directory DGRE 2012)

From an environmental viewpoint, studies evaluating the cost of water degradation in Tunisia (BM 2007) have estimated the cost of overexploiting water tables. Overall, the total cost of this phenomenon comes to $19 million, or 0.1% of Tunisia's GDP. Other authors previously indicated a higher cost, equal to 1.3% of GDP.

The south-west and the Kebili governorate in particular are most affected, with 43% of the national cost of degradation due to intensive exploitation of quasi-fossil groundwater.

Since the 2011 revolutionary movement, the public authorities are less present and less active in rural areas, and impacts on the public hydraulic domain are becoming more frequent, especially with the creation of deeper and deeper water abstraction points (Elloumi 2016).

It should be noted that overexploitation is likely to spread to all shallow but also deep aquifers in the future if effective regulatory and protective measures are not taken. Large-scale expansion of the phenomenon is likely to have a considerable long-term impact on socioeconomic conditions in the regions affected and on the country's food security.

Several studies and research projects[8] have been conducted or are underway to determine the most suitable approaches and methods to stem overexploitation phenomena in the different socioeconomic contexts encountered in public irrigated areas. However, political commitment remains timid in light of the complex and volatile social situation in the areas concerned (BM 2003).

2.5.6 Irrigation/Energy Balance

From technical and economic angles, energy will be a complex challenge for Tunisia in the future. As is the case everywhere, fossil fuel reserves are running out. State subsidies for energy are already dwindling. Energy will therefore weigh more heavily on the water exploitation balance sheet. Three main factors will directly increase the share of energy consumed for irrigation: (i) the deepening of wells and boreholes, especially in areas where groundwater resources are subject to intensive exploitation, and in the Saharan areas where resources are suffering from a loss of artesianism; (ii) modern irrigation techniques are often energy-intensive, and their widespread adoption will see an overall increase in energy for irrigation; and (iii) an improvement in agricultural intensity will also increase overall demand for irrigation water, leading to a rise in the energy used to pump water to satisfy this additional demand in regions in the North where resources are still available (MIT/ANME 2011).

When it comes to irrigation, water and energy always go hand in hand. It is therefore vital that their management is coordinated to respond to the environmental,

[8] See the research project on groundwater governance in Tunisia (El Haouaria-Cap Bon water table), IWMI-USAID/INRGREF2013–2016. http://gw-mena.iwmi.org

economic, and political challenges raised by the expected growth in demand for water and energy.

2.5.7 The Impact of Climate Change

According to recent studies on climate change-related impact on Tunisian agriculture, the main consequences of climate changes are likely to include an increase in water requirements resulting from an increase in temperature (between +0.7 and 1.5 °C depending on the region) and more unpredictable and declining precipitations (−4 to −12%), resulting in a reduction in available water resources.

In general, the vulnerability of water resources to climate change leads to a worsening of the imbalance by 2030, in particular a 28% decrease in groundwater availability in coastal aquifers, nonrenewable aquifers, and high salinity groundwater, as well as a 5% reduction in surface water. This reduction in quantity is associated with a decline in the quality of water resulting from the intensive exploitation of groundwater, most notably for coastal aquifers subject to a risk of seawater intrusion (MARH/GTZ 2007).

The direct impacts on the sector of irrigation usually occur at different levels. In the first place, agricultural water demand is much more sensitive to climate change than demand in other areas of use. Temperature increases generate a shift and a shortening of the period of crop growth. Increased needs for irrigation water are expected as a result of the increase in evapotranspiration under thermal action and the stimulation of photosynthesis. Overall, with the coupled decrease in available water resources and increase in crop needs, we must expect a decrease in the irrigable area.

To face climate change and its impacts, better management of supply and demand appears as an effective means of adaptation. Actions are already implemented at various levels of decisions and are focused in particular on a demand management policy that aims at reducing demand or at least slowing down its growth.

2.5.8 Ongoing Institutional Reforms

As for long-term measures, they address in particular issues likely to lead to deep reforms of the modes of administration and resource management (institutional, geographical framework of resource management, privatization, etc.). The recent study of the water sector has analyzed all of these questions but does not provide definitive answers (Cf. Sect. 3.1). It offers guidance in the form of recommendations or general management guidelines, and most of the crucial issues with regard to developing a new water policy remain open. New directions are likely to be given by a new study, currently in its start-up phase, ("Water Study 2050"), which is expected to address recent sectoral constraints.

Despite the difficulties experienced by the country after the revolutionary events of 2011, in particular the crisis of state authority and the weakening of its means of control to protect water resources, especially groundwater, against illicit abstraction in several regions, some progress is being made regarding:

- The development of a new Water Code, currently under debate at the national level, which aims in particular at planning for integrated resource management at the river basin level; strengthening the participation of the different stakeholders, empowering the National Commission of water; creating regulatory bodies; strengthening the role of the regions in water management; and providing a more effective regulatory framework for the protection of groundwater resources (DGRE/CNEA 2009).
- A change in the legal framework of the GDAs to give them real autonomy.
- The participation of civil society in water management, etc.

2.6 Conclusion

Considerable efforts have been made in Tunisia to promote irrigated agriculture, and technical progress in the past 30 years is unquestionable. However, results remain below potential and expected performance. Weaknesses in agricultural development (technical advice to irrigators, agricultural credit, value chains, etc.) are found in the majority of large-scale irrigated schemes. The degree of agricultural intensification in these areas and the productivity of water remain to be enhanced. Overall, the situation of agricultural development is not satisfactory for a country that has not yet reached an acceptable level of food security.

While a key sector of Tunisian agriculture and a central means of agricultural intensification, irrigation comes with its specific problems. It is an asset for ensuring a high level of coverage of future food needs but also a concern with regard to sustainable management of water and the durability of infrastructure.

Tunisia will face the need in the near future to rehabilitate and modernize part of its water infrastructure that has become obsolete, adopting a more radical water demand management policy and more efficient economic instruments, taking into account climate change and strengthening local and collective water management. Indeed, the building of new infrastructure becomes increasingly marginal, and the valorization of the huge investments already made a priority.

As a final remark, irrigation is still expanding in Tunisia, and the governmental policy has to support this process by better organizing the sector and making use of its potentialities but without stimulating the mobilization of new water resources. On the other hand, Tunisia is really turning to demand-side management but with varying degrees of success and enthusiasm and despite a powerful inertia of the supply-side model.

It is to be hoped that the sacrifices of the Revolution will provide a better basis for the development of the irrigated sector.

References

Abbas, K., et al. (2005). *Impacts des formules tarifaires binômes sur les systèmes irrigués de la Basse Vallée de la Medjerda*. Montpellier: IAM.

Achour, H. (2012). *Status and new developments on the use of brackish water for agricultural production in the Near East: Tunisia country report*. Cairo: FAO/Regional Office for Near East.

AfDB (BAfD). (2011). *Usages agricoles des eaux souterraines et initiatives de gestion au Maghreb: Défis et opportunités pour un usage durable des aquifères*. Note Technique.

AHT GROUP AG. (2009). *Identification et élimination des goulets d'étranglement pour l'utilisation des eaux usées dans le cadre de l'irrigation ou autres usages*: Rapport national –Tunisie, RG/2008-01/FTF, Pays Meda – Euro Med/BEI.

Bachta, M. S. (2006). *De l'eau aux dattes: aperçu sur la filière-datte tunisienne et perspectives d'interventions*, SIRMA-Deuxième atelier Régional, Marrakech-Maroc, 29–31 Maui 2006.

Bachta, M. S., et al. (2000). *Environnement institutionnel et relations physiques pour une gestion intégrée de l'eau dans le milieu semi aride: le cas tunisien*. In PHI-V/Documents Techniques en Hydrologie N° 51. Paris: UNESCO.

Banque Mondiale (BM) (2003). *Tunisie – Analyse de la performance environnementale*, Report N° TN 25566.

Banque Mondiale (BM) (2006). *Tunisie: examen de la politique agricole*, Rapport N° 35239-TN.

Banque Mondiale (BM) (2007). *Evaluation du coût de la Dégradation de l'eau*, Rapport N° 38856-TN.

Ben Boubaker, A. (2007). *Evaluation participative du PISEAU: Synthèse des résultats*. FAO/BM.

Ben Mustapha Jacox, A., & Jamin, J. Y., 1015) (2016) *Action collective analysée par ses acteurs: une association d'irrigants en Tunisie*. Nature Sciences Sociétés.

Besbes, M., et al. (2003). *Système aquifère du Sahara Septentrional, gestion commune d'un bassin transfrontalier*. La Houille Blanche, N°5.

Besbes, M., Chahed, J., & Hamdane, A., (2014). *La sécurité hydrique de la Tunisie-Gérer l'eau en conditions de pénurie*. Paris: Editions l'Harmattan.

Bessaoud, O. (2005). *Les organisations rurales au Maghreb: un essai d'évaluation de leur rôle dans le développement agricole et rural*. Séminaire SFER, 7-9 Novembre 2005.

Blue Plan (Plan Bleu) – UNEP. (2002). *Analysis of strategies and prospects for water in Tunisia, Sophia Antipolis/Marseille*.

Boussard, J. M. (1972). Economie de l'adoption de l'irrigation, *Bulletin Technique d'Information*-Juin 1972; Ministère de l'Agriculture, France.

Canesse, A. A. (2010). Gestion des ressources naturelles et système institutionnel de gouvernance en Tunisie, Maghreb-Machrek, N°202, Hiver 2009-2010.

Chahed, J., Hamdane, A., Besbes, M. (2008). A comprehensive water balance of Tunisia: Blue water, green water and virtual water. *Water International*, *33*, n°4.

Chemak, F., et al. (2010). *Farming system performance and water use efficiency in the Tunisian semi-arid region: Data development analysis approach*. In International Transactions in Operational Research. May 2010.

Cherif, A., & Kassah, A. (1995). *L'eau et l'agriculture irriguée en Tunisie*. Tunis: Publication de la Faculté des Lettres de la Manouba.

Chevalier, L. (1950). Le plan de modernisation de l'équipement en Afrique du Nord. *Population*, 50 ème année, N°1 (1950).

CIRAD. (2011). *Analyse comparative de l'expérience tunisienne et de plusieurs expériences internationales en matière d'associations d'usagers gestionnaires de périmètres irrigués*. Société du Canal de Provence- France.

CRUESI. (1970). *Recherche et formation en matière d'irrigation avec les eaux salées*. Rapport technique 1962–1969 – UNESCO/UNDP.

DGACTA (MA). (2005). *Examen et évaluation de la situation actuelle de salinisation des sols et préparation d'un plan d'actions et de lutte dans les périmètres irrigués de Tunisie*, Programme PISEAU.

Direction Générale des Etudes et du Développement Agricole (M.A.R.H). (2006). *Contribution à la revue de la politique du secteur agricole* -Partie: Agriculture irriguée- J. Blanken/FAO.
Direction Générale des Etudes et du Développement Agricole (MA). (1990–2012). *Résultats de l'enquête sur les périmètres irrigués-* Rapports annuels.
Direction Générale des Ressources en Eau – CNEA (M.A.R.H). (2009). *Etude de révision et d'amendement du Code des Eaux.*
Direction Générale des Ressources en Eau (M.A.R.H). (1999). *Etude du secteur de l'eau.* Bechtel/ SCET- Tunisie.
Direction Générale du Génie Rural (M.A). (1997). *Etude de la gestion et de la tarification de l'eau d'irrigation au niveau des périmètres irrigués.* Groupement CNEA/BRLi.
Direction Générale du Génie Rural et de l'Exploitation des Eaux (DGGREE). (2009). *Diagnostic de l'évolution des taux d'intensification dans les périmètres irrigués*: Etude des cas des gouvernorats de Kairouan, Sidi Bouzid et Kasserine, Rapports des phases I et II, KfW/Biche.
Direction Générale du Génie Rurale et de l'Exploitation des Eaux (DGGREE). (2012). *Evaluation de la situation actuelle des Groupements Hydrauliques* (in Arabic)
Direction Générale du Génie Rural et de l'Exploitation des Eaux (M.A.R.H). (2013). *Elaboration des approches et concepts pour la pérennisation de la gestion et de l'exploitation des systèmes d'eau* – Mission 1, Quatre rapports (C1, C2, C3, C4), KfW/IGIP.
Direction Générale du Génie Rural et de l'Exploitation des Eaux (DGGREE). (2015). *Rapports et études sur le secteur de l'irrigation.* Tunis.
El Amami, S. (1984). *Aménagements hydrauliques traditionnels en Tunisie.* Tunis: CRGR.
Elloumi, M. (2016). *La gouvernance des eaux souterraines en Tunisie*, IWMI Project Report-Groundwater governance in the Arab World- Report N°6. http://gw-mena.iwmi.org/outputs
FAO. (2007). *L'irrigation en Afrique en chiffres-* Enquête AQUASTAT 2005.
Faysse, N., et al. (2011). Agricultural use of groundwater and management initiatives in the Maghreb: Challenges and opportunities for sustainable aquifer exploitation (A. U. Ordu, J. Kolster, & N. Matondo-Fundani (Eds.), Economic Brief). Tunis: AfDB, 23 p.
Feuillette, S. (2001). *Vers une gestion de la demande sur une nappe en accès libre: exploration des interactions ressources-usages par les systèmes multi agents.* Application à la nappe de Kairouan, Tunisie Centrale, Thèse de Doctorat, Université Montpellier II- France.
Gauckler, P. (1897). *Enquêtes sur les installations hydrauliques romaines en Tunisie, Imp.* Tunis: La Rapide.
Gharbi, N. (2009). *Aménagements hydrauliques et améliorations de la gestion de l'eau dans les oasis*, CIRAD, Actes de l'atelier SIRMA, 25–27 février 2009- Tunisie.
Hamdane, A. (2002). *L'irrigation en Tunisie-* Institut Méditerranéen de l'Eau-Marseille.
Hamdane, A. (2006). *Suivi des progrès dans le domaine de l'eau et promotion de politiques de gestion de la demande*: Rapport national de Tunisie, Plan Bleu /CMDD.
Hamdane, A., & Bachta, M. (2015). *Etude sur l'intensification des périmètres irrigués en Tunisie*, Banque Mondiale/FAO.
Hassaïnia, J. (1991). *Irrigation et développement agricole: L'expérience tunisienne*, CIHEAM-Institut de Montpellier.
INECO. (2009). *Institutional framework and decision-making practices for water management in Tunisia: Towards the development of a strategy for improved groundwater management.* Tunis: International Center for Environmental Technologies.
Institut National de Recherches du Génie Rural, des Eaux et Forêts (INRGREF). (2002). *Cartographie de la bioclimatologie de la Tunisie.* Tunis: INRGREF.
ITALCONSULT. Ministère de l'Agriculture. (1993). Economie d'eau 2000- GMBH/GKW/Coyne &Belier/ CNEA.
ITES. (2011). *Etude stratégique, Eau 2050 en Tunisie*, Rapport de synthèse Juin 2011.
Kress, H. J. (1977). Eléments structuraux andalous dans la genèse de la géographie culturelle de la Tunisie. *Revue IBLA*, 1980, N°145 (Traduction).
Laghrissi, H. (2013). *Impact de la dynamique institutionnelle sur la gestion d'une nappe côtière souterraine*: Bsissi-Oued El Akarit Projet de fin d'études, Ecole Supérieure d'Agriculture Mograne.

Latiri, R. (2005). *Analyses des politiques hydrauliques – Cas de la Tunisie*. Actes du séminaire Euro Méditerranéen sur les instruments économiques et modernisation des périmètres irrigués, 21–22 novembre 2005, Sousse, Tunisie.

Lavenus, S., et al. (2016). *Gestion des ressources en eau souterraines comme bien communs*. AFD, note technique N°18, Juillet 2016.

Le Gal, P., et al. (2008). *Impacts de la chaîne d'approvisionnement export sur la valorisation de l'eau pour les dattes dans les oasis du Sud de la Tunisie, CIRAD, SIRMA*. Actes du 3ème atelier régional, Nabeul 4–7 juin 2007.

Lebdi, F. (2009). *Contraintes de l'agriculture irriguée aux opportunités du marché: cas de la Tunisie*. Notes d'analyse du CIHEAM N°51, Sept. 2009.

MARH-GTZ. (2007). *Stratégie nationale d'adaptation de l'agriculture tunisienne et des écosystèmes aux changements climatiques*. GOPA-Exaconsult.

Ministère de l'Agriculture. (1970). *Etude du plan directeur de l'utilisation des eaux du Nord.*

Ministère de l'Agriculture. (1998). *EAU XXI: Stratégie du secteur de l'eau en Tunisie à long terme 2030.*

Ministère de l'Agriculture, des Ressources Hydrauliques et de la Pêche (BPEH/DGGREE). (2016). *Etude d'impact (évaluation) du programme national d'économie d'eau en irrigation*, Phase I et Phase II- LB/SCET-Tunisie/OIE.

Ministère de l'Agriculture et des Ressources Hydrauliques (AFA) (2006) *La reforme agraire en Tunisie*, Conférence Internationale sur la reforme agraire et le Développement Rural (ICARRD)- Tunis, 2006.

Ministère de l'Environnement et du Développement Durable (MEDD). (2010). *Etude sur la gestion durable des systèmes oasiens*, GTZ/OTEDD.

Ministère de l'Environnement et du Développement Durable (MEDD)- ANPE. (2008). *Gestion durable des ressources en eau*, GTZ/OTED.

Ministère de l'Industrie et de la Technologie –ANME. (2011). *Etude du Développement et de la maîtrise de l'énergie dans le secteur de l'agriculture et de la pêche- Alcor 2010–2011.*

Office National de l'Assainissement (ONAS). (2015). *Rapport annuel d'activités de l'ONAS de 2014*. Tunis: ONAS.

Poncet, J. (1956). La mise en valeur de la basse vallée de la Medjerda et ses perspectives humaines. *Annales de Géographie*, t65, n°349, 199–222.

Sander, J., et al. (2008). *Water balance and evaluation of water saving investments in Tunisia.* BIRD/Waterwatch.

Schiffer, M. (1997). *Agriculture irriguée au Maghreb: limites et perspectives.* Berlin: Institut Allemand de Développement (IAD).

Treyer, S. (2002). *Analyse des stratégies et prospectives de l'eau en Tunisie*, 2 rapports, Plan Bleu- PNUE.

Watson, A. M. (1984). Agricultural innovation in the early Islamic world. *Journal d'agriculture traditionnelle et de botanique appliquée*, 31 ème année, bulletin n°3–4.

Xanthoulis, D. (2013). *Etude stratégique des formes de réutilisation des eaux usées en Tunisie*- Programme de Coopération FAO-Banque Mondiale (PISEAU II).

Websites

AQUASTAT – FAO water and agriculture information system: http://www.fao.org/nr/water/aquastat/main/index.stm

Institutional aspects of water management in Tunisia: http://www.semide.tn/bodyinstitution.htm

National Agricultural Observatory (ONAGRI): http://www.onagri.nat.tn/

Research on Journal Officiel de la Republique Tunisienne: http://www.cnudst.rnrt.tn/wwwisis/jort.06/form.htm

Statistical record of the Ministry of Agriculture (CNS): http://www.cnudst.rnrt.tn/wwwisis/jort.06/form.htm

Chapter 3
Morocco

François Molle, Oumaima Tanouti, and Nicolas Faysse

Abstract Morocco has well developed irrigation facilities that range from small scale communal systems based on springs, qanats or river diversions to groundwater-based individual initiatives and large scale public schemes. Water demand policies – e.g. water pricing, shift to drip irrigation, 'aquifer contracts' and other forms of participatory management – have shown little potential in curbing overexploitation of resources in many basins. Expansion of irrigated areas and the priority given to productivity have taken their toll on the environment, favored commercial agriculture, and contributed to a net depletion of groundwater estimated at 1 billion m^3/year. There is a need to better align agricultural development, water conservation, and environmental objectives.

After a short historical perspective, this chapter first reviews a number of trends in the irrigation sector (modernization, development of groundwater resources, wastewater reuse and desalination), before turning to regulatory and institutional issues, including participatory management, economic tools, privatization and an examination of the Plan Maroc Vert. The threats posed by climate change, water scarcity, and environmental degradation are then discussed.

Keywords Irrigation · Policy · Drip irrigation · Climate change · Water management · Morocco

---

F. Molle (✉)
UMR-Geau, Institute of Research for Development (IRD), University of Montpellier, Montpellier, France
e-mail: francois.molle@ird.fr

O. Tanouti
UMR-Geau, Institut de Recherche pour le Développement, University of Montpellier, Montpellier, France

Université de Paris Nanterre, Paris, France

N. Faysse
UMR-Geau, Cirad, University of Montpellier, Montpellier, France

Asian Institute of Technology, Bangkok, Thailand

F. Molle et al. (eds.), *Irrigation in the Mediterranean*, Global Issues in Water Policy 22, https://doi.org/10.1007/978-3-030-03698-0_3

3.1 Introduction

Because of the crucial importance of water resource in the development of the agricultural sector in Morocco, the history of irrigation policies during the twentieth and twenty-first centuries largely overlaps with that of water policy in general. Morocco is one of the countries in the world which offers the largest diversity of irrigation schemes, in terms of types of water source, technology, scale of development, and actors involved in their management. For at least one century it has presented two main facets, often described as the 'duality' of agriculture, with on the one hand plots belonging to the rulers and their entourage, colons, elites, national or international investors, and on the other a peasant agriculture. While the former is generally developed on large farms by absentee owners, capital- and technology-intensive, fully irrigated, and identified as the 'modern' side of Moroccan agriculture, the latter is made of small holdings generally using less capital-intensive irrigation and agricultural production techniques (not to mention its much larger rain fed component). This chapter first briefly describes the historical steps of irrigation development in Morocco and its present status, before zooming in on recent technical and institutional changes instigated by state policies, but also the boom in groundwater abstraction. It underlines how the status of irrigation is eventually tightly linked to the overall (over)development of water resources in the country, and possibly threatened by agricultural policies that do not sufficiently consider environmental and hydrological constraints.

3.2 Historical Perspective on Irrigation Development

3.2.1 Water and Land Use in Ancient Times

In Morocco water use ingenuity has been driven by the relative aridity of the land and the know-how brought or developed by its successive civilizations (Benzekri 2006). Romans have developed urban water systems and their attendant transfer canals, while in the ninth century the Idriss dynasty further developed water supply in sanitation in the city of Fez. Technical innovations in the domain of agriculture are often associated with the qanat systems-locally called *khettaras*- that were introduced in the twelfth century by the Almoravides, notably around Marrakech and in the Tafilalet. This technique was further developed by the Alaouites in the North, but also in oases (Benzekri 2006; El Faiz 2002). Believed to have originated in Persia and disseminated alongside Islamic expansion, khettaras tap superficial aquifers and act as 'artificial springs' serving communities, gardens and crops. They are notable both by the heavy investment in labour needed to excavate them (which often meant that either the state or a tribal collective would be undertaking their excavation) (Finet 2002; Pascon 1977), and by the sophisticated social rules that were developed to maintain them and manage the resource (El Faiz 2001).

The Almohade dynasty, while developing khettaras, turned to capturing superficial waters in *wadis* (generally intermittent stream) by means of *ougoug* (river weirs often made of earth, stones and branches) and earth diversion canals called *seguias*, that conveyed flood waters as well as low flows to irrigated lands. In the Marrakech plain (or *Haouz*), for example, farmers developed seguias between the ninth and the twelfth century, diverting the rivers coming from the Atlas mountain valleys to their alluvial fans in the main valley (Cressier 2006). In the mid twentieth century, they were irrigating 30,000 ha in the Haouz as well as serving Marrakech, diverting on average as much as 72% of the flow of the four main wadis feeding the Haouz (Pascon 1977). The 90 km long *Ya'qûbiyya* seguia linking the wadi Lakhdar basin to the Haouz plain is testimony of larger inter-basin transfers (El Faiz 2002). Accessing water was both a necessity and a manifestation of power, and the Soultaniya seguia illustrates how a 30 km long seguia could be placed upstream of all others and yet convey water to Marrakech across the land without being tampered with (Ennaji and Herzenni 1987; Cressier 2006). Other khettaras and seguias, and corresponding water rights, were controlled by zaouias (religious foundations) (Jolly 2002).

In the Middle and High Atlas, communities developed irrigation in the valleys by diverting local rivers. Water management was done collectively through the traditional organization of the *jmaa*. In general water rights are detached from land and can be lent or traded.

3.2.2 The French Protectorate (1912–1956)

Official colonization started in the Gharb plain in 1918, where large tracts of land were sold by the colonial administration to French entrepreneurs at "privileged conditions" (Le Coz 1968). This rich and fertile coastal plain characterized by semi-aquatic *merjas* (swampy land that needed to be reclaimed) "looked easy to settle due to the absence of permanent dwellers and was offering auspicious prospects for modern agriculture" (Le Coz 1968). Private colonization, whereby European settlers were buying land from Moroccan owners, started in the 1930s.

Colonial agriculture initially revolved around the cultivation of wheat due to the high prices on the French market reserved to this production as part of an indirect subsidy scheme (Amphoux 1933). But such a situation was precarious and many colons ended up indebted in the 1930s, prompting a search for diversification and cash crops. Between 1929 and 1933, several missions were sent to California to study agricultural expansion on the US Western coast. Because of their alleged climatic and geographic similarities, "Morocco was seen as resembling California at a time when the American West was a universal symbol of the modern world" (Bouderbala 1997). A new policy based on the development of irrigated agriculture and agricultural diversification came to be presented as the solution to the failure of wheat production under rain-fed conditions, and as the lever that would allow the expansion and the success of colonization across Morocco.

Irrigation was brought in colonial development schemes under several forms. State policies first considered establishing private investors in large-scale public schemes, such as in the Gharb. In 1933, the construction of El Kansera dam was completed, together with the Beht irrigation scheme devoted to citrus cultivation, the export of which was strongly encouraged by the state. It also considered using traditional irrigation schemes like in the Haouz, to govern through local tribal leaders or notables (*caids*) and landowners who, together with the colons, occupied the largest and most fertile properties, monopolizing input, capital, and modern technologies of production (Pascon 1977). Last, in the Tadla scheme, the administration attempted to modernize the traditional agro-pastoral space, with the introduction of irrigation, seen by Marthelot (1961) as "the only undertaking which, before the end of the World War II, had aimed without restriction at modernizing a purely Moroccan rural population". Préfol (1986) emphasized that the aim of the administration in the Tadla "was the development of the land by its population and to its benefit, the objective being the modernization of life and mentalities" and marvelled at the "wonders of irrigation in Morocco". Agricultural activities, however, were defined and directed by the administration (the Office), which distributed tools, fertilizers and seeds, and ensured ploughing by tractors and the distribution of irrigation water (Préfol 1986). Pascon (1977) emphasized the loss of autonomy and social cohesion associated with the shift from community-managed irrigation to state-run networks and technology.

3.2.3 The Post-Independence Period Until 2012

Unlike other newly independent countries which sought their prosperity in industrialization, Morocco placed a lot of hope in agriculture and irrigation, partly because of the key influence in the *Istiqlal*, the then ruling party, of agronomists and engineers trained in the French *Grandes Ecoles* (Ihazrir 2009; Bourdillon and Faris 2004), and also because many of Istiqlal members were large landlords (Mouhsine and Lakmahri 2014).

3.2.3.1 The Fate of Colonial Farms and Agrarian Reform

In the early 1960s, the government of Ibrahim Bouabid realized the failure of some early agricultural policies and programs such as the '*Ploughing Operation*', intensification programs based on improved seeds and fertilizers, etc. and decided to insert irrigation development in the new 5-year plan of 1960–1964, which embodied new hopes for a restructuring of the traditional agricultural sector, most especially through massive investment in irrigation (Lazarev 2012; Swearingen 1987). This spurred the creation of the ONI (*Office National de l'Irrigation*) in 1960, as a response to the prevailing fragmentation of state agencies dealing with irrigation (Lazarev 2012). The ONI "was endowed with the largest development budget in the country, and enjoyed a degree of decision-making autonomy which largely insulated it from other government bodies" (Lazarev 2012). But the ONI was not interested in hydraulic development alone and put emphasis on agricultural and social policies. It embraced in particular the idea of an agrarian reform, with measures such as the expropriation of all colonial land and a limitation in the size of farms, which could redress the perverse effects of inequity in the access to land. Since labor and water were abundant, the ONI chose to design and develop irrigation schemes based on gravity irrigation techniques. "The measures proposed for the five-year plan in order to carry out an agrarian reform were judged to be too revolutionary (…) and they met with the opposition of King Mohammed V, of the Istiqlal party, and of landed elites" (Swearingen 1987).

Consequently, in 1962 the plan was abandoned and the ONI was absorbed by a new centralized Office (OMVA: Office de Mise en Valeur Agricole), in charge of land development across the country but soon decentralized at the regional level. At the national level, around 400,000 ha of a total of roughly one million colonial land were acquired by Moroccan through different types of transactions, while "the state distributed 326,100 ha of land through the agrarian reform, the remaining (300,000 ha or 270,000 ha, according to different sources) came to constitute state property (*domaine de l'Etat*)" (Mahdi 2014; Bessaoud 2013; Bouderbala 1999a).

The failure to implement a fully fledged distribution of earlier colonial land reflected the interest of the political elite in acquiring land for themselves, but also the priority given to productivity and the quest for profitability, which came to challenge earlier priorities given to social and equity considerations. Policies further fuelled the duality of agriculture between, on the one hand, small peasants who received 5 ha plots and, on the other, an objective of intensification and profitability on state land (managed by two companies, SOGETA and SODEA) and large private landholdings which were largely preserved, especially in irrigated areas. At the end of the 1970s, for example, 1% of the landowners in the Haouz Plain controlled 38% of the land, while 62% of farmers owned only 12% of the land (Daoud 1981).

3.2.3.2 Dam Building (La "Politique des barrages")

Because of the sheer difficulty to reform agrarian structures, the administration put social reforms on the back burner and concentrated on technical solutions, with King Hassan II announcing in 1967 his famous policy to develop one million hectares of irrigated land, rekindling an old colonial promise. This policy also endorsed the recommendations of a World Bank mission dispatched in the wake of the economic crisis of the mid 1060s, which recommended the concentration of investments on regions suitable for irrigation and to the benefit of the "most advanced farmers" (Swearingen 1987). The Hydraulics Directorate was created in the same year to shepherd the project, including large dams, canals and land development for irrigation schemes, and in the 1965–1972 period 50% of public investments went to the agricultural sector. Capitalizing on the peak of the price of phosphate in 1974 (which was one of the main sources of income for the state), the irrigation programme was launched in 1975: in the 3-year plan of 1978–1980, 31% of public investments were devoted to rural and hydraulic infrastructures (Lazarev 2012).

But the hydraulic policy did not merely replace the agrarian reform. It was developed as an idea of its own. The new king, crowned in 1962, was bent upon making Morocco the 'California of Africa', with an agriculture based on an agro-export model (Ihazrir 2009), under the influence of the powerful corps of engineers (Akesbi 1988). "The monarchy [...] after the passing away of Mohamed V in 1961, made an alliance with rural elites which ensured the success of the constitutional referendum in December 1962 and the (more controversial) parliamentary elections of April 1963" (Leveau 1998). Agriculture in general and dams/irrigation in particular became a political trump card, whereby the support of rural elites was sought by extending personal benefits to them in exchange for rural stability, and were seen as essential to the stability of the regime (Akesbi and Guerraoui 1991; Leveau 1998; Regnier 1975). Rural areas were considered as a major reserve of votes in support of the regime, which could be mobilized to check other political forces, such as left parties and labour unions.

While in 1969 the trade balance showed a surplus of 750 million dirham (DH) (~75 million euros), in 1974 the surplus had been replaced by a deficit of over 750 million DH (Swearingen 1987). For technical and political reasons, however, investments were not discontinued. Subject to riots and failed *coups d'état* attempts the country was still fragile and public investments continued based on borrowed capital. Between 1967 and 1975, the irrigated area grew by 24,000 ha annually (Benhadi 1975). This, however, rapidly proved to be unsustainable and in 1978 Morocco engaged into fiscal austerity policies. In addition, between 1978 and 1981 the price of oil came to be multiplied by 2.7, worsening the state fiscal deficit, resulting in a hike of basic foodstuff prices, and eventually widespread riots. In response, the state implemented austerity measures and launched a 5 year (1981–1985) economic recovery plan. On top of that, Morocco was affected by a severe and prolonged drought (1981–1984) which depleted available water resources (during the 1980/1985 period the stored volume hardly exceeded 25% of the total capacity), and fuelled outmigration from rural areas to the cities. This was the – untimely – moment in which structural adjustment packages were imposed by the World Bank and the IMF, liberalizing the economy and rolling back the state well through the 1990s (Clément 1995).

A total of 12 dams were to be constructed between 1981 and 1985, but - with one exception- these dams were small compared to those constructed in the seventies. Yet, the initiation of the *Programme National de l'Irrigation* (PNI) in 1967, with the primary objective of bringing a total of 1 million hectares under irrigation before the year 2000, was reinforced by a 1986 declaration setting an objective of "one dam per year". Both goals were part of the Moroccan Government's longer-term strategy of "not one drop lost to the sea" (World Bank 1995) that would ten years later arouse the concerns of the World Bank that supply augmentation policies were reaching their limit (ibid.).

3.2.3.3 ORMVAs and the Agriculture Investment Code

Because of the considerable capital outlays mobilized by the Moroccan state to develop its irrigation sector, there was a need to ensure the profitability of these investments and the achievement of two major twin objectives: ensuring a degree of self-sufficiency in essential crops such like wheat, and promoting the export of cash crops. Guiding principles were encapsulated in the so-called '*Code des investissements agricoles*' issued in 1969, which principally concerned the irrigated sector. According to the Code, the state invests in large-scale irrigation schemes (managed by different ORMVAs: *Office Régional de Mise en Valeur Agricole*) and reserves itself the right to recover up to 40% of the investment costs from the beneficiaries (who do not need to be consulted prior to the investment) (Bouderbala et al. 1974; Bouderbala 1999b). The state is responsible for land reclamation and consolidation, land levelling and drainage. The beneficiaries must pay a lump sum of 1500 DH per ha per year and conform to the cropping patterns spelled out in the Code. In exchange, they may benefit from subsidized input, credit, and technical assistance. Collective land is allotted on the basis of 5 ha per beneficiary and land properties of less than 5 ha cannot be further parcelled. Land is earmarked for agriculture and cannot be turned into non-agricultural use.

Initially, the nine ORMVAs were not only responsible for water allocation and distribution but also determined cropping patterns, processed and marketed most industrial crops, including sugar, cereals, and cotton (Faysse et al. 2010). In addition, the State was strongly involved in the regulation of the market through prices setting and the implementation of development projects (Choukr-Allah 2004). Water distribution was rationalized based on cropping-patterns arranged according to pre-established patterns ('*trame A*' and '*trame B*') and fixed rotations (Pérennes 1993).

With time, this strong state involvement came under criticism. Fixed crops and water distribution patterns did not provide flexibility nor accommodated farmers' requirements or worries. Private initiative was reined in, recovery of service fees remained low, much maintenance was delayed, and the benefits expected by the state did not fully materialize (Benzekri 2006). The prohibition of land division below a threshold of 5 ha, meant to avoid land fragmentation, led to properties with multiple owners. Crop choice was eventually left to the farmers in 1985.

ORMVAs are in charge of constructing, maintaining (down to the tertiary level *included*), and managing infrastructures, but they also extend technical advice to farmers, while ensuring a water police function and recovering service fees. They are financially autonomous organizations placed under the Ministry of agriculture and administered by a board chaired by the Minister and mostly composed of representatives from the various administrations concerned.

3.2.3.4 Water Law 10-95

Hitherto largely centered on water resource development for irrigation, the Moroccan water policy acquired a new face in 1995, with the Water Law 10-95. This law – the elaboration of which started as early as the early 1980s, during the drought period (Chaouni 2005)-, signals a shift away from exclusive supply augmentation and emphasizes demand management policies. Following the Dublin principles and the principles of Integrated Water Resources Management, the law puts forth the user-pays and polluter-pays principles, and establishes decentralized River basin Organizations (*Agences de Basin Hydraulique*, ABH), while a National as well as Regional Environment Councils are created. The Water Law 10-95 also established the High Council for Water and Climate, which is supposed to define main water policy orientations at the national level, and provides for the coordination and the integration of water resources development and management policies at the level of each major River basin (Belghiti 2005).

The irrigation sector has been little affected by this law, and the prerogatives of the Ministry of Agriculture, in particular with regard to the ORMVAs, have been left largely untouched. Even in the recent Water Law 36-15 enacted in 2015, the Ministry of Agriculture has seemingly successfully negotiated its position (Tanouti 2017).

3.2.4 Current Status of Small-Scale and Large-Scale Irrigation Systems

The irrigated area in Morocco has been estimated at 1.46 million hectares, that is around 17% of the total agricultural area of the country (8.7 million ha), to which can be added 300,000 ha of occasionally irrigated land in communal schemes (Belghiti 2005) (Table 3.1). Large-scale public irrigation, divided into nine ORMVAs, represents roughly two thirds of the perennially irrigated land in collective schemes.

But this total area includes 441,430 ha of private irrigation, that is, irrigation developed based on groundwater and private wells. In reality, 626,610 ha belong to this category but 185,180 ha are located within large-scale public schemes and also use surface water. It must be noted that these figures are not very accurate because (i) the area effectively irrigated is lower than the area equipped, (ii) the area irrigated

Table 3.1 Irrigated areas (ha), at the end of 2010 (MAPM 2012, 2013)

	Area (ha), by technology			
Type of irrigation	Gravity	Sprinkler	Drip	Total
Large-scale irrigation	504,459	110,760	67,381	682,600
Medium and small-scale irrigation	256,209	2610	75,311	334,130
Private irrigation	317,571	16,951	106,908	441,430
Total	107,8239	130,321	249,600	1458,160

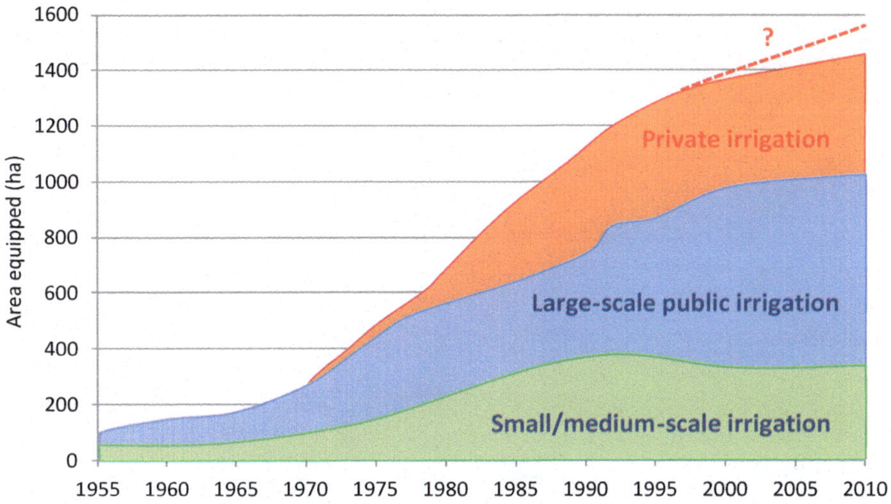

Fig. 3.1 Evolution of areas equipped for irrigation. (Source: Various official reports; the area under private irrigation has been drawn approximately, based on intensive phases of well drilling reported during the dry years of the early 1980s and early 1990s, and the stated current area of 441,430 ha)

by seguias depends on water availability each year, and (iii) there is no monitoring of groundwater-based irrigation. The 441,430 ha figure, as well as those of state irrigation, were already reported in 2004 (cited in Oubalkace 2007), and are still being used (see CESE 2014), showing that the continued expansion of this private irrigation is not well captured and not updated, and that private irrigation is probably underestimated (Fig. 3.1).

The nine ORMVAs are shown on Fig. 3.2, which also displays numerous dots showing small-and-medium scale irrigation schemes (PMH), mostly in or at the foot of the Atlas Mountains (and particularly in the mountain range between the Tensift and Souss river basins). Dark green dots show areas with intensive use of groundwater (principally coastal areas, the Saïss Plain, the Haouz and the Souss-Massa), but is also probably not fully up-to-date.

In general a provisional allocation is established each year by ORMVAs based on available supply (in the dams). Before each planned distribution period (typically every 2–4 weeks), individual farmers submit their request on whether they want to receive the quota they are entitled to. Then the ORMVA prepares the rotation schedule (starting and ending hours, and time derived from the ratio between target volume and flow rate (generally 30 l/s)), and then establishes the duration of supply to the corresponding tertiary. Farmers get a ticket with details about the supply to their fields planned for the following week, as does the ditch rider who distributes water within the tertiary. The system is somehow 'on demand' but requests are capped by the seasonal quota defined by ORMVAs and is *essentially a centralized system* (Plusquellec 2002). It must be noted, however, that the resulting volumetric

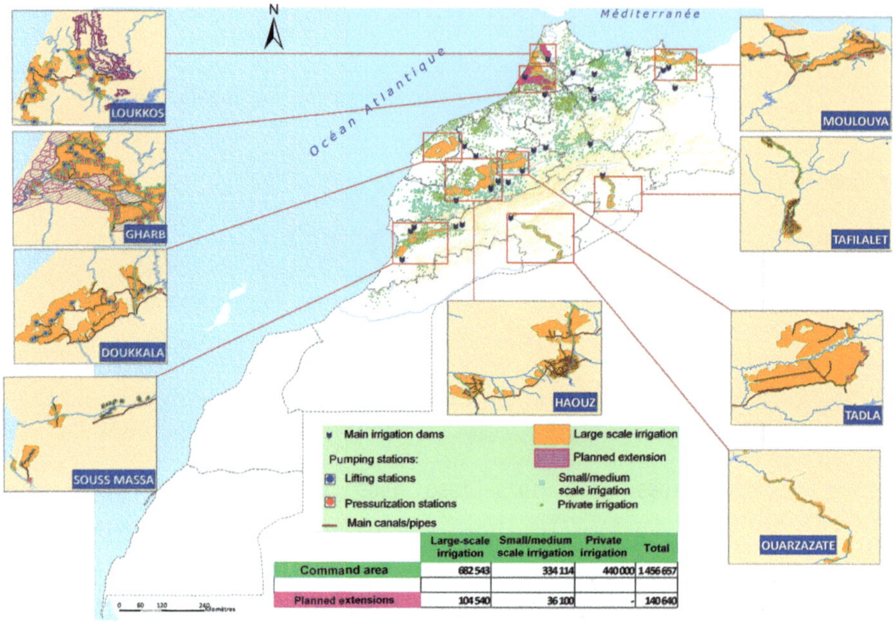

Fig. 3.2 Main irrigated areas in Morocco. (Adapted from MAPM 2012)

monitoring of water use at the farm level is technically rather exceptional, as far as large-scale gravity irrigation is concerned. Sectors irrigated by sprinkler can be found in the Gharb, Loukkos and Doukkala.

Because of uncertain/minimal deliveries and rigid rotations formerly adapted to centrally-imposed cropping patterns, diversification to crops such as vegetables (with frequent water requirements), has initially been constrained and this constraint has generally been bypassed by tapping groundwater (Kuper et al. 2012), or sometime through on-farm storage, either individual (investors) or collective (e.g. the Doukkala, see FAO 2012). Billing is generally done once (Haouz) or twice (Tadla) a year and the percentage of O&M fees collected is about 80% (Van Vuren et al. 2004), except in the areas under sprinkler irrigation in the Gharb (and Loukkos) schemes, where recovery is very low because it is difficult for the ORVMA to stop individual farmers accessing collective irrigation hydrants.

Maintenance of ORMVAs' distribution networks, although often delayed, were addressed with substantial and recurring outlays of public funds channeled through successive rehabilitation programs, including the Large-scale irrigation rehabilitation project (PRGI) and the two World Bank-funded project on large-scale irrigation (PAGI 1 and 2) (Belghiti 2004). This financial burden has been the main motivation behind the calls for more user participation and the establishment of WUAs in the 1990s (see Sect. 3.4.2), water pricing (see Sect. 3.4.3) and later for the search for models of privatization which could render ORMVAs financially self-sufficient (see Sect. 3.4.5).

If in the 1970s all state resources and attention were directed towards large dams and irrigation schemes, from the 1980s onward the government turned to medium and small-scale (in general communal) irrigation (PMH), recognizing the historical importance of such traditional systems and the opportunities for 'modernizing' or rehabilitating them, or constructing new ones. This was done, in particular, through two World Bank loans (PMH-1 and PMH-2) (El Gueddari and Arrifi 2009). A total of around 1500 schemes covering 334,130 ha, largely managed by water users themselves, source their water from small dams, springs, *khetaras*, tube wells, and wadi weirs for spate or continuous irrigation. Gravity irrigation is the most widespread method (74% in 2010), while sprinkler irrigation (10%) and most especially micro-irrigation (17%) are expanding with the recent *Plan Maroc Vert* (PMV) (SAM 2011).

Many khettaras and associated irrigation schemes have been particularly affected. In the Haouz, for example, the 500 khettaras have all dried up (in general because of the competition with wells) and/or been abandoned (due to very high maintenance costs or weakening of collective rules, and the urbanization process around Marrakech). In the oases, wells have often undermined khettaras systems (with some exceptions, like in Erfoud, near Ouarzazate, where well drilling has been reportedly been controlled by the community), but current irrigated areas have consistently increased with time. Out-migration and associated remittances have allowed investments back in irrigated agriculture, however unsustainable that may be (De Haas 2001).

3.3 Trends in the Irrigation Sector

3.3.1 The Modernization of Irrigation

Traditional gravity irrigation, and its practice in public schemes, has long been criticized for its alleged inefficiency and 'losses' (World Bank 1998). Moroccan policymakers have therefore extended substantial support to investments in micro-irrigation. Since 1982, taxes on the import of irrigation equipment or material have been reduced or cancelled (Laamari et al. 2011). In the 1990s, subsidies offered by the FDA (Agricultural Development Fund, established in 1986), averaged 17% of the investment costs in micro-irrigation. In 2002, a new decree raised the level of subsidies to 30–40% and extended this subsidy to all components of the project (including wells, pumps, and intermediate storage ponds). Shortly after, a National Program for the Development of Micro-Irrigation was launched with a target of 114,000 ha. After 5 years this program had only achieved 39% of the stated objective (El Gueddari and Arrifi 2009), with 75% of the conversion observed actually achieved through the existing FDA. It must be noted, however, that these statistics do not consider farmers who have adopted drip out of their own choice and without subsidies (e.g. because they were not owner of the land), like in the Coastal Chaouia.

Faced with the evidence that still few farmers had the capacity to invest, the government raised the rate of subsidies to 60% in 2006, and established in 2007 the National Program for Water Savings in Irrigation (PNEEI). Four years later, the PMV would finally raise the level of subsidies to 80% for large farms over 25 ha, and… 100% for small farmers. In less than 10 years the support to micro-irrigation shifted from a 17% to a 100% subsidy (though in practice, due to a maximal amount of subsidy defined for each component of the irrigation project, farmers often still have to pay part of the investment).

The PNEEI, subsumed into the PMV in 2008, adopted ambitious targets, seeking to achieve the conversion of 550,000 ha of land irrigated by gravity or sprinkler to drip irrigation, or in other terms to raise the percentage of micro-irrigation to 50% of the irrigated area in 15 years, at the cost of 37 billion DH. Seventy-two percent of this conversion would concern large-scale public irrigation (including some individual farm conversions and the modernization of the collective distribution of water through pressurized pipes), with the remaining devoted to private individual irrigation. Technical change is supposed to be accompanied by technical advice on irrigation practices, crop choice, and linkages with the agro-industry and export markets.

The expected outcomes of the program include water savings varying between 30% and 50% (with a total of around 1 billion m^3 (Bm^3) 'saved'), an increase in yields up to 100%, increases in job creation, rural incomes, and service fee recovery, and -rather optimistically- the reduction of the energy demand of the irrigation sector as well as a reduction of the overexploitation of aquifers (El Gueddari and Arrifi 2009). The PNEEI is believed to be "conducive to a revolution in Moroccan irrigated agriculture, not only with regard to irrigation water use efficiency, but also to productivity and competitiveness" (ibid.). There is, however, a worrying overlooked point in the PNEEI. The 70 page official main document of the PNEEI only has 2 paragraphs stating that water efficiency is scale sensitive and that losses at the plot level might not be losses when seen at the scale of the river basin. However, the rest of the document builds upon the ensuing statement "that it is at the level of the plot that water savings must be sought".

In Morocco, as elsewhere in the world, the rhetoric of water savings associated with micro-irrigation has been challenged by a growing body of evidence from the field, for example, the Haouz (Jobbins et al. 2015; Tanouti et al. 2016), the Souss (BRLi and Agroconcept 2013), or the Saïss (Benouniche et al. 2014). They emphasize the following issues:

- While the gross amount of water applied onto the plots is normally reduced by the technology, overirrigation with drip is common and studies in four different settings have shown plot-level efficiencies varying between 25% and 90% (Benouniche et al. 2014), due to farmers' lack of interest in delivering only the needed amount of water, a lack of know-how, poor system design, or technical problems such as clogging.

- With the shift to drip, farmers often choose to 'densify' their orchard or olive grove by adding (young) trees in the middle of old trees, adopting mixed-cropping (e.g. adding fruit tree lines between olive trees), or planting his plot anew with a higher density of trees (either olive or fruit). Vegetables and trees are clearly associated with drip and generally increase water requirements (and consumption) (Kuper et al. 2012).
- Even if the amount of water applied is reduced (and the same crop is maintained), it is uncertain whether the overall *net consumption* of water is reduced since reduction in non-beneficial soil evaporation is compensated by an increase in crop transpiration.
- Since drip systems are most frequently associated with individual wells, the water available is defined by the capacity of the well and is therefore not changed in the short term. With reduced application per unit of land, the farmer will be led to expand laterally to increase his irrigated area into plots that are either rainfed or left fallow because of insufficient supply, whether these plots are his own or rented from neighbors.

In other words, private benefits (increased yields and farm incomes) arise in tandem with a huge environmental cost: the depletion (or net consumption through evapotranspiration) of more water, which in the vast majority of the cases (most notably the southern deficit basins) translates into *further depletion of the aquifers*. Indeed, if supply of surface water is maintained, or even reduced, on the one hand, and water consumption enhanced, on the other, then the difference can *only come* from reduced return flows (to rivers or, mostly, aquifers).

This 'rebound effect', whereby drip irrigation translates into more water being consumed, while reducing the recharge of (already overexploited) aquifers, has now been well identified (Molle and Tanouti 2017). In Morocco the PMV or the PNEEI only have passing references to these problems and the risk of the "acceleration of the exploitation of aquifers" is said to be minimized "through the promotion of a rational practice of micro-irrigation and also through the establishment of aquifer contracts by the River basin agencies with which collaboration will be sought" (MAPM 2007a). Other parallel measures that this document promotes include an exhaustive inventory of wells, to be complemented by a survey on actual abstraction, awareness raising campaigns about the risks associated with overexploitation, and the installation of meters for all groundwater users, "a very tricky but totally indispensable measure to be implemented" (ibid.).

Last, the PMV/PNEEI carry with them the risk, identified by the Ministry (MAPM 2007a), of a poor technical or financial performance which would lead farmers to blame the new irrigation technology and revert to former practices. This risk is all the more real due to the fact that farmers are enticed by technology sellers to apply for these 100% subsidized projects, and also because market prices are uncertain and can slump due to local overproduction or external causes (e.g. the recent drop in the price of citrus or tomato).

3.3.2 *The Groundwater Rush*

3.3.2.1 Situation, Drivers, Regulations

From the 1970s onwards, but more strongly in the past 20 years, groundwater use for irrigation has developed quickly in Morocco. This has primarily been the case within the ORMVAs, where the failure to deliver adequate quantities of irrigation water has prompted farmers to tap the aquifer underneath their feet, which is largely replenished by the infiltration generated by the irrigation process. This trend was accentuated during the drought of the early 1980s. In the Tadla scheme, for example, the number of wells rocketed up from 900 in 1980 to 8735 in 1984 (Hammani et al. 2004). A second type of groundwater-based irrigation developed in the vicinity of ORMVAs' schemes (Haouz, Doukala, Gharb, Loukkos…), but also in partly-irrigated large plains like the Saïss, and generally uncultivated or rainfed areas (Guerdane, Chichaoua, etc). Small or big farmers sometimes follow a kind of mining strategy. Some investors from the Souss, for example, who had seen their groundwater resources exhausted, moved up north to the Chichaoua region (Tensift basin) to drill new wells, and then moved further afield to the Saïss Plain, when water levels dropped.

Because of the drought situation in the early 1980s (and later in the early 1990s) and the relief it provides, well drilling was initially subsidized by the state. In 1984, taxes on the import of drilling rigs and material were canceled and the drilling of wells, electrification, and irrigation material started to be subsidized (Laamari et al. 2011). The drilling of tubewells was boosted by the advent of Syrian drillers who spread over the whole Arab world with deep-drilling technical know-how (Quarouch et al. 2014). But it also benefited from a very lose enforcement of the law according to which water abstraction above a certain quantity requires an authorization. The 1995 Water Law reinforced the necessity to go through a licensing procedure for all kinds of water abstraction from the public domain (surface and groundwater). ORMVAs were in charge of licensing in their areas of responsibility. With the advent of the ABHs this responsibility shifted to the agencies but a negotiation took place whereby the ORMVAs remained responsible for drilling authorizations in the areas under their responsibility, while ABHs would grant water use permits. This double line of command introduced confusion and competition, and the possibility for the Ministry of agriculture to bypass restrictions associated with the overexploitation of aquifers and to promote the drilling of new wells as part of the PMV, against the will and the regulatory role of the Ministry of water and ABHs (Tanouti 2017).

Registration and licensing are part of the panoply of management tools. In Morocco a decree indicated in 1997 that water withdrawals points established before the enactment of the 1995 Law should be declared to the ABHs within a timeframe of 1 year after the publication of the decree. Because of the lack of responsiveness from users, two extensions of that registration period have been granted until October 2015, after which date no other extension of the delay would be granted and violators would face penalties of up to 2000 DH (BRLi and Agroconcept 2013; L'Economiste 2015). All wells drilled before April 2009 are in a position to be legalized. Wells drilled after this date are considered illegal, unless their owners have requested and received an authorization. The lack of interest shown by farmers can be explained by a burdensome procedure, and the (largely theoretical) condition to install a meter at their cost (the price of which varies between 1500 and 2000 DH for small farms and reaches 20,000 DH for large farms). In addition, the owner of the well must declare his consumption once a year and pay a water bill based on the rate of 0.02 DH/m^3 (L'Economiste 2015). It is only after well registration became a condition for accessing the generous subsidies of the PMV that farmers interested in the PMV showed willingness to officialize their well.

There is currently no reliable inventory of (agricultural) wells and boreholes in Morocco. In the Souss-Massa-Draâ basin, the system of authorizations is widely disregarded and out of an estimated 17,000 existing wells, only 2000 have been authorized (L'Economiste 2014). The ABH of Tensift has inventoried (circa 2010) 10,700 wells and boreholes and estimates the percentage of illegal wells at 31%, making up a total of around 14,000 wells and boreholes in the Haouz; but according to one official the real number is more likely to be around 25,000.

Calls by the Economic, Social and Environmental Council (CESE 2014) for the "installation of water meters on all tubewells of small, medium and large farms and combating illegal water abstraction for irrigation", or the intention of the PNEEI to install meters on all abstraction points, do not realize the difficulty of the task and, at this stage, are tantamount to wishful thinking. The installation of water meters is obligatory to get access to PMV subsidies, but field visits are limited and no follow-up is in order (BRLi and Agroconcept 2013). Sanctioning and the water police are reportedly weak to nonexistent (MAPM 2007a; CESE 2014; Del Vecchio 2013). Because of the strong and effective presence of the control of the Ministry of Interior at the local level (through *caids* and *moqaddems*), it can be inferred that slack policing reflects a lack of political will rather than a lack of means.

The 1995 Water Law also provides for mechanisms that should be activated in case of overexploitation of resources, such as safeguard or prohibition area delineation. However, Morocco has surprisingly never resorted to these plans of action, even though some ABHs have requested them. Given a groundwater overdraft estimated at 1 Bm^3 per year by the Minister of Water (Maroc.ma 2014), and the situation of overexploitation of all major aquifers in the country (Monitor Group 2008), it is hard to escape the conclusion that the political influence of interests associated with the expansion of groundwater-based agriculture has so far prevailed upon conservation policies.

However, current groundwater use is associated with growing negative externalities and clearly unsustainable. All the measures envisaged – registration, tariffs, micro-irrigation or, according to the Minister of Water, "the adoption of a new mode of governance that promotes participatory approaches, through 'aquifer contracts', involving all the stakeholders in decision-making" (Maroc.ma 2014) – are unlikely to stem the overdraft and not up to the challenges posed (Molle 2017). The main measure actually implemented –surface water supply augmentation through transfers– (e.g. Souss, coastal Chaouia, Saïss) will only provide some respite while fuelling more expansion of irrigated land.

3.3.2.2 The Groundwater Management Contracts

The experience in Morocco with aquifer comanagement, later called '*contrats de nappes*' (aquifer contracts), started in 2004 when the Souss Massa ABH carried out an awareness campaign about the new water law and proceeded to close illegal wells. This triggered social unrest and the Governor of the region suspended the decision and decided to approach the problem by creating a commission integrating representatives from 20 institutional partners. An agreement was signed in 2007 which included (among other things) 22 small dams and 5 large dams to be constructed by the state, and a regularization of 'illegal' wells, against a freezing of the expansion of irrigated areas for citrus and vegetables, a (subsidized) shift to drip irrigation, an increase of groundwater user fees, and a reinforcement of the water police (Faysse et al. 2012). Eight years after the signing of the agreement the situation has not really changed, as the contract has been undermined by a general laisser-faire attitude, the failure of the government to deliver on the supply augmentation projects and enforce regulations, and the event of a few good hydrologic years that have displaced the prevailing sense of urgency.

Although the aquifer contract in the Souss was not implemented, around 2011 the GIZ started supporting the ABH of Souss-Massa, Tensift and Tadla (GIZ 2011), with an overoptimistic planning of multi-stakeholder meetings expected to lead to an agreement on a convention after 1 year. These initiatives bumped into the technical, institutional and political complexity of the problem addressed and were not conclusive. In 2013 an inter-ministerial notification signed by the Ministries of Agriculture, of the Interior, and of Water, expressed political support to the establishment of aquifer contracts and clarified the procedure to be adopted.

In 2014, a national workshop on the management of groundwater resources put the aquifer contracts centre stage again (Tanouti 2017). The Minister for water acknowledged the failure of the control of groundwater abstraction, a problem which neither the law nor the 2009 National Water Strategy could successfully address, pointing to "a legal vacuum" with regard to aquifer contracts, and to some "reluctance among the partners of the Department in charge". Yet, the Ministry presented at the workshop a timetable of 15 major aquifer contracts, all slated to start before the end of 2016. The four components of these aquifer contracts, however, pointed to a rather top-down approach largely consisting of constraining measures (Molle 2017).

The 'aquifer contract' for the Saïss Plain, for example, mainly restates the measures spelled out in the basin masterplan (PDAIRE), does not declare the Saïss as a safeguard zone, and contains no provision for controlling the use of private wells and tubewells, or the participation of users, prompting Del Vecchio (2013) to comment that the aquifer contract is in reality "essentially a hydraulic and agricultural development policy characterized by a strong interventionism of the Moroccan state". The jury's still out as to whether this is a promising way forward but the failure of the Souss-Massa contract, the first aborted attempts in the Tadla or Haouz, and the current framing of the Saïss aquifer contract do not provide ground for much optimism. The very high number of farmers using groundwater for irrigation in an uncoordinated and largely informal manner outlines a challenging setting. Yet, in 2014 the government declared a policy to have such contracts established in all major aquifers (later reduced to three) in Morocco by 2016 (L'Economiste 2014), further to a loan conditionality set up by the World Bank.

3.3.3 *Reusing Wastewater for Irrigation*

The amount of wastewater generated by Moroccan cities has increased from 48 in 1960, to 500 Mm^3 in 2000, and 700 Mm^3 in 2010 (Aomar and Abdelmajid 2002; MAPM 2011). Morocco has been very slow to consider developing reuse as a means of increasing water supply. Farmers however often did not wait to make use of this resource and more than 7000 ha were reported to be irrigated with raw sewage in 2000 (ibid.), while about 546 Mm^3 of raw wastewater was discharged to the environment (60% of which to the sea). A few pilot projects have been developed in the 1990s near Agadir (aerated lagoon and filtration/percolation) or Ouarzazate (lagooning). Another project started in 2003 in Attaouia (El Kelâa des Sraghna). A national plan for the treatment of wastewater has been launched in 2005 with the aim of treating all wastewater by 2030. The National strategy of 2009 later set a lower target of 300 Mm^3 of treated wastewater to be reused by 2030 for the irrigation of golf courses, public gardens or crops.

One of these new projects is being implemented in Marrakech and is hitting the headline. The treatment station has a capacity to treat 110,000 m^3/day at the secondary level of treatment. Treated water is to be used for the irrigation of golf courses as well as to revive part of the famed palm tree grove of the city. The project has been advertised as epitomizing new environmental policies as well as conservation strategies in the face of water scarcity. While the analysis of the project's technical and financial difficulties point to a complex relationship between the state and private investors (Ennabih 2016), another aspect deserves emphasis: it appears that Marrakech's wastewater was already used by local farmers to irrigate an area of approximately 2000 ha. The project is therefore also tantamount to a reallocation of the resource from local farmers to golf courses (Tanouti and Molle 2013). Everywhere else in Morocco, wastewater is actually used by farmers, irrespective of regulations.

3.3.4 *Desalination*

Desalination is another option offered to Moroccan planners. Because of the cost of producing water it is more suitable for the production of potable water for coastal cities but it is also being contemplated as a means of substituting groundwater in the Chtouka area (in the Massa basin in the south), where the aquifer is being depleted. An elaborate Public-Private Partnership (PPP) is under planning, whereby an area of 12,000 ha would be supplied with desalinated seawater by a private company under the conditions that the larger landholders commit to buying this (expensive) water, and that the community of users helps freeze the expansion of wells. But desalinated water might also constitute a cost-effective solution in oasis areas, when coupled with solar energy and niche market opportunities (Benabderraziq 2017).

3.4 Institutional, Regulatory and Policy Frameworks

3.4.1 Formal Organization of the Irrigation Sector

From an institutional point of view, the creation of water/environmental administrations by the Water Law 10-95 came with the setting up of ABHs, and a reorganization of the water sector and the abolishment of the Regional Hydraulic Directorates (Directions Régionales de l'Hydraulique, or DRH) whose staff, however, has been largely transferred to the ABHs (Tanouti 2017).[1] The new structuring implied a substantial redistribution of roles and bureaucratic prerogatives. ABHs are supposed to take care of data collection, water abstraction authorizations, basin level planning (through the establishing of basin master plans called PDAIRE), and the yearly allocation of dam water. In theory, ABHs should regulate demand according to the available supply, integrate the different uses, including the environment, and promote participatory management. However, the central state clearly keeps a strong control over the ABHs and is predominant in its administration board, chaired by the minister in charge of water resources and consisting of 60% of members representing state agencies (Tardieu 2001). In addition, the establishment of ABHs has

[1] In the 1970s, the irrigation sector was focused on developing infrastructures and placed under the Ministry of Agriculture and Agrarian reform, with a Water Resource Development Directorate (*Direction de Mise en valeur*) that would include the ONI and the OMVA. It was later replaced by the Rural Engineering Administration, with two directorates responsible for the development of public irrigation infrastructure and the O&M of ORMVAs, respectively. But in parallel, the General Administration for Hydraulic Works – under the Ministry of Public Works–, and its regional branches (DRH), was responsible for other hydraulic infrastructures, notably dams. In 1994, 2 years after the Rio submit, a Secretary of State in charge of the environment is placed under the Ministry of Interior, but 1 year later the environment is put under the responsibility of a fully-fledged Ministry of the Environment. This Ministry will however be short-lived and subsumed 2 years later into the Ministry of Agriculture, Public Works, and Environment, under the powerful civil engineer Mziane Belfqih. One year later, in 1998, with the advent of a socialist government, the administration in charge of the environment was again shifted to a Ministry of Rural and Urban Planning, and Environment. In 2002 the ABHs were emancipated from this Ministry and placed under a new Ministry of Development Use Planning, Water and Environment. This new ministry placed water outside the Ministry of Agriculture, the Ministry of Interior and the Ministry of Public Works, but this relative independence was short-lived and in 2007 it was conflated with the Ministry of energy, mines, water and the environment, and placed under a Secretary of State in charge of Water and Environment. Last, in 2012, these two responsibilities were separated and each of them entrusted to a *Ministre délégué*.

been slow (7 years to establish them in most basins, and 14 years to extend ABH to the whole territory).

Expectedly, these new prerogatives located under the Ministry/Secretary of Environment have been seen by traditional central administrations, most especially the Ministry of Agriculture and its ORMVAs, as a "challenge to their power" (Tardieu 2001; Tanouti 2017). This is in particular true because the ABHs have taken over decisions about allocation of water at the basin level, which means that ORMVAs have become bulk water users among others and, although they still yield substantial negotiation power, do not decide anymore on the amounts of water to be released by the dams. Yet, the culture of planning and building infrastructure has largely been passed to the new structure, since the ABHs have frequently been sited in the quarters of the former DRH (de Miras and Le Tellier 2005), while the DRH staff have "accepted the transfer without renouncing to their past" (Tazi Sediq 2007). For cultural and political reasons, their 'regulatory' power intended remains weak.

Part of the budget of the ABHs comes from a water tax levied on industrial users and irrigators. The tax for using public domain resources has been fixed at 0,02 DH/m^3, and is passed on to the ABH by the ORMVAs which collect fees from farmers. ABH should also theoretically levy taxes from polluters but the polluter pays principle is not implemented yet. Theoretically, in line with the French model, the sums recovered should be transformed into subsidies to water conservation or treatment actions (Belghiti 2005).

Neither the irrigation sector nor the development of hydraulic infrastructure have been much challenged by the administration in charge of the environment which is currently, tellingly, lumped together with the sectors of mines, water and energy, as indicated earlier. In Sect. 3.5.1 we will return to commenting on sectoral policy contradictions and on how agriculture in general and irrigation in particular remain largely unaffected by environmental regulation. Despite a water law often lauded as a state-of-the-art legislation, some analysts consider that "Legislative and regulatory work and institutional development [have] led, respectively, to a compilation of fragmentary texts with a number of loopholes and inconsistencies, and to a juxtaposition of sectoral institutions and missions constituting a heavy and costly technical-administrative apparatus" (El Alaoui 2006).

3.4.2 Stakeholders' Participation: Conflicting Views and Practices

The large-scale irrigation schemes developed between the 1950s and the 1980s have been put under strong centralized management. After the adoption of structural adjustment policies in 1983 and economic liberalization, the government withdrew from agricultural and marketing activities, raised water fees, and – much to the insistence of the World Bank- started to explicitly call for the involvement of users in water management (van Vuren et al. 2004; Bergh 2007). The government introduced a Participatory Irrigation Management (PIM) policy and in 1990 passed Law 02-84 that specified the legal status of WUA and named them *Association d'Usagers de l'Eau Agricole* (AUEA). The 1992 Decree 2-84-106 that established the terms of agreement between the government and the AUEAs reads like a detailed list of the duties and incumbencies these associations have to comply with, with hardly any mention of the benefits they would get in return. The law specifies that an agreement must be signed with the state, expressing *"the commitment of the association to mobilize the funds needed to cover all the expenditures associated with the administration of the association, water distribution, and the maintenance and conservation of infrastructures (…) and the obligation to carry out regularly the maintenance works needed to keep all facilities in good working conditions"*.

An elected board implements the decisions taken by the general assembly, establishes budgets and submits them to the assembly, and can hire staff (Freitas 1996). The democratic nature of elections is affected by the high rate of illiteracy of farmers, with more than 70% of them found illiterate in the case of the Tadla scheme (van Vuren et al. 2004). A degree of control over WUAs is exerted through the nomination of the seventh board member (dubbed '*the seventh man*') by the Administration, as a representative of the Ministry of Agriculture, while Article 15 stipulates that other representatives of the government can have advisory roles if they wish so. The official reason for this dispositive is to facilitate communication with the government, but in practice, it is often used as a means of imposing governmental decisions. WUAs are also constrained by their stated attributions: some WUAs started to supply water to farmers located outside the official irrigation system, while charging them higher fees, but this practice remains informal and even illegal (Errahj et al. 2009). The lack of political will to empower farmers (Faysse et al. 2010) is well illustrated by the authorities' indecisiveness to grant WUAs financial independence (van Vuren et al. 2004): in the Tadla scheme WUAs were first established with budgets constituted by the 20% of collected fees handed back to them. However, this transfer was later cancelled because it was found to be in contradiction with the Code of Agricultural Investments (Doukkali 2005).

"As had been the case in the past, the government took the initiative, defined the rules of the game in its own way, and maintained the right of oversight of the associations operations" (Belghiti 2005). These policies did not yield significant changes in governance and most of the 408 WUAs officially found in large-scale irrigation schemes remain apathetic or non-existent (Aloussi and Anbari 2012), especially in the Doukkala, Gharb and Loukkos schemes (Faysse et al. 2010). In the Tadla, Van Vuren et al. (2004) identified a lack of political will for an effective transfer of irrigation management, and noted that "the staff from the ORMVAT does not want to lose its position as irrigation manager".[2]

[2] Likewise, Bergh (2007) identified a problem with the "purely technical and depoliticized view of participation that dominates the attitudes of Moroccan civil servants… The deeper origin of such attitudes lies in the central government's reluctance to open up the spaces that are necessary for a more political sense of agency to develop. The latter would in turn allow participation to unfold as a truly transformative power for rural development".

From the 2000s onwards, the state started an ambitious policy to shift from surface and sprinkler irrigation in large-scale schemes to drip irrigation. The PAGI project, the *Programme d'Amélioration de la Grande Hydraulique* funded by the World Bank, was conditional upon the establishment of WUAs. This was done on paper but most of these WUAs remained dysfunctional or non-existent, partly because of the non-implication of farmers in the design of both infrastructural and institutional interventions,[3] partly because of the lack of interest from managers unwilling to relinquish their roles and prerogatives (Mahdi 2012; Kadiri and Mahdi 2009). ORMVAs attempted to "revitalize" the WUAs, so that they could play a key role in the coordination between farmers and the irrigation office during the modernization process, however the lack of means of WUAs limited their capacity to play such a role (Hadioui et al. 2014). The case of the N'fis scheme, near Marrakech, showed that farmers' commitment to participate in water management through WUAs was very weak. Farmers did not ask for the establishment of the associations, were not involved in their formation, and in the manner through which they were grouped under a particular WUA (Keita 2006; Raki and Ruf 2004; Valony 2006). More recently, however, some WUAs in the Tadla scheme, which had been inactive for years, were revived when a younger and more informed generation became involved in irrigation management (Faysse et al. 2010).

WUAs have been more successful in small-scale irrigation schemes that were already managed by farmers (Bekkari and Castillo 2011). In the Aït Hakim and Aït Bouguemez valleys, for example, sophisticated and flexible water sharing arrangements were formalized, with different upstream and downstream irrigation rules, and also different village-level rules (Keita 2006; Romagny and Riaux 2007). In many cases, however, traditional management did not require the establishment of WUAs and WUAs were merely pre-conditions for the World Bank (Programmes SMIS 1 and 2) and other donors to have identified beneficiary groups before releasing funding (Riaux 2006). Yet Bergh (2007) showed that lack of openness in the preparation of development plans and planning phases, as well as project implementation, was a main impediment to the successful execution of IMT/PIM programmes.

But new institutions may also be a window of opportunity for some local actors to change the *status quo*. Kadiri et al. (2008) showed also that the competitive recruitment of young graduates played an important role in the improvement of management in the *Moyen Sébou* irrigation scheme in Morocco. They simplified administrative procedures, adapted the rules in place when needed, and reacted more quickly to emergencies (breakdowns of canals, pumps).

[3] Belghiti (2005) acknowledges that "the integration of human dimension in the design of the projects makes them sometimes more difficult and costly in time and effort because it is complicated in term of choice of standards design and of process of implementing".

3.4.3 Economic Issues and Tools

According to the 1969 Investment Code, the contribution of beneficiaries to irrigation investment costs was established at 1500 DH/ha, and later raised in 1984 to 30% of the average irrigation investment costs. Until 1997, this contribution was waived for farms under 5 ha, and for the first 5 ha of farms under 20 ha, while payment was staggered over 17 years, with a 4% interest rate. In 1997 the contribution was raised to 40% for all farms, and the interest rate fixed at 6% (Belghiti 2004). Although eventually moderate, the percentage of investment costs recovered is rather high when seen comparatively at the world level (Molle and Berkoff 2007).

Since the establishment of the nine ORMVAs, a major concern of the state has been their physical and financial sustainability. Delayed maintenance prompted two large World Bank loans (PAGI 1 and 2, in the 1983–2000 period) to avoid the degradation of the investments. The World Bank developed at 'Bulk water pricing study' as part of it 1998–2004 'Morocco Water Resources Management Project', that clearly spelled out the costs of the water service and the ability to pay of farmers (Belghiti 2004). Morocco was obliged to raise its level of cost recovery several times, first as part of the Structural Adjustment Plan during 1980–1984, then of the World Bank (PAGI) project during the 1997–2006 period, and finally of an European grant conditionality (FAS-EAU), after 2007 (MAPM 2007a; Belghiti 2004).

The financial contribution of farmers might not be sufficient to cover all O&M costs but it is quite substantial, especially when compared with other countries in the MENA region, notably in the ORMVAs which have water lifting costs. The rate of recovery of the service fee is also high in gravity irrigation schemes (much less so for sprinkler irrigation), since water supply is often on demand and ORMVAs can – technically- easily exclude farmers who do not pay their fees. ORMVAs combine three mandates (studies and development, O&M, extension services) and the service fees (around 0.35 DH/m^3) actually cross-subsidize other functions that should be covered by the state budget. More recently, management costs have come to also include the ABHs, as "the main part of their budget has to come from water users" (World Bank 2009). In any case, raising prices to the level of covering management costs has proved to be problematic. For Doukkali (2005), "the inability to complete water pricing reforms, clearly suggests how political costs have overshadowed the real socio-economic and resource costs".

Irrigation water pricing has also long been advocated by the World Bank and mainstream organizations, based on the idea that to cope with water scarcity, "full recovery of the cost of irrigation water and the integrated management of the resource would lead to significant gains in efficiency and encourage a shift to crops requiring less water" (OED 1998). ORMVAs hardly cover half of irrigation needs and there is therefore no 'wastage' from which one could expect farmers to reduce their 'demand'. The marginal value of this water in terms of production is way higher than its cost to the farmers, and this would still be the case even if the full O&M costs were recovered from them. It is well recognized that too high prices would also push farmers to shift to groundwater (ORMVAs' water is priced at

0.35 dh/m^3 against 0.02 dh/m^3, for a lump volume of 3000 m^3/ha), when aquifers are not too deep or when groundwater is not saline (Monitor Group 2008; Belghiti 2004). This is what many already do, not for reducing costs but because groundwater is available on-demand and escapes the rigidity of irrigation rotations.

Water pricing has also resurfaced, however timidly, as one of the measures listed to help combat the overexploitation of aquifers by individual users. At the moment an extremely limited percentage of farms pay for the groundwater they abstract, as mandated by the Law. The National Water Strategy briefly mentions the need to "adopt a water pricing system based on volumetric metering" (Royaume du Maroc 2011). The Monitor Group (2008) considers that water pricing is a 'lever' with a "high level of influence" on demand, and proposes to increase the price of groundwater, and also to establish a premium (higher) price for water abstracted in overexploited aquifers.

At the same time, policies are silent about the more politically sensitive issue of energy subsidies, notably about how domestic gas (that came to be cheaper than diesel) is now largely used for groundwater pumping. By envisioning a volumetric pricing system, policy-makers also make it conditional upon the identification of existing wells and the thorny prerequisite that the use of each farmer should be volumetrically known.

3.4.4 The "Green Morocco Plan" and Irrigation

The Green Morocco Plan (*Plan Maroc Vert*, or PMV) is a sector-wide comprehensive agricultural development plan launched in 2008 with ambitious objectives (a contribution to GDP of 174 Billion DH, the creation of 1.15 million jobs by 2020, the tripling of the income of 3 million people in rural areas). It rests on two pillars: the accelerated development of a modern and competitive agriculture, with regard to both production and food processing; and the support to smallholder agriculture through the implementation or professionalization of 545 projects of small farms in poorer rural areas. These two pillars are to be strengthened by a number of cross-sector reforms concerning water and land tenure, and an original mechanism called 'aggregation', whereby small and fragmented landholdings are expected to access technology and markets through an association with more advanced and capital-intensive actors. With regard to water and irrigation, ongoing projects from the Ministry of Agriculture have been subsumed into the PMV, conceived as an overarching programme (Belghiti 2010). This concerned not only the PNEEI but also the "Programme to bridge the gap between dams and irrigated areas", that aimed at completing the development of irrigated areas originally planned under the main dams on the country (140,000 ha over 10 years); and the PPP "for the delegation of public irrigation service" (see next section).

Section 3.3.1 has already discussed the impact of the PNEEI (now under PMV) on water resources, mostly aquifers, because of the overall likely increase in water consumption associated with the conversion to drip irrigation. The PMV set out to shift the level of support to the conversion to drip irrigation to 100%, to offset any reluctance from farmers due to financial considerations.[4] But the PMV does not only aim at the conversion of gravity irrigation into drip. It also includes the expansion of irrigated agriculture. This is constrained by the need for farmers or investors to show a legal land title and well permits if they want to benefit from the subsidies. While farmers had largely ignored previous regularization campaigns, in the 2010s the windfall subsidies of the PMV made the number of applications soar and the authorizations for abstracting groundwater were initially given with few exceptions. To bypass the growing reluctance of ABHs to give permits, the PMV offered loopholes to farmers so that drip irrigation projects and their subsidies could go ahead. The PMV allowed farmers to apply for reconversion subsidies if they signed a letter certifying that they were using the well before 2005 (Saïss), or exhibited an administrative document showing they have initiated an administrative request to regularize an allegedly old well (Haouz) (del Vecchio 2013; Tanouti 2017). Illegal, but also legal, well drilling is still possible either through personal arrangements with local authorities, or through the intervention of powerful people in the government or political system. There is clear evidence of investors using influence to get well authorizations for developing plantations on (collective tribal) land leased by the Ministry of Interior, on the public domain of the SODEA/SOGETA[5] parceled out to investors, or acquired on the market.[6] In the Saïss Plain, Fofack (2012) has also found that farmers could sometimes go through either the *caïdat* (Ministry of interior) who can mediate the request through the ABH, or through the sellers of drip irrigation systems, who can take care of the application process and possible arrangements with the Ministry of Agriculture. These parallel lines of authorities undermine, if not nullify, the responsibility of the ABHs for sustainable water use.

In other words, the PMV is not always subsidizing conversion to drip irrigation but is also planning expansion of groundwater-based private irrigation. The expected additional water demand is estimated at 1 Bm^3, according to the Monitor Group (2008). To which must be added the likely increase in water consumption due to the

[4] Such a level of subsidy has generated a situation where irrigation equipment salesmen and design companies enter directly in contact with farmers or water user associations to propose them turnkey projects in which they take care not only of the implementation of the project, but also of the administrative procedure to receive the subsidies and request drilling authorizations (BRLi and AgroConcept 2012).

[5] Public bodies in charge of managing land recovered from colons after independence (and not redistributed).

[6] The intricacy of Moroccan land tenure has also often been seen as an obstacle to development (World Bank 2008). In 2005, a new law (so-called 'de la main levée') allowed members of Agrarian reform cooperatives to obtain a private title for their plot of land, and therefore to trade it on the land market (Valette et al. 2013). Collective lands under customary ownership (terres collectives) have also long been coveted by investors to develop new tree plantations, such as in the Haouz or the Souss (BRLi and Agroconcept 2013), and these lands are attributed by and under the control of the Ministry of Interior, which lies outside of the prerogatives of line ministries. They are also facing a process of privatization (with 300,000 ha of collective land about to be registered in the name of right holders) (Belghazi 2016). Other types of traditional collective land tenures (habous, guich) are also under pressure to be privatized (maroc.ma 2015).

intensification and expansion associated with drip irrigation. Investments to raise agricultural production, and land and water productivity probably have a positive impacts on wealth generation, incomes and jobs (although this conclusion is subject to debate)[7]; but they run counter to hydrological realities and belie the official window-dressing discourses that underline the integration of sectoral policies and objectives.

3.4.5 Privatization and PPPs

The financing of both the O&M of existing ORMVAs and the development of new irrigation infrastructure has long been seen as a recurring burden (Belghiti 2004). Morocco has been very receptive to ideas of privatization of services in general and has also largely operated a transfer of technical water expertise from the state to private companies. In the 2000s the World Bank, and later the French Development Agency (AFD), assisted Morocco in carrying out feasibility studies about the autonomy of the branch of ORMVAs responsible for water delivery services, through private or parastatal entities. These studies have not identified a viable economic model for such institutional change, largely because of the impossibility to generate enough profit to attract private companies.

But the idea of private sector involvement in funding and management has been shifted to new projects promoted under the seemingly fashionable label of PPP and targeting high-value agriculture because of its capacity to pay for part of the investment and for high delivery costs. Such an approach was first initiated in the Guerdane area in the Souss region (Houdret 2012; Houdret and Bonnet 2016). A private company built a pipe to transfer water from a dam to a 10,000 ha where groundwater depletion had strongly impacted the profitability of citrus farms and dried up orchards. The investment of €70 million has been shouldered by the state (48%), the company (44%) and the users (8%), pointing to a still considerable level of subsidy. This project de facto limited access to water to capital-intensive farmers and was not immune to financial problems (e.g. the drop in the profitability of citrus and the wave of defaulting on water fees in 2015) (Houdret and Bonnet 2016); not to mention than in a closed basin such a transfer also amounts to a spatial reallocation of the resource.

Another project is starting in the Coastal Chaouia region (between Azemmour and Bir Jdid): an area of 3200 ha cultivated by 600 farmers will receive 15 Mm^3/year from the Oum Er-Rbia river and the delivery service has been entrusted to a private company; 90% of the €37 million needed for the infrastructure will come from the state (Albayane 2013). Another project is being completed in the Massa region (south of Agadir), where a desalination plant will be built to irrigate market crops and citrus for export. Yet another project concerns the transfer of 100 Mm^3 of

[7] Discussing the merits of the PMV lies beyond the scope of this chapter. For a critique see Akesbi (2012, 2014), Bentaleb (2015), and Salem and Ait Benhamou (2016).

water annually from the Mdez dam on the Sebou River to the Saïss Plain, where the aquifer is being depleted. All these projects still include large outlays of public money for the investments and their profitability, from the perspective of the managing operator, is linked to the possibility to charge water fees that are higher than in ORMVAs and fully cover delivery costs. This clearly targets the production of cash crops by a capital-intensive agriculture favored by the government.

3.5 Threats and Future Prospects

3.5.1 Production vs. Environmental Conservation: Sectoral Contradictions

At the beginning of the 2000s, the water strategy stressed "the necessity to integrate the diverse programs and policies on development, water and the environment, and to ensure their consistency" (ONDH 2004). Later, in 2008, Morocco gave itself an "ambitious and innovative [updated] National Water Strategy (…), with a roadmap for water management and precise quantitative objectives up to 2030", that "has been harmonized and made consistent with that of agriculture documented, in the PMV", according to the CESE (2014).

Behind conventional political window-dressing, it is now becoming increasingly clear that Morocco's adherence, up to these days, to the longer-term strategy of "not one drop lost to the sea" (World Bank 1995) is now proving incompatible with environmental sustainability. As early as 1995, the World Bank (1995) clearly showed little enthusiasm for the irrigation expansion policy of the government, underlining that "The Government's objectives of expanding irrigated areas, in the context of increasingly scarce water supplies, rising costs and less protected output prices, poses a severe challenge for irrigated agriculture". Twenty years later Morocco is still intent to develop 140,000 ha of what it calls a 'catching up' (*rattrapage*) with development plans, slowly taking the last excess river basin – the Sébou – towards full commitment of its water resources.

But beyond the water sector itself, the worsening status of groundwater resources has much to do with the contradictory and antagonistic policies of the Ministry of Agriculture. As demonstrated earlier, support to agricultural intensification through drip irrigation, and to expansion by subsidizing investors is detrimental to the water status of the country. More generally, it is the environment which is being impacted by the stressed water regime (seawater intrusion, degraded wetlands, polluted rivers, dried-up springs, etc.) (MEMEE 2010; Akesbi 2014). The culture of the ABHs, the decentralized 'regulators' under the Ministry of Water, is inherited from the former Regional Hydraulic Directorates and is only slowly being directed towards environmental sustainability. The limited power entrusted to this Ministry is indicative of the current political priorities of the Moroccan state.

Agricultural policies are also coupled with energy policies but the PMV has been largely elaborated without consideration of its impact on the energy sector, despite the impact of the conversion to drip irrigation on the country's €10 billion energy bill (96% of Morocco's energy is imported), 13% of which on account of agriculture (Doukkali and Lejars 2015). In 2011, total subsidies involved in energy consumption by agriculture amounted to €0.75 billion (ibid.).

The CESE (2014) recently issued a disquieting report emphasizing the absence of an operational regulatory body in the water sector (the inter-ministerial Committee being non-operational for years), failures to properly implement the 1995 Water Law (e.g. with regard to the polluter-pay principle, safeguard and prohibition zones on overexploited aquifers, sanctioning, etc), insufficient coordination between department concerned with water, ABHs lacking means, autonomy and independence, and the *Conseil Supérieur de l'Eau et du Climat* (CSEC) having unclear roles and prerogatives.

3.5.2 Regional Water Deficits and What They Mean for Irrigation; National Planning

Irrigated agriculture, although limited in terms of area when compared with rainfed agriculture, is a major contributor to the agricultural GDP as well as national exports. Its share of total water use is around 84%. In a context where most river basins (except the Loukkos and part of the Sébou) are overexploited, this suffices to understand how the future of irrigation is tightly associated with the future of water resource development and status.

It is becoming increasingly clear that much of the participatory gloss put on the National Water Strategy, the decentralization of management at the level of river basins, the elaboration of the PDAIREs, or the expected aquifer contracts do not fundamentally alter a policy that is still largely based on state-planned supply management (subsidies to well drilling by the PMV, remaining marginal dam sites to be developed, north-south transfer, desalination, etc). The reality is a desiccation of the country gradually expanding northward and 'downward' (with at least 1 Bm^3 of groundwater depletion each year), under continued and unchecked water resources development by both the government and individual farmers or investors.

In Morocco, the process of basin closure started in the Souss-Massa in the 1980s, moved North to the Tensift and the Oum er-Rbia, and is underway in the Sebou,[8] where the remaining surplus water will soon be committed (100,000 ha expansion in the Gharb, irrigation development in the middle-Sébou, transfers to the Saïss

[8] The PDAIRE, under optimistic hypotheses and disregarding climate change, foresees an average discharge to the sea in 2030 of 500 Mm^3, roughly 10% of average available resources in the basin. While the upper Sebou basin is increasingly overcommitted, northern catchments in the Rif region contribute a lot of runoff (largely stored in the Wahda dam) but these resources cannot be transferred at acceptable costs.

Plain, increasing urban needs in Fès and Meknes, water transfers to Casablanca and further south, etc). Although exhibiting a deficit of 840 Mm^3/year (ABHOER 2012), the Oum er-Rbia basin is strikingly *still* witnessing an expansion of irrigation, like in Khenifra, or in the coastal Chaouia area…

As alluded to earlier, all major aquifers are now overexploited, with a drop in water levels, a depletion of at least 1 Bm^3/year, and negative impacts on springs and river baseflow (Maroc.ma 2014). A situation that prompts GIZ (2014) to stress that with no action "the exhaustion of the aquifer in the short term is inevitable, with severe socio-economic as well as ecological consequences". Yet, as one magazine observes, "The drop in the cost of well drilling, pumping and irrigation equipment, as well as subsidies allocated by the state have encouraged the farmers to invest in wells", and with the lack of rain in 2015–2016 "professionals of the well drilling and irrigation equipment sectors see with satisfaction their activities on the rise" (Finances Hebdo 2016). Thirty years ago (in 1974) the masterplan for the Souss basin stated specifically that should the private sector continue to disregard planting or pumping bans, it should in the future fully bear the most disastrous consequences (ABHSM 2010).

Regulation of demand in order to keep it in line with available resources is the responsibility of the ABHs but these lack funds, are understaffed and with budget deficits. They have been involved in data collection, contracting out technical studies, licensing of well drilling, or yearly water allocation, but in their planning tasks and have had to compromise with the ORMVAs (representing agriculture), the Ministry of Interior and private interests (Tanouti 2017). The PMV makes light of environmental limitations and even provides a loophole for farmers to drill subsidized wells in overexploited aquifers, in contradiction with attempts by the ABHs to regulate (over)exploitation.

The overextension of irrigated farming makes it very vulnerable to climatic vagaries. In case of prolonged drought, its supply will have to be dramatically curtailed. Impacts will be all the more serious because the percentage of tree crops is growing and this may result in the loss of sunk capital. Whether because of climate irregularity or exhaustion of groundwater resources, irrigated agriculture, currently merely seen as an engine of growth, is poised to be impacted by failing water supply.

The overall bleak water situation of Morocco does not mean that no agricultural expansion or intensification is possible. However, identifying what can be done and where without further impairing the water balance requires the development of a more elaborate knowledge of hydrology, with more attention to return flows, changes in stocks, and surface/groundwater interactions.

3.5.3 Climate Change

Studies carried out in the late 2000s to update River basin masterplans found a decrease of "regularized volumes" by around 30%, compared with earlier masterplans (done around 30 years earlier). This difference is even sharper at the level of river basins such as Bouregreg, Oum Er Rbiâ and Sébou (46%, 45% et 38% respectively) (El Gueddari and Arrifi 2009). The annual potential of renewable water resources, hitherto estimated at 30 Bm^3, is now considered to be 22 Bm^3 (MAPM 2007b).

Even though a lot of uncertainty remains, climate change predictions all point to a continued trend in the occurrence of extreme events and in rainfall reduction, and "the country's water resources, both superficial and groundwater, are expected to further slump by around 15 to 20 percent by 2030" (MAPM 2007b). Aoubouazza et al. (2013) forecast a drop in water availability between 14% and 23% by 2050, and Driouech (2010) a decline in rainfall by 20% and a rise in temperature between 2 and 6 °C by the end of the century.

Yet, surprisingly, PDAIREs do not consider these expected reductions and plan water resources on the basis of current availability. The PDAIRE for the Tensift bears no mention of climate change; add-on studies have been conducted after the publication of the PDAIREs to study the impact of climate change in some basins, like the Oum Er Rbia where the supply-demand deficit estimated by the PDAIRE at around 560 Mm^3 for 2030, was found to be rather around 1500 Mm^3, with the observed climate trends and forecasts available to date (Hydraumet 2013)[9]!

In other words, not only is Morocco currently still increasing resource mobilization and stretching out its irrigation area, but its planning -against all evidence- is disregarding the coming increase in temperature and evapotranspiration, and decrease in rainfall and runoff. Irrigated agriculture stands out to be the most impacted.

3.5.4 *What Future for Irrigated Agriculture?*

While the future of irrigated agriculture is primarily threatened by the status of water resources, it also has to face several other challenges. The financing of reconversion to drip and of expansion heavily draws on financial resources. €3.6 billion are slated for the PNEEI and Doukkali and Lejars (2015) estimate subsidies at €11,000 per ha, on top of the subsidies allocated to other agricultural investments. Mobilizing ever marginal and costly water, transferring water from the North to the South, desalination, etc. are supply-augmentation policies largely made inescapable by the continued expansion of irrigation. They form a large part of the National Water Plan, which will require a total investment of $27 billion (Telquel.ma 2014).

Irrigated agriculture is also potentially threatened by a risk of deepening socio-economic differentiation which could result from policies such as the PMV, or the PPP projects, which systematically favor a model of capital intensive agriculture

[9] "The water supply and demand balance for 2030 (the current PDAIRE) is optimistic and does not sufficiently integrate climate risk and its possible effect on the potential water of the area to this date: the supply-demand deficit estimated around 560 Mm^3 for 2030 should be rather around 1500 Mm^3, with the observed climate hydro trend and forecasts available already today" (Hydraumet 2013).

linked to export markets. The PMV is seen by some experts as perpetuating the productivist and technicist policy models that Morocco has implemented, from colonial time throughout to current days, to tackle a situation that is known as the 'duality of agriculture' (Akesbi 2012, 2014; Ihazrir 2009). While the first pillar of the PMV is devoted to boosting 'modern' agriculture, the second 'pillar' is supposed to help small landholders transition towards a 'modern' agriculture (Faysse 2015), grossly overlooking the social reality of the countryside according to Akesbi (2014).

Another process of social differentiation is being observed in areas which critically depend on groundwater. In the Tadla scheme (Kuper et al. 2012), the Souss basin (BRLi and Agroconcept 2013) or the Saïss Plain (Ameur et al. 2017), the necessity to access groundwater at ever increasing depth and costs clearly favors investors growing cash crops to the detriment of small holders without the capacity to deepen their wells or shoulder these costs. More generally interventions on the water cycle, whether through technological change such as drip irrigation or wastewater treatment, like in Marrakech, often somewhat invisibly redistribute access to water, with an impact on weaker constitutions (Molle 2012).

Last, the future of irrigated agriculture in its different forms is linked to the evolution of factor prices (e.g. the cost of energy or labor) as well as market conditions. The competitiveness of Morocco on international markets is vulnerable to competition by other countries in the Mediterranean and elsewhere, to health or environmental standards, and to unpredictable changes in market prices (Akesbi 2011, 2012).

For the past 15 years, several factors concomitantly contributed to an increase in agricultural production in Morocco: (i) the development of intensive groundwater-based irrigation practices in areas previously cropped with rainfed crops or used as pasture; (ii) the development of drip irrigation which led to sizable increases in yields, especially for vegetable production; (iii) the many incentives for agricultural investments extended by the Green Morocco Plan. However, there has not been a parallel increase in the export of agricultural products. As a consequence, there is an increased competition on domestic markets, especially in the fruit and vegetable sector. For instance, in the Middle Atlas mountains, many farmers have since around 2006 started to invest in intensive apple and plump trees, using boreholes and drip irrigation systems, on land previously devoted to pastures. Sellika and Faysse (2015) and Sellika et al. (2015) showed that the most probable scenario in the ten coming years (and already apparent in 2017) is a strong decrease in the benefits of apple and plump farmers, a scenario that might affect fruit and vegetable growers using groundwater in many regions of the country.

3.6 Conclusions

As mentioned in the introduction, Morocco can boast substantial achievements in the water sector and these achievements are well underlined in the literature. The CESE (2014) stresses that Morocco is considered as a "regional and continental model in terms of water management", with an "institutional governance model" and legislative framework considered as "exemplary", and a national water strategy that is "coherent and ambitious". According to Doukkali (2005), "from a historical perspective, the institutional reforms undertaken in Morocco, especially since the new law of 1995, are truly remarkable". Building on a strong historical basis of communal irrigation systems and hydraulic know-how, the state investments in dams and large-scale public irrigation have tripled the irrigated area of the country, before individual farmers and investors drilled wells to develop groundwater-based irrigation, expanding Morocco's aggregate irrigated area to around 1.5 million ha.

Morocco's success with supply management has not been paralleled with corresponding achievements in terms of demand management, and Morocco's adherence, up to these days, to the longer-term strategy of full utilization of its resources (World Bank 1995) is now proving incompatible with environmental sustainability and hydrologic variability. While threats have been periodically well identified they have often resulted in ad hoc declarative postures translated into strategies and even laws. But the gap between policy intentions and reality on the ground is now exposing a lack of political will to enforce regulations, in general, and control the expansion of irrigated agriculture, in particular. The overbuilding of public irrigation schemes (whose needs are covered only at 60% on average, and much less in southern basins) has already put most other river basins 'in the red' (Bouregreg, Oum er-Rbia, Tensift, Souss-Massa, Tafilalet), and groundwater-based agriculture has compounded the situation and provoked an annual deficit of 1 Bm^3 (and probably much more than that in reality).

Yet expansion of irrigation is still being promoted by the PMV in the name of modernization, food security, poverty alleviation and other objectives. Morocco tends to be caught up in the ubiquitous (global) incantation that we need "to do more (irrigation) with less (water) in a sustainable way (no overexploitation)", wishfully assuming that the circle can be squared (Molle and Tanouti 2017). Expansion is expected to be compensated by an improvement in the performance of irrigated agriculture, including water savings, intensification and higher water economic productivity. This makes light or even completely ignores that, in most cases, expected savings at the scheme or plot levels are illusory (because 'losses' largely return to aquifers that are already overexploited, with notable exceptions such as the Doukkala where they go to sinks, like saline aquifers).

Policies enlisted to solve this contradiction are proving ineffective or even counterproductive. The technical fix of drip irrigation comes with enhanced intensification and diversification to cash crops that both tend to result in higher depletion of water. Irrigation modernization also has a bearing on energy use (Raïs et al. 2016) and recent policies aimed at reducing the use of subsidized butane through the promotion of solar energy[10] are extremely worrying in terms of groundwater use. Agricultural water pricing has been proved to be important for cost recovery but ineffective for saving water (especially for groundwater use where wells are not

---

[10] http://leconomiste.com/article/1016688-pompage-solaire-de-nouveaux-mecanismes-financiers-pour-l-irrigation

metered in general). Institutional changes, such as the establishment of WUAs or aquifer contracts have not delivered on their promises. The proposed toolbox and the language of alleged 'risk mitigation' measures are distracting us away from the evidence that the country is squeezing itself dry by increasing water resources development and depletion; even more so because government's current subsidies are achieving just the opposite. Such dynamics belie the official discourse that underlines the integration of sectoral policies and objectives.

The development of Moroccan agriculture is however a key national objective. Irrespective of whether current public investments are cost-effective or not, the conversion to drip irrigation has likely benefits in terms of income, land and water productivity. These benefits, however, are largely limited to those who are able to access groundwater individually. Large-scale scheme are being gradually upgraded to provide pressurized water (but this has yet to be tested on the mid-term), while small schemes have problems of land tenure and fragmentation and would require heavy investments. These benefits may even warrant some moderate increase in water consumption, but (1) they should not be drummed up as a solution to water scarcity; (2) they should be geographically concentrated and allowed only in areas deemed compatible with the current availability of water resources. Morocco's knowledge of its water resource, despite a respectable data collection network, is now clearly insufficient to address the challenges of managing closed basins and overexploited aquifers.

Irrigated agriculture is witnessing a number of challenges that will become more acute in coming years. From a social point of view, various dynamics of social differentiations are at work (fostered by export markets standards, price vagaries, growing water abstraction costs, etc), while climate change will both substantially reduce available supply and include more extreme events which will be increasingly hard to weather in basins where resources are already overexploited. Morocco's is currently reaping short-term political and economic benefits at a high a financial cost (PMV), but at the cost of a growing long-term vulnerability to social and climatic upheavals which have also, history has it, unsettled its political stability several times in the past.

References

ABHOER. (2012). *Résumé du PDAIRE*. Projet de Plan Directeur d'Aménagement Intégré des Ressources en Eau du Bassin de l'Oum Er-Rbia et des bassins côtiers atlantiques.

ABHSM Agence du Bassin Hydraulique du Souss Massa. (2010). *Les changements climatiques. À la recherche d'un équilibre hydrologique dans le bassin du Souss Massa*. WWF Marseille.

Akesbi, N. (1988). Dépendance alimentaire et vérité des prix: mythes et réalités. In *La crise et l'endettement du tiers monde* (p. 541). Etudes Maghrébines.

Akesbi, N. (2011). La nouvelle stratégie agricole du Maroc annonce-t-elle l'insécurité alimentaire du pays ? *Confluences Méditerranée, 78*(3), 93–105.

Akesbi, N. (2012). Une nouvelle stratégie pour l'agriculture marocaine: Le Plan Maroc Vert. *New Medit, 2*, 12–23.

Akesbi, N. (2014). Le Maghreb face aux nouveaux enjeux mondiaux. *Les investissements verts dans l'agriculture au Maroc* (Notes de l'IFRI). Paris: IFRI.

Akesbi, N., & Guerraoui, D. (1991). *Enjeux agricoles*. Casablanca: Le Fennec.

Albayane. (2013). *Azemmour-Bir Jdid: Doukkala des Eaux délégataire du projet d'irrigation*. 23-12-2013. http://www.maghress.com/fr/albayane/120507

Aloussi, M. N., & Anbari, A. (2012). *Gestion participative de l'irrigation au Maroc*. Powerpoint.

Ameur, F., Amichi, H., Kuper, M., & Hammani, A. (2017). Specifying the differentiated contribution of farmers to groundwater depletion in two irrigated areas in North Africa. *Hydrogeology Journal, 25*(6), 1579–1591.

Amphoux, M. (1933). L'évolution de l'agriculture européenne au Maroc. *Annales de Géographie, 236*(42), 175–185.

Aomar, J., & Abdelmajid, K. (2002). *Wastewater reuse in Morocco*. Morocco: Ministry of Agriculture, Rural Development and Waters and Forests.

Aoubouazza, M., Rajel, R., & Essafi, R. (2013). Impact of extreme climate events on water resources and agriculture and biodiversity in Morocco. *Journal of Climatology and Weather Forecasting* 2013, *1*(2), 8.

Bekkari, L., & Del Castillo, I. Y. (2011). L'appropriation du modèle d'association d'usagers de l'eau par une communauté villageoise du Moyen Atlas au Maroc. *Cahiers Agricultures, 20*(1), 73–77.

Belghazi, A. (2016). Les terres collectives irriguées seront immatriculées gratuitement. *Medias24*, 14 Mars 2016.

Belghiti, M. (2004). *Valorisation de l'eau et tarification dans les périmètres de «grande hydraulique» au Maroc*. Revue H.T.E. N° 130 – Septembre/Décembre 2004.

Belghiti, M. (2005). *Farmers participation in the management of public irrigation schemes in Morocco*. FAO Report.

Belghiti, M. (2010). *Place de l'irrigation dans la stratégie du plan Maroc Vert*. Powerpoint.

Benabderraziq, H. (2017). Pers. Com.

Benhadi, A. (1975). La politique marocaine des barrages. *Annuaire de l'Afrique du Nord, 14*(1976), 275–294.

Benouniche, M., Kuper, M., & Hammani, A. (2014). Mener le goutte à goutte à l'économie d'eau: ambition réaliste ou poursuite d'une chimère ? *Alternatives Rurales, 2*, 36–47.

Bentaleb, H. (2015). Le Plan Maroc Vert creuse les sillons de l'inégalité. *Liberation*, Samedi 17 Octobre 2015.

Benzekri, E. M. (2006). *50 ans de politique de l'eau au Maroc*. https://fr.scribd.com/document/241156065/50-Ans-de-Politique-de-l-Eau

Bergh, S. I. (2007). *Public administration capacity to implement 'participatory' development policy – Case studies from Morocco*. Paper presented at the Panel 17 States at work: African public services in comparative perspective at AEGIS Second European Conference on African Studies, Leiden.

Bessaoud, O. (2013). *Les réformes agraires post-coloniales au Maghreb: un processus inachevé*.

Bouderbala, N. (1997). La modernisation et la gestion du foncier au Maroc. In *La modernisation des agricultures rnéditerranénnes* (Options Méditerranéennes, Série A./n029).

Bouderbala, N. (1999a). Les systèmes de propriété foncière au Maghreb. Le cas du Maroc. In A. M. Jouve & N. Bouderbala (Eds.), *Politiques foncières et aménagement des structures agricoles dans les pays méditerranéens: à la mémoire de Pierre Coulomb*. Montpellier: CIHEAM.

Bouderbala, N. (1999b). L'aménagement des grands périmètres irrigués: l'expérience marocaine. In A. M. Jouve & N. Bouderbala (Eds.), *Politiques foncières et aménagement des structures agricoles dans les pays méditerranéens: à la mémoire de Pierre Coulomb* (pp. 171–184). Montpellier: CIHEAM.

Bouderbala, N., Chraibi, M., Pascon, P. (1974). La question agraire au Maroc. *Bulletin économique et social du Maroc,* p 423

Bourdillon, J., & Faris, M. (2004). *Les ingénieurs des Ponts au Maroc, la promo 59, l'enthousiasme et l'amitié des années 60, l'héritage*. http://bit.ly/2AfiDu8.

BRLi, & Agroconcept. (2013). *Gestion de la demande en eau dans le bassin méditerranéen – Exemple du Maroc – Cas d'étude du Souss Massa*. AFD et Plan Bleu.
CESE. (2014). *La gouvernance par la gestion intégrée des ressources en eau au Maroc: Levier fondamental de développement durable*.
Chaouni, M. (2005). *La loi sur l'eau et le droit à l'eau. Une interprétation de la réglementation de l'eau à l'usage des utilisateurs et des gestionnaires des ressources en eau* (p. 222). Rabat: Imprimerie Al Maârif Al Jadida.
Choukr-Allah, R. (2004). Participatory irrigation management in Morocco. *Options Méditerranéennes Series B, 48*, 209–217.
Clément, J. F. (1995). Les effets sociaux du programme d'ajustement structurel marocain. *Politique Etrangère, 4*, 1003–1013.
Cressier, P. (2006). Géométrie des réseaux et marqueurs des territoires. *Mélanges de la Casa de Velázquez, 2*(36), 39–59.
Daoud, Z. (1981). Agrarian capitalism and the Moroccan crisis. *MERIP Reports: Land and Labor, 99*, 27–33.
De Haas, H. (2001). *Migration and agricultural transformations in the oases of Morocco and Tunisia*. Utrecht: KNAG.
De Miras, C., & Le Tellier, J. (2005). *Gouvernance urbaine et accès à l'eau potable au Maroc: Partenariat Public-Privé à Casablanca et Tanger-Tétouan*. Paris: L'Harmattan.
Del Vecchio, K. (2013). *Une politique contractuelle sans contrôle ? La régulation des ressources en eau souterraine dans la plaine du Saiss au Maroc*. Mémoire de Master. Université Lumière Lyon 2.
Doukkali, M. R. (2005). Water institutional reforms in Morocco. *Water Policy, 7*, 71–88.
Doukkali, M. R., & Lejars, C. (2015). Energy cost of irrigation policy in Morocco: A social accounting matrix assessment. *International Journal of Water Resources Development, 31*(3), 422–435.
Driouech, F. (2010). *Distribution des précipitations hivernales sur le Maroc dans le cadre d'un changement climatique: descente d'échelle et incertitudes*. Thèse de doctorat, Université de Toulouse.
El Alaoui, M. (2006). Développement de l'agriculture irriguée, dispositif juridique et institutionnel et stratégie de gestion de l'eau au Maroc. In S. Bouarfa, M. Kuper, & A. Debbarh (Eds.), *L'avenir de l'agriculture irriguée en Méditerranée. Nouveaux arrangements institutionnels pour une gestion de la demande en eau*. Cahors: Actes du séminaire Wademed.
El Faiz, M. (2001). La Grande Hydraulique dans le Haouz de Marrakech: fascination technologique et émergence du pouvoir des ingénieurs. In E. Gobe (Ed.), *Les ingénieurs maghrébins dans les systèmes de formation. Systèmes de formation, filières coloniales et pratiques professionnelles* (pp. 213–225). Professionnalités contemporaines. Actes de la réunion intermédiaire du programme Ingénieurs et société au Maghreb. Rabat.
El Faiz, M. (2002). *Marrakech patrimoine en péril*. Actes du Sud/Eddif.
El Gueddari, A. B. S., & Arrifi, M. (2009). L'agriculture irriguée au Maroc face à la rareté des ressources en eau? In *L'Etat des ressources en eau au Maghreb en 2009, UNESCO, Rabat*.
Ennabih, A. (2016). *Mythe et réalité de la politique des eaux usées de Marrakech: entre intérêts privés et légitimation publique*. Mémoire de Master 2, Université de Lyon.
Ennaji, M., & Herzenni, A. (1987). *L'irrigation des terres makhzen dans le Haouz de Marrakech sous le règne de Hassan premier* (pp. 68–69). ANAFID.
Errahj, M., Kuper, M., Faysse, N., & Djebbara, M. (2009). Finding a way tolegality, local coordination modes and public policies 296. *Natures Sciences Sociétés, 58*, 358–369.
FAO. (2012). *Le passage à l'irrigation localisée collective. Les résultats d'une expérience dans le périmètre des Doukkala*. Rome: FAO.
Faysse, N. (2015). The rationale of the Green Morocco Plan: Missing links between goals and implementation. *The Journal of North African Studies, 20*(4), 622–634.
Faysse, N., Errahj, M., Kuper, M., & Mahdi, M. (2010). Learning to voice? The evolving roles of family farmers in the coordination of large-scale irrigation schemes in Morocco. *Water Alternatives, 3*(1), 48–67.

Faysse, N., El Amrani, M., El Aydi, S., & Lahlou, A. (2012). Formulation and implementation of policies to deal with groundwater overuse in Morocco: Which supporting coalitions? *Irrigation and Drainage, 61*(1), 126–134.

Finances Hebdo (Maroc). (2016). Irrigation: fort intérêt pour le forage des puits. 15 Février 2016.

Finet, A. (2002). *Diagnostic des systèmes de production du périmètre irrigué du N'Fis (Maroc). Un aménagement aux résultats contradictoires. Mémoire pour l'obtention du Diplôme d'Ingénieur en Agriculture.* Centre National d'Etudes Agronomiques des Régions Chaudes.

Fofack Tsabou, R. L. (2012). *Analyse des règles d'accès à l'eau souterraine dans un contexte de mutations de l'agriculture et des politiques publiques au Maroc: le cas de l'aquifère du Saiss.* Master: Science Politique et action publique. Université de Lille 2.

Freitas, R. J. (1996). *Design for the implementation of water users associations in the Tadla irrigation perimeter.* Report No 41. USAID.

GIZ. (2011). *Élaboration du contrat de la nappe du Haouz-Mejjate. Analyse de la gestion actuelle de la nappe.* Février 2011.

GIZ. (2014). *Elaboration d'une convention pour la gestion intégrée des ressources en eau dans le bassin du Haouz-Mejjate.* Note d'information pour la Wilaya Marrakech-Tensift-Al Haouz.

Hadioui, M., Faysse, N., & Kemmoun, H. (2014). Participation des agriculteurs à la conception d'un projet de reconversion à l'irrigation localisée dans le périmètre du Tadla. *Alternatives Rurales, 2,* 96–109.

Hammani, A., Kuper, M., Debbarh, A., Bouarfa, S., Badraoui, M., & Bellouti, A. (2004). Evolution de l'exploitation des eaux souterraines dans le périmètre irrigue du Tadla. *Revue HTE, 130,* 39–43.

Houdret, A. (2012). The water connection: Irrigation, water grabbing and politics in southern Morocco. *Water Alternatives, 5*(2), 284–303.

Houdret, A., & Bonnet, S. (2016). Le premier partenariat public-privé pour l'irrigation au Maroc: durable pour tous? *Cahiers de l'Agriculture, 25*(2).

Hydraumet. (2013). *Stratégie d'adaptation de la gestion des ressources en eau des bassins de la région hydraulique de l'Oum Er Rbia.* Rapport pour la Banque Mondiale et l'ABHOER.

Ihazrir, A. (2009). La politique hydraulique marocaine à l'épreuve: sécheresse et crise des identités rurales. In H. Ayeb, & T. Ruf (dir.), *Eaux, pauvreté et crises sociales.* IRD Éditions.

Jobbins, G., Kalpakian, J., Chriyaa, A., Legrouri, A., & El Mzouri, E. H. (2015). To what end? Drip irrigation and the water–energy–food nexus in Morocco. *International Journal of Water Resources Development, 31,* 393–406.

Jolly, G. (2002). La gestion des périmètres irrigués Méthodologie de diagnostic Cas d'un transfert de gestion: les associations d'usagers du périmètre du N'Fis (office du Haouz, Maroc). In P. Garin, P. Y. Le Gal, & T. Ruf (Eds.), *Atelier du PCSI (Programme Commun Systèmes Irrigués) sur la gestion des périmètres Irrigués collectifs, Montpellier, France* (pp. 25–45).

Kadiri, Z., & Mahdi, M. (2009). Eau: une coordination à plusieurs cycles. *CESEM Economia, 16,* 45–51.

Kadiri, Z., Kuper, M., Faysse, N., & Errahj, M. (2008). Transformation d'une innovation institutionnelle: l'exemple des associations d'usagers des eaux agricoles dans le Moyen Sebou au Maroc. In T. Hartani, A. Douaoui, & M. Kuper (Eds.), *Economies d'eau en systèmes irrigués au Maghreb. Actes du quatrième atelier régional du projet Sirma, Mostaganem, Algérie, 26–28 mai 2008.* Montpellier: Cirad.

Keita, B. (2006). Gestion sociale de l'eau et projet de modernisation hydraulique dans une vallée du haut atlas au Maroc: les Aït Hakim (des Aït Bouguemez). In A. Richard, P. Caron, J. Y. Jamin, & T. Ruf (Eds.), *Coordinations hydrauliques et justices sociales. Actes du séminaire, novembre 2004, Montpellier, France.* Montpellier: Cirad.

Kuper, M., Hammani, A., Chohin, A., Garin, P., & Saaf, M. (2012). When groundwater takes over: Linking 40 years of agricultural and groundwater dynamics in a large-scale irrigation scheme in Morocco. *Irrigation and Drainage, 61*(1), 45–53.

L'Economiste. (2014). *Souss-Massa-Draâ: L'anarchie des forages de puits.* 07/11/2014.

Laamari, A., Boughlala, M., Herzenni, A., Karrou, M., & Bahri, A. (2011). *Water policies in Morocco – Current situation and future perspectives.* ICARDA.

Lazarev, G. (2012). *Les politiques agraires au Maroc. 1956–2006*. Un témoignage engagé. Economie Critique.
Le Coz, J. (1968). Le troisième âge agraire du Maroc. *Annales de Géographie, 77*(422), 385–413.
L'Economiste. (2015). *Marrakech: Le tourisme continue à tirer les investissements*. Edition N°: 4477 Le 06/03/2015.
Leveau, R. (1998). Réussir la transition démocratique au Maroc. La monarchie, acteur central du système politique. *Le Monde Diplomatique*.
Mahdi, M. (2014). Devenir du foncier agricole au Maroc. Un cas d'accaparement des terres. *New Medit, 4*, 2–10.
MAPM (Ministère de l'agriculture et de la Pêche Maritime). (2007a). *Programme national d'économie d'eau en irrigation*. Document principal. 10/7/2007.
MAPM (Ministère de l'Agriculture et de la Pêche Maritime). (2007b). *Programme national d'économie d'eau en irrigation*. Note de synthèse.
MAPM (Ministère de l'agriculture et de la Pêche Maritime). (2011). *Projet de renforcement des capacités sur l'utilisation sans danger des Eaux Usées en Agriculture*. Rapport National du Maroc. December 2011.
MAPM (Ministère de l'Agriculture et de la Pêche Maritime). (2012). *Place de l'eau dans le Plan Maroc Vert*. Powerpoint.
MAPM (Ministère de l'agriculture et des pêches maritimes). (2013). *Situation de l'agriculture Marocaine en 2012*. Dossier N°10.
Maroc.ma. (2014). *Les eaux souterraines jouent un rôle très important dans le développement socioéconomique du Maroc*. 16/9/2014. www.maroc.ma/fr/actualites/les-eaux-souterraines-jouent-un-role-tres-important-dans-le-developpement-socioeconomique
Maroc.ma. (2015). *Assises nationales sur "la politique foncière de l'Etat et son rôle dans le développement économique et social"*. 8/12/2015. www.maroc.ma/fr/actualites/ouverture-skhirat-des-assises-nationales-sur-la-politique-fonciere-de-letat
Marthelot, P. (1961). Histoire et réalité de la modernisation du monde rural au Maroc. *Tiers-Monde, 2*(6), 137–168.
MEMEE. (2010). *Rapport National sur l'État de l'Environnement*. Ministère de l'Energie, des Mines, de l'Eau et de l'Environnement.
Molle, F. (2012). La gestion de l'eau et les apports d'une approche par la political ecology. In D. Gautier & T. A. Benjaminsen (Eds.), *L'approche political ecology: Pouvoir, savoir, environnement* (pp. 219–240). Paris: Quae.
Molle, F. (2017). *Conflicting policies: Agricultural intensification vs. water conservation in Morocco* (G-Eau Working Paper). http://bit.ly/2j220bA
Molle, F., & Berkoff, J. (Eds.). (2007). *Irrigation water pricing: The gap between theory and practice*. Wallingford, UK: CABI.
Molle, F., & Tanouti, O. (2017). Squaring the circle: Agricultural intensification vs. water conservation in Morocco. *Agricultural Water Management, 192*, 170–179.
Monitor Group. (2008). *Etude de Mise à Jour de la Stratégie Nationale de l'Eau et des Plans d'Action à Court, Moyen et Long Termes pour le Développement du Secteur de l'Eau au Maroc*.
Mouhsine, R., & Lakmahri, S. (2014). *L'impossible réforme agraire*. Zamane, 28/3/2014.
OED. (1998). OED Précis no. 152 – Maroc: un potentiel encore inexploité.
ONDH (Observatoire National de Développement Humain). (2004). *Stratégie Nationale de l'Eau*.
Oubalkace, M. (2007). *Stratégie méditerranéenne pour le développement durable – Suivi des progrès dans le domaine de l'eau et promotion de politiques de gestion de la demande*. Rapport final (Royaume du Maroc). Commission Méditerranéenne du Développement Durable.
Pascon, P. (1977). *Le Haouz de Marrakech*. Rabat-Tanger, Éditions marocaines et internationales.
Pérennes, J. J. (1993). *L'eau et les hommes au Maghreb: contribution à une politique de l'eau en Méditerranée*. Karthala Editions.
Plusquellec, H. (2002). *How design, management and policy affect the performance of irrigation projects – emerging modernization procedures and design standards*. Bangkok: FAO.
Préfol, P. (1986). *Prodige de l'irrigation au Maroc. Le développement exemplaire du Tadla, 1936–1985*. Paris: Nouvelles Éditions latines.

Quarouch, H., Kuper, M., Abdellaoui, E. H., & Bouarfa, S. (2014). Eaux souterraines, sources de dignité et ressources sociales: cas d'agriculteurs dans la plaine du Saïss au Maroc. *Cahiers Agricultures, 23*(3), 158–165.

Raïs, I., Faysse, N., & Lejars, C. (2016). Impacts d'un changement de politiques énergétiques sur les exploitations irriguées: éclairage sur la base d'un échantillon d'exploitations dans le Saiss (Maroc). *Alternatives Rurales, 4*, 90–102.

Raki, M., & Ruf, T. (2004). La participation des usagers de l'eau des périmètres irrigués méditerranéens. Regards croisés sur deux modèles marocain et égyptien. In A. Richard, P. Caron, J. Y. Jamin, & T. Ruf (Eds.), *Coordinations hydrauliques et justices sociales*. Montpellier.

Regnier, J. J. (1975) Monarchie et forces politiques au Maroc. In: *Introduction à l'Afrique du Nord contemporaine* (pp. 341–358). Aix-en-Provence: Institut de recherches et d'études sur le monde arabe et musulman.

Riaux, J. (2006). La gestion participative de l'irrigation: exemple d'intervention publique dans une vallée du Haut Atlas marocain, les Aït Bou Guemez. In L. Auclair, C. Aspe, & P. Baudot (Eds.), *A l'heure du développement durable, le retour des paysans?* (pp. 259–274). Paris: Edisud.

Romagny, B., & Riaux, J. (2007). La gestion communautaire de l'eau agricole à l'épreuve des politiques participatives: regards croisés Tunisie/Maroc. *Hydrological Sciences-Journal des Sciences Hydrologiques, 52*(6), 1179–1196.

Royaume du Maroc. (2011). *Stratégie nationale de l'eau au Maroc* [de 2009].

Salem, A., & Ait Benhamou, Z. (2016). Plan Maroc Vert, le grand mirage. *Le Desk*. 2/10/2016. https://ledesk.ma/

SAM. (2011). *Situation de l'agriculture Marocaine No 10*. Maroc: Ministère de l'Agriculture et des Pêches Maritimes.

Sellika, I. E., & Faysse, N. (2015). Perspectives de production et de commercialisation de la pomme au Maroc à l'horizon 2025. *Alternatives Rurales, 3*, 102–116.

Sellika, I. E., Faysse, N., Bousadni, A., & Ait El Mekki, A. (2015). Perspectives de production et de commercialisation des prunes et des pruneaux au Maroc à l'horizon 2025. *Alternatives Rurales, 3*, 117–130.

Swearingen. (1987). Terre, politique et pouvoir au Maroc. *Revue de l'Occident Musulman et de la Méditerranée, 45*, 41–54.

Tanouti, O. (2017). *La Gestion Intégrée des Ressources en eau à l'épreuve du bassin versant – Le cas du bassin du Tensift au Maroc*. Thèse de Doctorat. Université de Paris Nanterre.

Tanouti, O., & Molle, F. (2013). Réappropriations de l'eau dans les bassins versants surexploités. *Etudes Rurales, 192*, 79–96.

Tanouti, O., Molle, F., Leduc, C. (2016). *Analyse du statut actuel des ressources en eau, de leur gouvernance et des réformes de politiques publiques*. Livrable D2.5. Projet AMETHYST – Rapport sur la gouvernance de l'eau et des réformes de politiques publiques.

Tardieu, H. (2001). *L'Agence de Bassin Oum Er R'bia: le chemin vers une gestion de l'eau globale et décentralisée*. Article pour l'Agence Française de Développement.

Tazi Sediq, H. (2007). *Du droit de l'eau au droit à l'eau au Maroc et ailleurs*. La Croisée des Chemins.

Telquel.com. (2014). *Charafat Afilal détaille son plan national de l'eau*. https://telquel.ma/2014/09/23/charafat-afilal-detaille-plan-national-leau_1416910

Valony, M. J. (2006). Des communautés d'irrigants aux associations d'usagers de l'eau. Un modèle mondial face aux différentes configurations socio-territoriales. In P. Caron, J. Y. Jamin, A. Richard, & T. Ruf (Eds.), *Coordinations hydrauliques et justices sociales*. Montpellier: Cirad.

Van Vuren, G., Papin, C., & El Haouari, N. (2004). *Participatory Irrigation Management: Comparing theory with practice a case study of the Beni Amir irrigation scheme in Morocco* (Projet INCO-WADEMED). Rabat: Actes du Séminaire Modernisation de l'Agriculture Irriguée.

Valette, É., Chéry, J. P., Debolini, M., Azodjilande, J., François, M., & El Amrani, M. (2013). Urbanisation en périphérie de Meknès (Maroc) et devenir des terres agricoles: L'exemple de la coopérative agraire Naïji. *Cahiers Agricultures, 22*(6), 535–543.

World Bank. (1995). *Water sector review. Kingdom of Morocco*. Washington, DC: The World Bank.

World Bank. (1998). *Projet de gestion des ressources en eau rapport d'évaluation. Groupe du développement rural, de l'eau et de l'environnement*. Bureau Régional Moyen-Orient et Afrique du Nord. Rapport 15760.

World Bank. (2008). *Marchés fonciers pour la croissance économique au Maroc. Volume I – Héritage et Structures Foncières au Maroc. Les contraintes structurelles et institutionnelles à l'émergence d'un marché efficient du foncier au Maroc*. Maroc: Banque Mondiale.

World Bank. (2009). Project performance assessment report. Morocco Water Resources Management Project. Report No. 48732.

Chapter 4
Spain

Marta García-Mollá, Carles Sanchis-Ibor, Llorenç Avellà-Reus, José Albiac, Daniel Isidoro, and Sergio Lecina

Abstract The persistent application of a supply-side management model has led Spain to a critical situation, placing irrigation at the center of national political debates. This chapter describes how, to overcome this hydro-political crisis, the administration and the farmers' associations have supported three non-conflicting strategies: introducing water markets to give flexibility to the allocation of water for irrigation, stimulating treated wastewater reuse, and modernizing irrigation systems, particularly through the massive transformation of gravity schemes into drip irrigation. This chapter analyzes these strategies and also provides a critical view on groundwater development for irrigation. Finally, agricultural and environmental policy debates and the expected impact of climate change on the irrigation sector are considered.

Keywords Irrigation policies · Water management · Technological change · Irrigation institutions

4.1 The Hydraulic Mission: Irrigation in Spain in the Twentieth Century

The Iberian Peninsula has been viewed on numerous occasions as a continent in miniature due to its variety of climates and landscapes (De Terán 1949). The territory shared by Spain and Portugal exhibits significant contrasts in terms of temperature and rainfall, which unevenly shape agricultural production and the availability and

M. García-Mollá (✉) · C. Sanchis-Ibor · L. Avellà-Reus
Valencian Centre for Irrigation Studies (CVER), Universitat Politècnica de València, València, Spain
e-mail: mgarmo@esp.upv.es

J. Albiac · D. Isidoro · S. Lecina
Agrifood Research and Technology Centre of Aragon (CITA), Aragon Government, Aragon, Spain

F. Molle et al. (eds.), *Irrigation in the Mediterranean*, Global Issues in Water Policy 22, https://doi.org/10.1007/978-3-030-03698-0_4

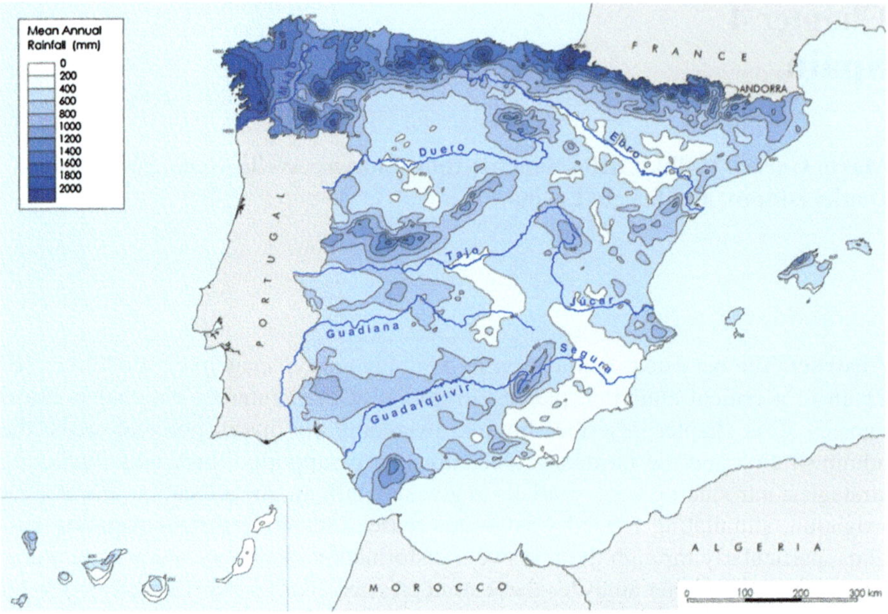

Fig. 4.1 Latitudinal gradient and longitudinal asymmetry of rainfall distribution. (Data from Centro Nacional de Información Geográfica and Agencia Española de Meteorología)

use of water for irrigation (Fig. 4.1 and Table 4.1). Contrasted geographical and historical scenarios have created different irrigation landscapes in the Iberian Peninsula, from the supplementary irrigation of cereals in the central plateau to the intensive citriculture of the Mediterranean façade.

At the beginning of the twentieth century, Jean Brunhes (1902) coined the terms *wet Iberia* and *dry Iberia* to divide the peninsula (Fig. 4.2). Due to the differing conditions of precipitation, the use of irrigation was necessary to guarantee the viability of particular crops or increase their production. An isohyet of 600 mm established the dividing line between these two domains. More than 1 million hectares of irrigated land previously existed to the south of this line: gardens planted during the Islamic period, feudal irrigation ditches, and canals designed after the Enlightenment period. Irrigation, essential for the national food supply, was also the basis for an incipient export-oriented agriculture consolidated at the beginning of the twentieth century (Fig. 4.3).

The distinction between the two Spains, dry and wet, was popularized in the early decades of the twentieth century through educational handbooks, geographical essays, and speeches on water policy. These terms became commonplace and served both to explain the distribution of irrigation on the peninsula and to justify State plans and projects for the expansion of irrigated lands (Swyngedouw 1999).

Table 4.1 Irrigation profiles by region

	Mean farm size (ha)	Main crop	Secondary crops	Water resources %			Irrigation technique %		
				Surface	Ground-water	Others	Gravity	Drip	Sprinklers
Andalucia	7.8	Olive groves	Fruit trees	71.25	27.59	1.16	18.2	72.8	9.0
Cantabric regions	0.6	Grass	Fodder	–	–	–	61.2	7.4	31.4
Duero Valley	10.6	Cereals	Alfalfa, vineyards	81.98	17.66	0.36	32.1	5.4	62.5
Ebro Valley and Catalonia	8.6	Cereals	Fruit trees	93.18	6.53	0.29	48.7	22.7	28.6
Islands	1.9	Fruit trees	Vegetables	–	–	–	12.5	55.6	31.9
Tajo and Guadiana Valleys	12.0	Cereals	Fruit trees	67.02	32.64	0.34	17.4	51.0	31.6
Valencia and Murcia	3.5	Citrus	Fruit trees, vegetables, rice	58.07	34.75	7.17	26.2	73.0	0.9
Total	6.2			76.92	21.50	1.59	27.9	47.8	24.4

Source: INE (2012); MAGRAMA (2013). Cantabric region encompasses Galicia, Asturias, Cantabria, and Basque Country; Duero Valley is Castilla y León; Ebro Valley includes Aragón, Navarra, and La Rioja, and Tajo and Guadiana valleys encompass Castilla la Mancha and Extremadura

4.1.1 "Regenerationism" and Irrigation Policy at the Beginning of the Twentieth Century

At the beginning of the twentieth century, the impact of the loss of overseas colonies (1898) was the catalyst for "Regenerationism," an intellectual movement that proposed an ambitious project to modernize Spain. Led by Joaquín Costa, Regenerationism viewed irrigation as the central axis of the socioeconomic regeneration of the country. The domestication of peninsular waters and their mobilization for economic development became the principal components of national reconstruction discourse. This essentially agrarian proposal also aimed at frustrating contemporary revolutionary movements (Ortega Cantero 1979, 1999; Gómez Mendoza 1992; Gil Olcina 2002).

The *General Plan for the Irrigation of Canals and Reservoirs* of 1902, known as the *Gasset Plan,* was an early manifestation of the change in water policy under the auspices of Regenerationist ideology (Ortega Cantero 1995). This simple listing of

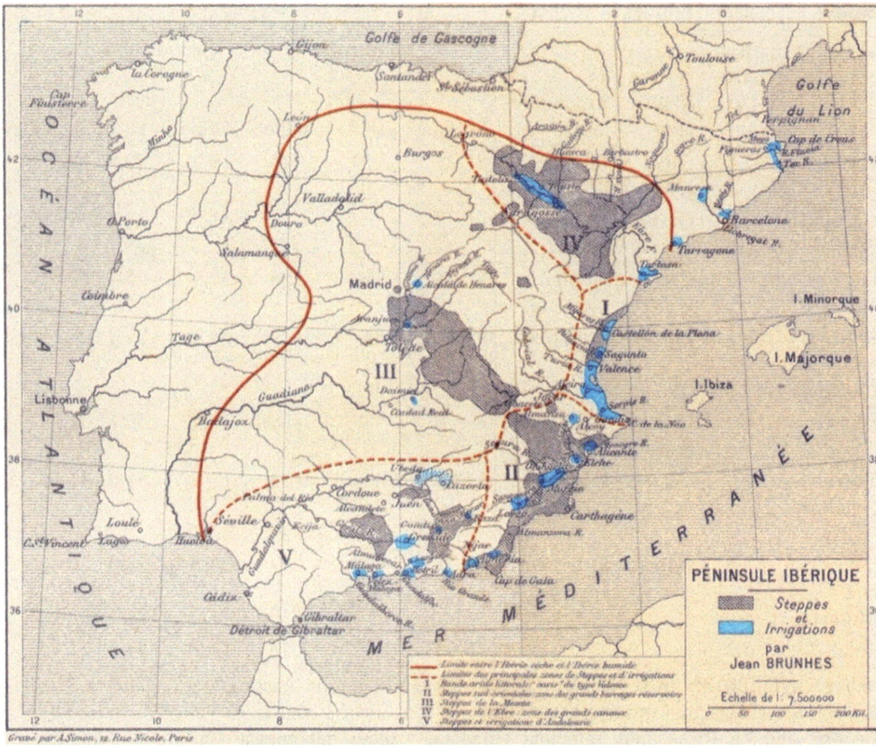

Fig. 4.2 Dry and wet Spain, according to Jean Brunhes (1902)

water projects, 222 reservoirs and 112 irrigation canals, intended to provide irrigation water to 1.18 million hectares but only managed to transform 150,000 ha. The lack of financial support by the state, the opposition of landowners and hydroelectric companies, and the neglect of the Mediterranean coast were some of the causes of its failure.

In light of the failure of this Plan, the first three decades of the century were marked by a substantial increase in irrigated areas due to the proliferation of wells and pumps. In the Mediterranean basins especially, the spread of this technology enabled a significant mobilization of groundwater for irrigation. This development occurred at the margins of the State, given that subterranean resources were considered private property according to the prevailing water legislation. In the Mediterranean regions, the introduction of pumping technology increased the ability to drain wetlands and to lift river waters to areas contiguous to traditional irrigation sites, particularly in places such as the Segura basin, where the first reservoirs built generated a (ephemeral) windfall in the availability of resources. These irrigation extensions were financed by large landowners, small groups of medium-sized farmers who formed private partnerships for the joint exploitation of the new resources, and by large companies with national and foreign capital, such as the Riegos de Levante companies in Alicante province or REVA in Valencia, attached to hydropower projects.

Fig. 4.3 Current distribution of irrigated lands. (Modified from *Ministerio de Agricultura, Alimentación y Medio Ambiente*)

Along with its commitment to the water planning mission, Regenerationism had a strong interest in administrative decentralization and in user participation in water management. In 1926 the *Syndicated Hydrographic Confederations* were created as autonomous organizations at the basin level. The *Confederations*, governed by an assembly of users, hold responsibilities for water planning and management (Cano García 1992; Molle 2009). These institutions were instigated by Regenerationist engineers in the large basins of the peninsula (Ebro, Duero, and Guadalquivir); but in the small basins of the Mediterranean (Segura, Júcar, Turia), the Confederations were lobied by the *sindicatos de riego* (irrigation syndicates), Water Users Associations (WUAs) later known as *comunidades de regantes (irrigation communities)* (Melgarejo 1995; Mateu Bellés 2011; Sanchis-Ibor 2012). In these regions there was a successful tradition of collective irrigation management that had been recognized and consolidated by the water laws of 1866 and 1879.

At the end of this period, Manuel Lorenzo Pardo, instigator of the Confederation of the Ebro, was responsible for drafting the National Water Works Plan in 1933, which, unlike the 1902 Plan, was based on hydrological and agrarian economy studies (Romero 1995). The Plan decisively advocated the expansion of irrigation on the Mediterranean coast through various interbasin water transfers. The government of the Second Republic promoted this Regenerationist project as a means of combating rural poverty, thus allowing the postponement of more drastic agrarian reforms.

4.1.2 The Hydraulic Mission During the Dictatorship

After the Civil War (1936–1939), the totalitarian regime of Franco took over the hydraulic mission of Regenerationism, although abandoning some of the fundamental principles of the ideology of Joaquín Costa. For the dictatorship, the Regenerationist project of modernization was useful inasmuch as it proposed, through public works and strong State control, the development of an ambitious plan to expand irrigation and increase agricultural wealth. However, the Franco regime blocked users from being able to participate in water management and planning. In 1942, a ministerial order suspended the autonomy and participatory structures of the Hydrographic Confederations and established stricter governmental control and a technocratic structure (Mateu Bellés 2011). Totalitarian delirium led the Franco administration to plan the dissolution of the historical *comunidades de regantes* to integrate them into the administrative structures of the State. This initiative was finally stopped due to the opposition of local conservative elites which, led by the irrigators of the Júcar, formed the national federation of irrigators in 1956 (D'Amaro 2014).

The Regenerationist project was furthered through a new plan of water works known as the Peña Plan, which was approved in 1939. This plan was a slightly revised version of the Lorenzo Pardo Plan of 1933 and was assigned a new name to gain political and legal legitimacy in the context of the authoritarian state. This document served as a road map to allow Spanish civil engineers to implement this transformational project over the course of more than four decades. Due to the execution of this plan, between 1940 and 1970, the capacity of Spanish reservoirs increased by a factor of 10, while irrigated land expanded from 1.4 million ha to 2.2 million ha (Fig. 4.4).

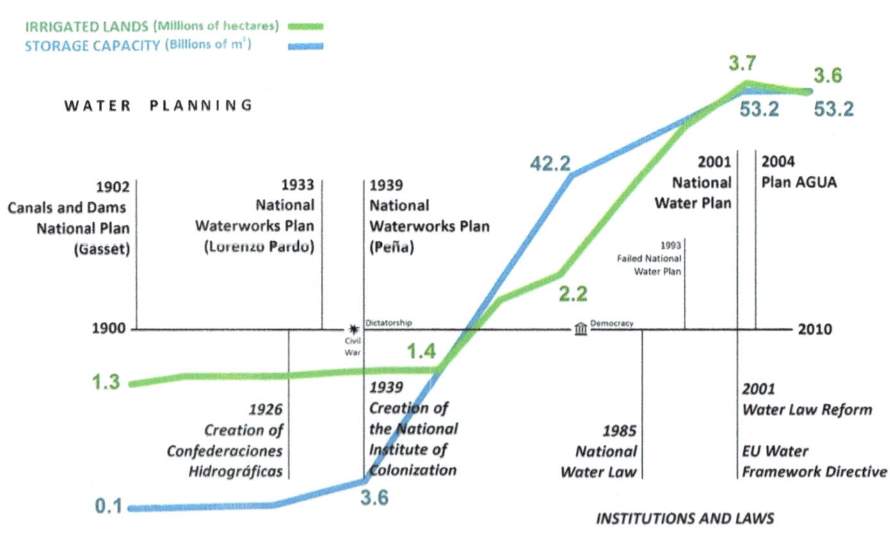

Fig. 4.4 Twentieth century timeline for water policies and development

The National Institute of Colonization (NIC) was created in 1939 to carry out this hydraulic mission. The NIC directed the expansion of irrigation and created all types of rural infrastructure, ranging from irrigation canals to population settlements on newly irrigated lands. The Badajoz Plan in the Guadiana River basin, which created 100,000 irrigated hectares, is the clearest example of this internal colonization. In the 1960s, NIC efforts shifted to promoting the use of groundwater by drilling well fields that enabled the expansion of irrigation in areas located from the reach of large irrigation canals. The development of irrigation with groundwater was not limited to the actions of the State, and substantial development was driven by small and large agricultural landowners, especially on the Mediterranean side.

The most significant component of Lorenzo Pardo's original project was the Tajo-Segura Water Transfer. It was designed to carry waters from the Atlantic basins to the Mediterranean to expand irrigation in the provinces of Alicante and Murcia (Fig. 4.4). In 1967, when the Minister of Public Works announced the initiative in Murcia, the aim of the project was to transfer 1000 mm^3 each year. The transfer was planned to improve irrigation in 147,000 ha with insufficient water supply and to extend irrigation to 122,000 ha of dry land. However, the average volume historically transferred has been only 348 mm^3, just a little more than a third of the canal capacity. Consequently, expansion of the irrigation areas was achieved with considerably less water resources than planned, contributing to an important structural deficit (Avellà 2003; Sanchis-Ibor et al. 2011).

4.1.3 The Unsustainablity and Impacts of Supply-Side Policies

Spain provides an excellent example of continued application of supply-oriented water management (Carles and García-Mollá 2003). Policies to expand water infrastructure were implemented consistently and included public financing of large water projects, negligible prices for water, scant control of actual use, very high consumption, and lack of control of water quality. In technological terms, the twentieth century was the century of concrete. Between 1900 and 2000, 1137 reservoirs were constructed with a total capacity of 56.4 billion cubic meters (Fig. 4.4). Cement was used profusely in the construction of new irrigation canals, which increased the irrigated area by more than 2 million ha, and in lining a large portion of traditional irrigation channels, beginning in the 1950s. The use of new technologies to collect and pump groundwater also played an important role in the development of this model (Fig. 4.5).

The Spanish hydraulic bureaucracy achieved the water goals of Regenerationism and its socioeconomic objectives. Additional objectives were also achieved: the security of urban supplies in a context of intensive urbanization; increased food security; and improved social conditions in rural areas, primarily due to the expansion of irrigation. However, the implementation of these policies led to water resource degradation in the basins of many regions and the overcommitment

Fig. 4.5 Main interbasin transfers planned in Spain
Note: The Tajo-Segura transfer and the abandoned Ebro-Southeast transfer project, connecting several basins to boost the agricultural expansion of the Mediterranean seaboard. The other two smaller transfers were built in the twenty-first century. The Negratín-Almanzora has operated successfully during the last 15 years, but the Júcar-Vinalopó is underutilized and socially conflict-prone due to the Jucar basin closure.

of basin resources and ecosystems degradation, definitively creating situations that have been defined as *basin closure* in other regions of the world (Molle et al. 2010).

In the closing decades of the 20th century, numerous regions saw increases in demand (agricultural, industrial, water supply, etc.) to levels that were significantly higher than supply. Obtaining new resources – at the time and place desired – became more complex and faced growing marginal costs, which hindered public financing of hydraulic works. The permanence of old water rights increasingly generated conflicts between irrigation and new users. Freshwater quality deteriorated, with severe impacts on wetlands and riparian ecosystems. The first notable scarcities ocurred during the droughts of the 1970s, when serious supply problems occurred in numerous cities, particularly in Andalusia, with severe hourly restrictions in the distribution of drinking water. In some basins, such as the Segura River, the overexploitation of river resources caused a water shortage for irrigation canals, particularly in the lower part of the river, causing the bed to run dry during several months of the year.

The intensive and uncontrolled extraction of groundwater led to a paradigmatic "tragedy of the commons" situation in numerous aquifers (Sevilla et al. 2010). Many of the aquifers located in the southern half of the country were officially declared overexploited by the administration and, in some of these areas, such as the wetlands of Las Tablas de Daimiel National Park, serious environmental effects were observed (Hernández-Mora and López-Gunn 2001). In addition, most coastal aquifers in the Mediterranean had problems with marine intrusion, which in many cases led to the salinization of water resources for irrigation (Gómez Gómez et al. 2003).

Lack of control over water resource quality also seriously impacted the natural environment and agricultural production. Irrigation ditches were the cause of a large portion of this contamination, as they had also been used for wastewater disposal. This connection, which had been very useful for preindustrial agriculture lacking nutrients, led to enormous complications after the chaotic urban-industrial development of the 1960s and 1970s. Numerous peri-urban irrigation canals began to receive huge quantities of pollutants, exceeding the tolerance thresholds of associated aquatic ecosystems and leading to the eutrophication of Ramsar sites such as L'Albufera de València (Vicente and Miracle 1992; Verdú et al. 1999). In the Segura River, water reached alarming concentrations of mercury, nickel, tin, ammoniac, and phenols, while biological oxygen demand exceeded legal levels (oxygen injectors had to be installed in the river in some urban areas to reduce the stench). Watercourse pollutants, coming from traditional irrigation drainage systems, were disseminated through irrigation canals, increasing soil and water salinity across the entire floodplain (Sanchis-Ibor et al. 2011).

4.2 The Hydraulic Transition: Renewing Institutions, Policies, and Technologies at the Turn of the Century

4.2.1 The Advent of Democracy: A New Framework for Irrigation

With the advent of democracy in Spain, the Constitution of 1978 created a new State structure in which power was shared between the central State administration and the regions (*autonomous communities*). Thus, during the 1980s, the water administration of the young democracy had to face a double challenge. On the one hand, it was necessary to develop a model of demand management due to the critical situation generated by the continued application of the supply-side model during the twentieth century. On the other hand, the new decentralized structure of the State complicated the management at the level of the hydrographic basin that had previously been carried out by the state administration through the Hydrographic Confederations (Fig. 4.6). The Water Law of 1985 addressed these challenges, prioritizing the hydrographic basin over the regional subdivisions, and creating a legal framework directed toward improved efficiency of water resource management.

Fig. 4.6 The current nine Hydrographic Confederations are represented in orange with black boundaries. White boundaries divide the regions (autonomous communities). In yellow, internal basins transferred to regional governments: *1*. Galicia; *2*. Basque Country; *3*. Catalonia; *4*. Andalucía

The significant changes in water policy implemented with the Water Law required deep restructuring of the administrative bodies, which took many years.

According to the Water Law of 1985, public activities and private water uses had to be planned through the *National Hydrologic Plan* (Plan Hidrológico Nacional, PHN) and the hydrologic plans for each basin. The Preliminary National Hydrological Plan was presented in 1993. The Plan intended to connect all of the basins in the Iberian Peninsula with a water transfer capacity of 4000 mm^3 in order to promote irrigation consolidation and expansion (Gil Olcina 2002). The proposed huge investments and enormous volume of water transfer generated considerable controversy and caused the withdrawal of the Plan. However, at the same time, the National Irrigation Plan (1995) advocated for the creation of 180,000 ha of irrigated lands in the Ebro and Duero basins, and the various basin hydrologic plans stipulated the expansion of irrigation over more than 1 million ha.

In 2001, the conservative government introduced a new Hydrological Plan with the same spirit as the previous one but with a lower volume of works. The most controversial idea in this plan was the proposal to build a water transfer from the Ebro to the regions of the southeast, which were characterized by structural water deficits (see Fig. 4.6). In addition, the Appendix II stipulated the construction of more than one hundred new reservoirs. For these reasons, this Hydrological Plan was also the target of numerous critiques by social, political, and environmental organizations.

In the same year the Water Framework Directive (WFD) imposed a shift in water policy in Spain, emphasizing the eco-social value of water. The transposition of the WFD was very opportune. The WFD prioritized environmental functions over productive functions, subordinating quantitative measures to the goal of guaranteeing good quality of water bodies. This was an enormous challenge for an irrigation sector trained in the development of water supply and an administration created to serve these objectives. The reactions triggered by this regulation were very significant. Some authors and institutions have adopted a critical position, understanding that the WFD spirit and design are not adapted to the Mediterranean, which has extremely demanding agricultural requirements and a hars climate. In contrast, others have viewed the WFD as an external aid and stimulus that is necessary for correcting the direction of Spanish water policies, to overcome the inertia of earlier models and combat social resistance.

4.2.2 How to Meet the Irrigation Demand? Water Transfers Versus Seawater Desalination

The successive shifts in Spanish water planning highlight two fundamental problems mentioned earlier: (i) the difficult transition between two models of management, which was evident in some plans that continued to insist on the massive mobilization of water resources to satisfy irrigation demand through water transfers or through desalinization, and (ii) the additional complexity incurred by the presence of a decentralized political power and an atomization of feelings of identity in the various regions or nationalities of the State (López-Gunn 2009).

The hydrological plans unleashed a political storm in which territorial and socio-hydraulic interests clashed. Regarding the Ebro water transfer, a neoliberal discourse stressing interregional "solidarity" was constructed and permeated deeply into the receiving regions and conservative sectors of society; water was to be disconnected from the territory and placed at the disposition of the users who could make the most productive use of the resource. The regions providing water and left-wing political parties opposed this action; they argued the need for constructing a "new culture of water" based on sustainability, with emphasis on efficiency and environmental principles, before economic development. In both territories, the construction of these discourses effectively linked the debate to questions of identity and political allegiance. Dissent was considered no less than a betrayal of the interests of these regional homelands. Sizeable mass mobilizations formed, along with direct impacts on election results, with crushing majorities for the conservatives in Murcia and Valencia, and victories of the left in Aragon and Catalonia.

In 2004, the Socialist Party won the elections after a campaign partially focused on the Ebro Transfer. The water transfer was canceled (RD Law 2/2004) and replaced by the AGUA Program (Spanish acronym for Actions for Management and Use of Water), whose primary objective was also to increase the supply of water but through desalinization. The alternative offered in 2004 by the Socialist government also did not significantly depart from the supply-oriented model (Swyngedouw 2013). The AGUA Program intended to increase the availability of water resources through the construction and improvement of desalination plants (31 plants with an investment of 1945 million euros to generate an annual supply of 713 Mm^3). The program also sought to increase the technical efficiency of irrigation by changing gravity-based irrigation to micro-irrigation, lining canals, and drainage ditches with concrete and reducing losses in reservoirs and ponds. The goal was to increase water supply by 1063 Mm^3 in regions in the southeast and by smaller quantities in other regions, at an estimated cost of 3.9 billion euros.

The majority of desalination projects were undertaken without establishing prior commitments with irrigation communities to purchase or use the water produced. The larger part of the infrastructure constructed ended up considerably underutilized, resulting in a substantial increase in unit costs (€/m^3) of desalinated water. The few studies conducted to date have shown excessively high costs for agricultural use (to which one would likely need to add costs for pumping or discharge). According to Lechuga (2007), the cost would be 0.45 €/m^3, whereas Martínez Vicente (2009) estimated prices between 0.53 €/$m^3$and 0.69 €/m^3. According to Lapuente (2012), the costs of seven desalinization plants (Aguilas, Valdelentiscos, S. Pedro del Pinatar I and II, Torrevieja, and Alicante I and II) would be 0.75 €/m^3 for 80% output and 1.2 €/m^3 for 40% output.

Now that many years have passed since the AGUA Plan was put into effect, the use of desalinated water for irrigation continues to be irrelevant, and today is primarily destined for urban supply in various coastal zones of the country. Only in the intensive horticulture of Murcia and Eastern Almeria is desalinated water regularly used for irrigation, and it has been recently introduced in the Western Almeria area. However, the administration and the farmers' associations have a growing interest in the mobilization of this resource for irrigation. During the last dry period (2015–2016), the government subsidized the price of this resource in some areas of Murcia and Alicante to avoid water shortages (finally distributed at 0.30 €/m^3) and has forced the use of desalinated water in some irrigated districts that were supplied through exceptional interbasin transfers in previous drought events (Morote et al. 2017). Moreover, in Escombreras (Murcia Region), farmers are planning a new desalinization plant that will be based on photovoltaic energy, constructed by a private company and fully paid by the irrigators (Sanchis-Ibor et al. 2018).

4.2.3 Wastewater Reuse in Irrigation

The failure of the interregional water transfer plans and the underutilization of the AGUA Program infrastructures stimulated the use of treated wastewater. The efforts developed by public sanitation and water treatment agencies over recent decades have created a new resource for agricultural purposes, in addition to improving the health of aquatic ecosystems. The success of these policies is due to various factors: (i) they contribute significantly to reducing the environmental externalities of water management, solving some problems and reducing the impacts caused in past decades by supply-oriented policies; (ii) they achieve the goal of improving efficiency at the basin scale and managing demand; (iii) they are low cost for agricultural users, which in the case of the Valencia region is estimated to be approximately 0.22 €/m^3 for direct users (in fact, there are many examples of concessions at considerably lower prices, as well as of indirect users who benefit from return flows to rivers and aquifers); and (iv) they do not lead to territorial or political conflicts but instead to broad social consensus.

Since 2007, a specific legal framework has regulated both the applications and quality of wastewater being reutilized, and the Hydrographic Confederations have delivered administrative concessions to users. In most cases, farmers do not make direct use of these resources but mix them with other water sources to reduce water salinity and to ensure a better irrigation quality. In other cases, small desalination plants treat this water for agricultural purposes.

To estimate precisely the exact scope of agricultural reuse is difficult because the administration does not collect disaggregated data and computes these resources with other non-conventional resources. According to Melgarejo (2009), 261 Mm^3 of water was reused for irrigation in Spain in 2008, of which approximately 60% (197 Mm^3) was used in the Murcia and Valencia regions. This only amounts to 1.6% of the national agricultural demand but equates to 4.7% in the regions of Murcia and Valencia.

Outside the irrigation sector, between 2001 and 2010, treated wastewater used for agriculture, industrial, or environmental purposes increased from 230 Mm^3/year to 450 Mm^3/year. Recently, the Ministry of the Environment launched an ambitious plan to increase wastewater reuse between 2010 and 2015, hampered by the financial crisis. The plan projected to increase reutilization for agricultural purposes by 123.9 Mm^3, of which 91.3 would be reused in the regions of Murcia and Valencia.

The potential of this non-conventional water resource is still important, because in many regions reutilization lags behind compared with the coastal zones of the Mediterranean. Furthermore, it is necessary to increase quality control to avoid accidents regarding food safety. Making the reuse of these waters possible would also necessitate training plans for farmers and strict control of the quality of water to be reused (including online access for users and the use of random sampling).

4.2.4 Technology and Efficiency: Modernizing Irrigation Systems to Reduce Agricultural Water Demand

Throughout the last decades, the modernization of irrigation has long been promoted by public administrations to reduce water consumption in the agricultural sector. Agriculture was the primary recipient of water resources in Spain (more than 70% of total water use) and an important portion of the irrigated area had outdated structures. Traditional surface irrigation amounted to 59% of the total irrigated area, and 71% of this area used structures more than 25 years old. As a consequence, land and water productivities were poor, labor requirements were high, working conditions were hard, and negative environmental impacts were significant (MAGRAMA 2002). In the 1980s and the beginning of the 1990s, the public promotion of irrigation modernization was moderate and mainly focused on the concrete lining of channels and the rehabilitation of historical irrigation systems. Sprinklers expanded over the meadows of Castilla and Andalusia, frequently introduced in newly irrigated lands. This technology reached a peak of 905,000 ha in 1999, after three decades of continuous growth.

The severe drought of 1994–1995 and the failure of the 1993 National Hydrologic Plan created a new scenario, characterized by the involvement of the public administration in the dissemination of water-saving technologies and also by the deployment of drip irrigation. The first step in this direction was made by the regional governments of Andalusia (Decree 97/1995) and Valencia (Decree 13/1995). Later, the National Irrigation Plan (2002) and the Plan for Sustainable Irrigation (2008–2013) aimed to convert more than 2 million ha to micro-irrigation, with a total investment of more than 5000 M€, of which public investment represented approximately 3400 M€ (Naranjo 2010).

The need to increase the reliability of water supply has been the major driving force behind the incorporation of these technologies (Escribano 2006; Corominas 2008; Alcón et al. 2009), together with improving water productivity (Gallego Bono 1996; Gómez Limón et al. 2007; Gil Meseguer 2010; Gómez Espín et al. 2007a, b). The introduction of water-saving technologies has generally been more intensive in regions with a more pronounced imbalance between resources availability and use and more rapid in areas using groundwater where farmers aim to reduce pumping costs. In addition, irrigators have deeply valued the comfort and reduction in labor costs. The possibility of adopting irrigation on-demand systems – with the consequent end of night irrigation – the introduction of fertigation, and the reduction in land preparation works prior to irrigation, has led to a notable reduction in work hours and a greater compatibility with private life and other economic activities. Consequently, an alliance has formed between users, administrations, professional associations, and private companies that benefited from the installation and maintenance of new infrastructures.

This alliance disseminated a powerful discourse on irrigation modernization, supported by the water savings estimated in scientific studies on irrigation efficiency, the attractiveness of the notion of modernization, and the need to restore the image of some irrigation systems often unfairly labeled as wasteful. The modernization discourse had clear political articulation in some regions, where efficiency and productivity of irrigation became a banner for those demanding the transfer of water resources, such as the Ebro water transfer planned in 2001. The zones receiving the transferred resources had the moral obligation to demonstrate that they were using every last drop of water before hoping to obtain new resources. Thus, irrigators from Valencia and Murcia regions have been continuously politically active to demand the Ebro water transfer and have been working to improve the efficiency of their water use (Sanchis-Ibor et al. 2016).

Consequently, an unprecedented technological shift in Spanish irrigation occurred. In 1978, only 3500 ha used drip systems, half of which was in the Canary Islands (Medina 1988). By 2015 1,727,341 ha had drip networks installed, making up 44% of the total irrigated lands, mainly in the Mediterranean regions. Only India has a larger area equipped with microirrigation systems (ICID 2016). In turn, sprinkler irrigation, which is not used on the Mediterranean coast, has not increased during the last 15 years. Recently, the financial crisis has slowed public works and private initiatives. Gravity irrigation decreased annually by 3.94% in the period 2002–2008 and by a further 1.39% during 2009–2015. The use of pressurized irrigation increased by 6.61% and 2.06% during these periods, respectively (Fig. 4.7).

The irrigation modernization plans aimed at enhancing water conveyance, delivery, and application. At irrigation district and plot scales, the modernization process has been mainly focused on switching from gravity to drip or sprinkler systems. Networks of ditches were replaced by pressurized pipe networks. Small reservoirs were constructed at the head of the new networks in order to ensure on-demand water delivery to the plots. Pumping stations were often needed to extract water from aquifers and/or meet the pressure requirements of the new drip or sprinkler irrigation systems. Only in some cases, land consolidation processes were conducted before modernization in order to increase plot size.

There have been some exceptions to this general modernization procedure: historical irrigated areas such as the *huertas* of Murcia and Valencia. These lands present specific features: (i) high-value crops (i.e., orchards and vegetables), (ii) very small land plots, and (iii) soils suitable for gravity irrigation. As a consequence, these systems have been modernized by improving their gravity irrigation struc-

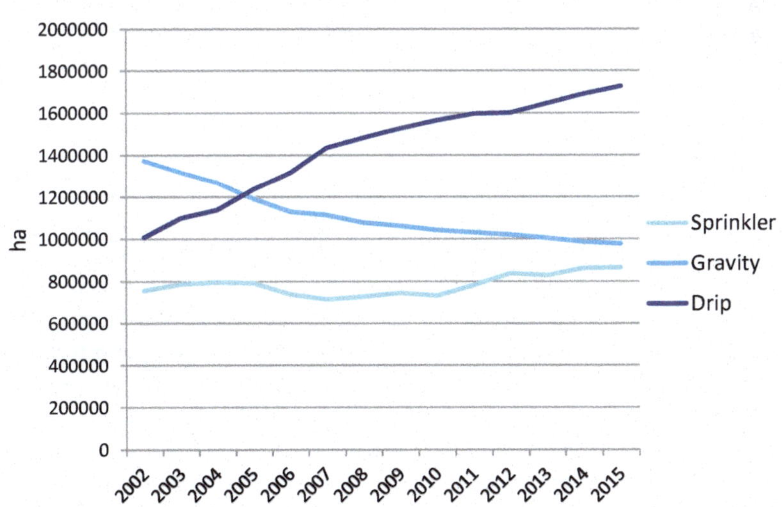

Fig. 4.7 Recent evolution of irrigation techniques in Spain (ha). (Source: ESYRCE (MAGRAMA))

tures. Ditches have been massively lined or replaced by low pressure pipes to increase their water delivery capacity and facilitate their operation and maintenance. In coastal wetlands, traditional rice farming has been improved by installing pumping stations to reuse return flows (García-Mollá et al. 2013).

In addition to the improvement of irrigation structures, the enhancement of management tools and procedures has also been addressed. Telecontrol has been widely installed to monitor and operate irrigation structures in real time from conveyance canals and in-line reservoirs to water delivery networks and drip and sprinkler irrigation systems (Damas et al. 2001). Consequently, labor requirements have been reduced and working conditions have been improved in farms and irrigation districts. Furthermore, the telecontrol systems have been combined in some cases with specialized software based on Geographic Information Systems (GIS) to enhance the operation and maintenance of irrigation structures, and the administrative tasks of irrigation districts (Playán et al. 2007).

However, different studies have questioned the validity of pressurized systems as a method to achieve reduction in water consumption and have particularly highlighted the presence of unexpected effects on water and energy consumption (López-Gunn et al. 2012b, c). With respect to saving water, some authors (Playán 2002; Lecina et al. 2010a, b; Playán and Mateos 2006; Perry et al. 2009; Cots 2011; Sese-Minguez et al. 2017; Corominas and Cuevas-Navas 2017) have documented various scenarios in which the introduction of water-saving technologies could have an effect opposite to that intended, the so-called Jevons paradox or rebound effect. First, the modernization of irrigation channels could cause an intensification of crops toward varieties that are more productive and more water demanding, thus increasing evapotranspiration and water consumption.

Second, the water saved locally may often be redirected for meeting new demands. When the obtained water savings envisaged through modernization projects are viewed as water resources freed (or dry water savings), the administration can cause the *rebound effect* by reallocating water to new users. These sort of "paper water" transfers have been seen as a technical solution in "closed" basins such as the Júcar River, where the Basin Plan of 1998 stipulated the allocation of the water resources saved through the modernization of traditional irrigation systems to the Júcar-Vinalopó transfer project, for irrigation and urban supply purposes.

Third, water savings can be used to expand the area irrigated; thus, consumption at the scale of the basin would increase, although in some cases, its economic use would be enhanced. In several regions, public subsidies have been used to revive old irrigation expansion aspirations or to launch new irrigation plans. This is the case of the Andalusian, Valencian, or Castilla La Mancha olive groves or almond trees, where drip technology installation has become a necessary condition for the transformation of traditional rain-fed areas into new groundwater irrigated lands. As a result of this process – and due to the low requirements of irrigated olive groves – in regions such as Andalusia, the average amount of irrigation water required per hectare has decreased, but the total amount of water consumed has increased (Gómez-Limón et al. 2013).

However, in some areas where irrigation expansion has not taken place and no intensification or crop changes have happened, substantial reductions of water withdrawals have been observed. Thus, Sanchis-Ibor et al. (2017), comparing the average decrease in water use in 77 modernized and not modernized WUAs in the Valencia region, attributed to the introduction of drip irrigation a 26% mean decrease in water applied between 1998 and 2009. Borrego-Marín and Berbel (2017) observed similar results (in water use and water costs) in nine irrigation schemes in Andalusia. As Dumont et al. (2014) have stated, the rebound effect is not a necessary consequence of drip irrigation installation; it can be avoided if the irrigated area and crop varieties remain under control and accompanying measures, such as water rights revisions, are implemented.

With respect to energy, Hardy and Garrido (2010) argued that the modernization of irrigation in Spain has brought with it an increase in farm consumption of electricity, against a backdrop of rising energy prices. Rodríguez-Díaz et al. (2012a, b) also observed this increase in some Andalusian districts, whereas others (Gómez Espín et al. 2007a, b) detected significant savings. These results are not contradictory. The impact of these costs could vary regionally, and it is frequently related to the origin of water resources. Sanchis-Ibor et al. (2017) identified a different behavior between groundwater and surface water-based drip irrigation areas in the Valencia region. Significant increases in water costs are found in surface irrigated areas, however in groundwater areas, the decrease of abstractions after drip installation leads to a slight decrease in water costs. This calls for putting energy efficiency considerations at the center of any modernization policy.

Researchers identified a gap between the potential and actual performance of the new structures (Zapata et al. 2009; Moreno et al. 2010; Salvador et al. 2011; Rodríguez-Díaz et al. 2012a; Rocamora et al. 2013; Stambouli et al. 2014; among others). While the most modern irrigation technology is adopted, in many cases it is not harnessed to maximize the economic profitability of the farms. Telecontrol systems are a paradigmatic case. This technology is mainly used to monitor on-demand water use and facilitate administrative tasks such as water billing. Nevertheless telecontrol systems can also be used to automate irrigation and optimize water use and energy consumption. This is particularly important to cope with rising water and energy prices induced by the Water Framework Directive and the liberalization of the electricity market.

The technological shift is also generating some positive environmental externalities. In the Ebro River basin, the modernization of the irrigation system of La Violada Irrigation District (converting a traditional gravity irrigation system with fixed irrigation turns into an on-demand sprinkler irrigation system) has resulted in a higher water-use efficiency and an increase in yield but also an improvement in N use efficiency and therefore reduction of off-site pollution induced by irrigated land (Jiménez-Aguirre et al. 2014; Barros et al. 2012; Isidoro et al. 2006a, b).

4.2.5 Groundwater Exploitation and Aquifers Governance

Following the massive mobilization of groundwater for irrigation throughout the twentieth century, several authors (Llamas and Garrido 2007; Sahuquillo 2009; Llamas et al. 2015) have advocated a reasoned use of groundwater in Spain as an alternative to surface water transfers or seawater desalination. According to these works, and under a proper institutional framework and a correct social and environmental assessment, water mining practices and conjunctive use of groundwater and surface water resources – through artificial recharge or through alternate conjunctive use – would be instrumental in meeting present and future water demands in several regions. Optimal solutions are, however, extremely site-dependent, both in ecological and socioeconomical terms (Molinero et al. 2013).

This option faces important obstacles, some of them resulting from the historical division between the management of surface water (public) and groundwater (private), the so-called hydroschizophrenia (Llamas 1975). The Water Law of 1985 changed this legal framework, declaring all water sources to be public regardless of their nature; however, groundwater uses established prior to January 1, 1986, preserved their private status, temporarily or in perpetuity.

Since then, the administration has launched various programs to register private groundwater use. In 1994, the ARYCA program made an inventory of private wells, with poor results. In fact, at the end of 2001, the Segura Hydrographic Confederation estimated that there were approximately 10,000 ha of extralegal irrigation in the basin. In 1995, a study estimated that 20,350 wells were used for irrigation in the area overseen by the Segura Hydrographic Confederation. This number differs markedly from the 4500 that had been declared to the Confederation in 2001, of which 2574 were registered, providing evidence of a relative impunity derived from the absence of oversight and enforcement mechanisms (Avellà and García-Mollá 2008).

The second program, called ALBERCA, has registered more than half million of wells (49.000 per year) since 2001. Despite this effort, and the advances obtained in several regions, Llamas et al. (2015) estimate that one million of existing wells have not been included yet in this national inventory and report a "hydrological insubordination" in several aquifers, where illegal wells have been recently drilled. The lack of funds and personnel seems to be hindering administrative control, negatively affecting the severely overexploited aquifers.

The 1986 Water Law created additional instruments to provide better public control of groundwater, such as the declaration of "overexploited aquifers." Under this legal provision, the *Confederaciones Hidrográficas* have to order the creation of a Water Users Association including all the individual and collective users of the area and to approve an exploitation plan to achieve aquifer recovery. According to the law, this exploitation plan must regulate the regime of extractions and can establish the replacement of individual wells with wells controlled by the new collective institution created for the management of the aquifer.

This legal measure has been successful in some cases but insufficient in others. The region of La Mancha is a good example. On one hand, in the Eastern Mancha, irrigated lands increased from 20,000 ha to 100,000 ha in 30 years, causing severe aquifer depletion and decreasing natural discharge in the Júcar River. In 1995, water users were compelled by the administration to create an association for the collective management of groundwater resources. The support of local users led to an agreement between the aquifer irrigation association, the central government, and the Júcar basin authority to implement sustainable management. The agreement was based on the recognition of water uses prior to 1985 (when groundwater was not public domain), the regularization of uses that started between 1986 and 1997, and a process of characterization of uses and control of extraction. This process is based on monitoring by remote sensing and individual cultivation plans provided by each farmer. The key for this system to work is that the farmers themselves are involved in the enforcement and control process. The efforts of the Water Users Associations, together with the support of the basin authority and the state government, have resulted in a reduction in abstraction during the 2000s – from 400 hm^3 to 320 hm^3 – also induced by crop changes (López-Gunn and Rica 2012; Sanz et al. 2015).

On the other hand, in the Western Mancha, the degradation of the wetlands of the Tablas de Daimiel National Park because of aquifer depletion required additional measures to achieve a better water governance (Hernández-Mora and López-Gunn 2003; López-Gunn et al. 2012a). Overexploitation began in the mid-1970s when new irrigation systems and cultivating techniques were introduced, under the private regime of groundwater extractions prior to the 1985 Water Law. In 1991, the basin agency imposed strict quotas on extractions from existing wells and prohibited the creation of new wells (Varela-Ortega 2007). This created serious social opposition and sparked the creation of illegal wells, thus increasing extractions. In 1993, an Income Compensation Plan was adopted that guaranteed farmer income in compensation for the losses due to water restrictions. This program was quite successful. However, the high cost to public funds (Rosell 2001) was unsustainable and in 2003 the ICP was modified. Water quotas were reduced well below those included in the previous plan, while compensation payments barely covered income losses. As a result, most farmers abandoned the program (Varela-Ortega 2007).

Subsequently, the Special Plan for the Upper Guadiana River was passed in 2007 to reduce the overutilization of the aquifer. The Plan proposed an investment of 5.5 billion euros to reduce overexploitation by 220 Mm^3. Basically, the plan included the transformation of private water rights into concessions of public water, through the purchase of water utilization rights and its allocation to new users or environmental purposes. The plan was only partially developed due to the financial crisis and was widely criticized due to the lack of transparency and the uncertain efficacy of the measures implemented. However, after 2011, the coincidence of several wet years and the decrease of abstractions – linked to changes in the Common Agricultural Policy – the wetlands ecosystem has significantly recovered (Rodríguez Cabellos 2014). This recent groundwater recharge masks the complexity of managing groundwater overabstraction and the relative powerlessness of state policies hitherto (Closas et al. 2017).

The significant government focus on groundwater quantity problems contrasts with a lack of interest in groundwater quality (Esteban and Albiac 2012). The legislation was only modified after 2012 to consider the "water bodies in risk of not achieving a good quantitative or chemical state." According to the basin agencies, only 392 of the 730 groundwater bodies are now in good state. This situation is particularly alarming in the Guadiana and Segura basins, with the 75% and the 68% of the groundwater bodies, respectively, affected by pollution (De Stefano et al. 2012).

4.2.6 Making Allocation Flexible: Changes in Rights, Markets, and Prices

Traditionally, economic instruments such as water markets and water pricing have rarely been used for water allocation in Spain. Water trading has only taken place occasionally, mostly through informal markets. The approach to water allocation has been mainly institutional, based on the public water concession regime and the cooperation of stakeholders with basin authorities. However, in the last two decades, these neoliberal measures have achieved a significant consensus among experts, perceived as a new formula to face problems caused by scarcity. However, the scope of these policies has so far remained limited.

The current system of water pricing is based on public and private tariffs that are essentially what users pay for the use of the water. Public prices include regulation tariffs (resource management fee) paid by beneficiaries of works for the regulation of surface or groundwater conducted by the State and the water-use fee paid by users of other public works. The tariffs and fees are calculated and billed annually. The prices for private irrigation, which users pay to the Water Users Associations, include the public prices cited above plus those belonging to each entity, costs of amortization of the investments they have made, expenses for the maintenance of their own facilities, energy costs for obtaining water (if applicable), costs of personnel and administration, and other costs, subtracting subsidies received.

Tariffs and fees vary widely among Spanish irrigators as a function of the origin of the resource, its size, the method of water application, etc. They have never been used by the administration as a means of reducing demand. This idea was proposed to the National Water Council in 2006 but was widely rejected by regional governments and Water Users Associations representatives. The use of small penalties as a mechanism to stimulate the adoption of water-saving technologies has only been documented within some irrigation schemes in Valencia that were shifting from gravity irrigation to drip irrigation systems (García-Mollá et al. 2014).

With regard to water allocation, the Water Law of 1985 did not permit the sale of water nor the existence of markets but enabled the revision of irrigation water concessions that were excessively generous. Water remained strictly limited to the uses and places indicated in the concession document, without the possibility of changing them. Thus, if the size of irrigable area or other conditions stipulated in the concession changed, the basin authority could reduce the volumes authorized. Further, the basin hydrological plans allow the government to cut down or cancel the concessions of public waters in situations of extreme drought or overexploitation. The Hydrological Planning Regulation of 2007 established that the calculation of authorized volumes should account for the efficiency of irrigation systems, for the possibility of reusing water, and also for any water savings resulting from changes in land area or crop type, adoption of new irrigation techniques, or infrastructure improvement.

The Watermaster Offices of the basin authorities are responsible for reviewing the characteristics of officially registered water uses, especially in the Mediterranean regions, including the irrigated area of each group of crops, technical efficiencies, net and gross maximum endowments (m^3/ha-year), and maximum instantaneous flow (l/s) for each intake. However, Serrano (2008) emphasized that revisions and changes in the due date of public allocations only began to take effect in 2004, primarily in the confederations of the north, despite the broad changes in land use and irrigation technology that have occurred in the last two decades. Currently, these reviews are slowly progressing in basins in the south, due to the lack of personnel and financial support, despite the fact that in Guadalquivir, Segura, and Júcar, significant shortfalls exist between the volumes granted by the administration and actual river resources (Del Moral 2011).

The reform of 1999, incorporated in 2001 in the Water Law, introduced significant changes in water rights in order to relax resource allocation and seek greater efficiency. Since then, the legislation allows the temporary sale of water-use rights. Two instruments are used to relax concessions, the *transfer of rights* and *centers for exchanging rights*, whose regulation has been improved with slight legislative modifications in 2003 and 2013.

Traditionally, *informal markets*, without authorization or oversight, have existed in Spain for decades and were described for the first time in the Canary Islands (Aguilera-Klink and Sánchez García 2002). There is a great diversity and typology of agreements outside of public administration, some originating centuries ago and most linked to irrigation (Hernández-Mora and De Stefano 2013). These agreements generally appear in contexts in which there is significant pressure on water resources, and the administration has tolerated them by viewing them as flexible local solutions that are sustainable and well fitted to their respective contexts (Gómez-Limón and Calatrava 2016). Since 2001, with the reform of the Water Law, some of these informal markets have become legally regulated; however, numerous exchanges remain entirely informal in nature.

Temporary Rights Transfers were introduced into the law in 2001. These exchanges are exclusively temporary between concessionaries or owners of water-use rights via contracts between interested parties that must be authorized by the Hydrographic Confederation. The price of the resources is agreed upon between the two parties, although the administration may impose a maximum value to prevent speculation. By 2011, the *Temporary Rights Transfers* in Spain had only exchanged 196 Mm^3, and prices ranged from 0.15 €/m^3 to 0.30 €/m^3 for intrabasin transfer contracts and from 0.18 €/m^3 to 0.25 €/m^3 for interbasin contracts (Palomo-Hierro et al. 2015).

The amount of exchanges after 2011 has been negligible. In almost all the cases, irrigation in Almería and Murcia (Southeast Spain) has been the destination of the transferred water.

The *Centers for Water-Use Rights Exchange*, known informally as water banks, after California, enable the temporary or permanent exchange of rights and are organized through the publication of *Public Offers for the Acquisition of Rights*. The basin authority must publish the offer of acquisition, describing the volume, prices, technical characteristics and criteria, and term (temporary or permanent), and approve the exchanges once the bids of the parties are received. The *Centers for Water-Use Rights Exchange* only have managed the exchange of 169 Mm^3, and the bulk of it (77.9 Mm^3) through public offer of acquisition at the Júcar Basin Authority, published during the 2006 to 2008 drought at prices ranging from 0.13 €/m^3 to 0.26 €/m^3 (Ferrer and Garijo 2013). More than 80% of this transaction volume was acquired by basin organizations for environmental purposes (Palomo-Hierro et al. 2015), mainly to solve the impact of drought on rivers and wetlands.

In Spain, both the use of contracts for ceding rights and exchange centers is very limited, and these mechanisms have essentially been activated during periods of drought only. Most experts have positively assessed the implementation of these measures, although criticizing the lack of transparency of these operations and the lack of competence in the markets (Gómez-Limón and Calatrava 2016). Despite the positive overall balance, Hernández-Mora and Del Moral (2015, 2016) have identified some cases of water transfers to the Segura basin with negative impacts on some ecosystems of the Tajo basin. They have analyzed this experience critically in the context of a general "marketization" process of water management in Spain and have highlighted the necessity of developing a better social, environmental, and economic assessment of these exchanges.

4.3 Future Challenges

4.3.1 Agrarian and Economic Challenges: The Role of the New CAP

The mean net margin per hectare for irrigated lands is 4.4 higher than in rain-fed areas (Gómez-Limón 2014; MIMAM 2007), but despite this difference in competitiveness, the irrigation sector also faces substantial economic difficulties. Input prices have recently shown substantial volatility and a general rising trend, causing a significant variability in the income of agricultural producers. Technology has been the basis for the recent improvements in efficiency and productivity but also the cause of the rise in the agricultural inputs. In numerous irrigation holdings, modernization processes have led to an increase in energy consumption, which has taken place in a context of rising tariffs. According to Fenacore (2014), the recent reforms of the electricity tariffs have resulted in an 80% increase of the agricultural energy costs between 2008 and 2014 (from 389 M€ to 700 M€), and energy costs currently make up 40% of the irrigation costs.

The viability of irrigated agriculture is also threatened by the low remuneration of agricultural productions in the national and international markets (Atance Muñiz 2013). Increased market prices have not been reflected in farmers' incomes. The instability of the value chain has compounded the uncertainty of an agriculture that is increasingly market-oriented. For these reasons, agricultural producers demand policies that provide a more equitable distribution of the added value in this chain (García Alvarez-Coque and Martínez Gómez 2014).

In this context, the CAP plays a determinant role for several productions of the irrigation sector. The midterm review (2003) decoupled subsidies from farmers' production decisions through a system of direct payments. The last reform (CAP 2014–2020) maintains this payment system and has introduced additional mechanisms to increase equity (modulation) and sustainability (eco-conditionality and greening). It is too early to assess the impact of the new CAP, but several authors suggest that the Spanish irrigation sector is not going to be affected by this new regulation. The pressure exerted by the Spanish government and the agricultural trade unions during the last negotiation process succesfully pushed to maintain the status quo in the internal distribution of the European subsidies, so that the irrigation farmers continue to receive an economic support significantly higher than the rain-fed producers. New measures such as improved modulation or greening are not expected to have significant effects (Colino et al. 2014). Consequently, European subsidies still benefit the most productive agricultural (irrigated) areas, to the detriment of marginal agriculture and rural poorer areas.

The CAP is crucial to support irrigation modernization projects, which have been hampered by the financial crisis of the last decade. The second pillar of the CAP (rural development) still prioritizes the policies of irrigation modernization in the Regional Development Plans (RDP), even though the budget will be reduced. The new CAP has introduced various conditions for the financing of these irrigation investments. New modernization projects must be considered in the new Basin Water Plans, and investments must demonstrate a minimum water-saving potential between 5% and 25% or a reduction of more than 50% in effective water use (if the affected water bodies are not in good state). Some authors (Gómez-Limón 2014) expect that these demanding requirements will reduce the number of new modernization projects.

Nevertheless, the new regulations still allow for the funding of the transformation of rain-fed areas into new irrigation systems or the enlargement of the existing networks, but in both cases conditions are much more strict. Financial support is only permitted for water bodies in good condition. Recent experiences (Sanchis-Ibor et al. 2011; Corominas and Cuevas-Navas 2017; Closas et al. 2017) show that the Spanish administration has not been capable of controlling irrigation expansion processes, and irrigation is still increasing in numerous districts of the country (while decreasing in others). In the coming years, in a likely context of declining water resources, the application of these measures to control the expansion of irrigated lands will become a key challenge for the Spanish administration.

4.3.2 The Adaptation to Climate Change Scenarios

The main difficulty that Spanish irrigation will face is likely to be climate change (CC) and the adaptive strategies that these changes will require. National and international studies quantifying the effects of CC all foresee an increase in temperature and reduction in rainfall (as well as other factors, such as hours of sunlight, wind, types of soil, etc.) that will cause reductions in the total amount of water resources available. In 2030, the annual reduction of total precipitation in Spain will vary between 5% and 14% according to predictions. The largest reduction will affect the Canary Islands and to a lesser extent the Balearics and the basins of the South: Guadalquivir, Guadiana, southern Andalusia, Segura, and Júcar (Moreno 2005).

The effects of CC have previously been mentioned in the White Book of Water (1998) and in some current Hydrological Plans, but their consequences in practical terms, such as a modification of concessions, reserves, etc., are essentially nonexistent. For example, the Júcar Hydrological Plan approved in July 2014 states that the total runoff for the period 1980/1981–2008/2009 was reduced by 9.5% with respect to the complete period 1940/1941–2008/2009. This is only partially due to CC, because precipitations were only reduced by approximately 2.8% and the mean temperature increased by 1.4%. Other environmental changes have played a key role in this decrease in resources, such as the reforestation of the mountainous portion of the basin and the overexploitation of some aquifers, which have reduced river baseflow.

The administration admits that climate change "will necessarily involve the revision and redefinition of water policies" and emphasizes the need to adapt regulatory elements and "management norms oriented to the real balance of resources and demands" (Moreno 2005). Agreeing with these declarations of principles is easy, but disappointingly there has been little progress in the materialization of adaptation mechanisms. The National Strategy of Adaptation to CC was conceived as a general framework for coordination among administrations, but not as an action plan. For the agricultural sector, it barely defined a few actions of monitoring and R + D (OECC 2006).

The main guidelines for adapting irrigation to CC were published 6 years later by the Ministry of Agriculture, Food and the Environment (CEDEX 2012) and can be grouped in three axes: technology, economy, and training. The technological axis focused on furthering the irrigation modernization policy. It recommended the replacement of gravity irrigation with drip or sprinkler systems but also increasing the efficiency of the existing systems through a second-generation modernization phase. This involves implementing systems for monitoring humidity at plot level and for automating irrigation to optimize water use and energy consumption. The need for more R + D investment and measures to make public subsidies conditional upon efficiency objectives have also been highlighted. Additionally, this document recommends the promotion of controlled deficit irrigation (RDI) and conservation agriculture techniques (CA) – such as reducing tilling and introducing straw mulches – to reduce agricultural water use.

The second axis included economic instruments to reduce agricultural water demand, recommending strengthening the application of the full-cost recovery procedures recommended by the WFD. The implementation of progressive water tariffs with consumption brackets was also highlighted as a useful tool to stimulate water saving in irrigation. Moreover the flexibilization of the concessional regime and the promotion of water markets were also mentioned as essential mechanisms to achieve a more efficient allocation of irrigation water. Finally, the formation and promotion of a water-saving culture among farmers was recommended (CEDEX 2012), through training actions for technicians and users, and by means of the creation of advising services and the publication of agricultural best practices.

Many of these recommendations to adapt to climate change are tasks that are underway in various basins, due to the need to correct previously existing shortfalls that occurred during droughts, such as in 1994–1995 and 2006–2008. Thus, the adaptation to CC clearly overlaps with and prolongs the policies developed during the last years to improve demand management, under the pressure of basin closure processes. Unfortunately, since 2008, the financial crisis has considerably slowed down the implementation of these guidelines for irrigation water demand management.

4.3.3 Lasting Policy Debates

It must be emphasized that supply-oriented measures, such as interbasin transfers or the use of non-conventional resources, were not considered in the above-mentioned document (CEDEX 2012). Nevertheless, these measures are still strongly advocated by farmers' associations and have been pivotal strategies in the recent water policies. In this area, the only current consensus is the promotion of the agricultural use of recycled wastewater. This measure is not politically conflicting and reconciles the agricultural and environmental lobbies and discourses. The financial crisis has almost paralyzed the execution of the National Plan for Wastewater Reuse, but their guidelines and projects are still in force and valid to address scenarios that will be much more demanding. The use of desalinated water or interbasin transfers are still controversial due to their politicization in the past. However, the potential of the existing desalination plants is clearly missed, and it could play a major role if water resources availability severely decreases as a result of CC. In this case, should policies be adopted that allow (or oblige) urban water suppliers to meet their own needs by using desalinized waters in order to release other water resources for agriculture?

This is one of the numerous dilemmas that could be faced in the coming years if the supply of irrigation systems is affected. Most of debates, undoubtedly, will arise from the anticipated reform of the concessional regime: Should a much stricter review of water concessions be conducted and cost-efficiency criteria be applied? How can efficiency be guaranteed but equity in water use still be considered? What other measures can be adopted to relax the concession regime? Could the relaxation of the concession regime resuscitate old interbasin transfer projects?

These challenging questions are even more troubling if the difficulties of application of the WFD are considered. In recent years, a degree of slackness has been observed with respect to the mechanisms for environment quality control, likely for political and economic reasons but also partially due to administrative inertia. Quality objectives were attained in some aquatic ecosystems but are very far from being achieved in the majority of cases. Deadlines for fulfillment have been extended in the most problematic places and important problems regarding diffuse contamination and other impacts caused by irrigation channels remain.

One of the weak points of the Directive is the mechanism of disproportionate costs, which allows for exceptions with regard to fulfilling environmental objectives for particular water systems until 2027, when reversing environmental impacts are very difficult and costly. This mechanism has postponed compliance to a more demanding scenario. Or has it left a door open to non-compliance? As Bueno Hernández (2007) stated, "one has the impression that [WFD] application offers serious doubts that the law will address everything, that all these problems will be solved through good planning…on paper, but reality is quite different…some objectives are so rigorous that they will lead to a situation where it is assumed that they are impossible to fulfill." It is worth asking whether the Spanish government has arrived at this same conclusion.

4.4 Conclusions

Spain is a paradigmatic example of a country with prolonged and intensive application of supply-side management policies, evident principally in the public promotion of the expansion of irrigation. Throughout the twentieth century, technology and institutions served this political end, a hydraulic mission in the form of reservoirs, wells, and irrigation canals that eventually tripled the area of irrigated land. The continuing application of this model resulted in notable success in terms of production and agricultural income but also in significant environmental externalities which brought water resource exploitation to its limit or beyond. At the end of the last century, numerous aquatic ecosystems underwent serious environmental crises, and Spain faced severe watershortages during drought events.

The exhaustion of this model and the territorial reorganization associated with the return to democracy has allowed a genuine transition toward more sustainable water management models. This transition has not been devoid of conflict between partisans of sustainable management of water resources and those social sectors, institutions, and economic agents that are resistant to change. The emphasis on the mobilization of water resources of the national water plans of 1993, 2001, and 2004 is the clearest indication of this inertia, which reflects significant economic interests, electoral motivations stemming from the complex territorial puzzle, and the idea, deeply rooted in Spanish society, that irrigation plays an essential role in national economic development (despite its modest contribution to GDP).

Over the course of the last three decades, a growing social demand for policies aimed at sustainability in the management of natural resources has become evident, particularly with regard to improving and protecting aquatic ecosystems. However, in situations of severe economic crisis, such as that experienced in Spain since 2008, it becomes particularly difficult to justify the preeminence of the eco-environmental dimension of water to the population, in light of its value as an economic good.

At these crossroads, where contradictory actions and initiatives frequently clash, resource-saving technologies have been able to generate important synergies and create one of the few areas of consensus in water policy. These technologies have been equally attractive to individuals who would like to reduce their use of water in agriculture and to those who aspire to continue to expand irrigated areas. Pressurized systems have been developed and now play a leading role in terms of surface area. There has also been considerable advance in the reuse of wastewater, and mechanisms to improve flexibility of resource allocation have been used in some systems that traditionally were very rigid.

Without underestimating the difficulties arising from the complex political climate generated by territorial issues, the primary challenge confronted by irrigation agriculture in Spain is CC, which intensifies the problems and weaknesses that arose from recent basin closure processes. The estimates of the reduction in water resources considered by river basin organizations cast serious doubt on the sustainability of current irrigation systems, given that demand for water resources in many regions has been increasing over the last decade. Therefore, matching water supply and demand could require important changes in the pattern of irrigated crops in the future, and a stabilization, geographical redistribution, or even reduction in irrigated area could take place.

References

Aguilera-Klink, F., & Sánchez García, J. (2002). *Los mercados del agua en Tenerife*. Bilbao: Bakeaz.

Alcón, F., Arcas, N., De Miguel, M. D., & Fernández-Zamundio, M. A. (2009). Adopción de tecnologías ahorradoras de agua en agricultura. In J. A. Gómez Limón et al. (Eds.), *La economía de agua de riego en España. Una perspectiva regional*. Almería: Fundación Cajamar.

Atance Muñiz, I. (2013). *Sostenibilidad económica de la agricultura española. La sostenibilidad de la agricultura española*. Almería: Cajamar Caja Rural.

Avellà, L. (2003). Efectos estructurales de los trasvases en las cuencas receptoras: La experiencia del trasvase Tajo-Segura. In L. Del Moral & P. Arrojo (Eds.), *La directiva marco del agua: Realidades y futuros* (pp. 443–462). Zaragoza: Fundación Nueva Cultura del Agua.

Avellà, L., & García-Mollá, M. (2008). Institutional factors and technology adoption in irrigated farming in Spain: Impacts on water consumption. In A. Dinar & J. Albiac (Eds.), *The management of water quality and irrigation technologies* (pp. 197–226). London: Earthscan.

Barros, R., Isidoro, D., & Aragüés, R. (2012). Irrigation management nitrogen fertilization and nitrogen losses in the irrigation return flows of La Violada irrigation district (Spain). *Agriculture, Ecosystems & Environment, 155*, 161–171.

Borrego-Marín, M. M., & Berbel, J. (2017). Impacto de la modernización de regadíos sobre el uso del agua y otras variables socioeconómicas en comunidades de regantes de Andalucía

Occidental. In J. Berbel & C. Gutiérrez-Martín (Eds.), *Efectos de la modernización de regadíos en España*. Cajamar Caja Rural: Almería.
Brunhes, J. (1902). *L'irrigation. Ses conditions géographiques ses modes et son organisation dans la Péninsule ibérique et dans l'Afrique du Nord*. Paris: C Naud.
Bueno Hernández, F. (2007). La Directiva Marco del Agua. *Ingeniería y territorio, 80*, 4–11.
Cano García, G. (1992). Confederaciones Hidrográficas. In A. Gil Olcina & A. Morales Gil (Eds.), *Hitos históricos de los regadíos españoles* (pp. 309–355). Madrid: Ministerio de Agricultura Pesca y Alimentación.
Carles, J., & García-Mollá, M. (2003). La coherencia de las instituciones y los modelos de uso del agua. In J. Albiac Murillo (Ed.), *Los instrumentos económicos en la gestión del agua en la agricultura* (pp. 115–127). Madrid: Mundiprensa.
CEDEX. (2012). *Estudio de los impactos del cambio climático en los recursos hídricos y las masas de agua. Efectos potenciales del cambio climático en las demandas de agua y estrategias de adaptación*. Madrid: Centro de Estudios Hidrográficos y Experimentales and Ministerio de Agricultura Alimentación y Medio Ambiente.
Closas, A., Molle, F., & Hernández-Mora, N. (2017). Sticks and carrots to manage groundwater over-abstraction in La Mancha Spain. *Agricultural Water Management, 194*, 113–124.
Colino, J., Martínez-Carrasco, F., & Martínez-Paz, J. M. (2014). *El impacto de la PAC renovada sobre el sector agrario de la Región de Murcia*. Murcia: CES Región de Murcia.
Corominas, J. (2008). *¿Modernización o reconversión de regadíos? Dimensiones socio-económicas ambientales y territoriales*. VI Congreso Ibérico sobre Planificación y Gestión del Agua, Vitoria, Fundación Nueva Cultura del Agua.
Corominas, J., & Cuevas-Navas, R. (2017). Análisis crítico de la modernización de regadíos Pensando el futuro: ¿cómo será el nuevo paradigma? In J. Berbel & C. Gutiérrez-Martín (Eds.), *Efectos de la modernización de regadíos en España*. Almería: Cajamar Caja Rural.
Cots, L. (2011). *Desarrollo y calibración de un modelo de simulación de recursos hídricos aplicado a la cuenca del río Corb dentro de la zona regable de los canales de Urgell (Lleida)*. PhD thesis, Universitat de Lleida.
D'Amaro, F. (2014). Instituciones y conflicto en el franquismo La reticencia de los regantes a la integración en los sindicatos (1944-1957). In C. Sanchis-Ibor, G. Palau-Salvador, I. Mangue, & L. P. Martínez-Sanmartín (Eds.), *Irrigation society landscape tribute to Thomas F Glick*. València: Universitat Politècnica de València. https://doi.org/10.4995/ISL2014.2014.
Damas, M., Prados, A. M., Gómez, F., & Olivares, G. (2001). Hidrobus® system: Fieldbus for integrated management of extensive areas of irrigated land. *Microprocessors and Microsystems, 25*, 177–184.
De Stefano, L., Martínez-Cortina, L., & Chico, D. (2012). An overview of groundwater resources in Spain. In M. De Stefano & R. Llamas (Eds.), *Water agriculture and the environment in Spain: Can we square the circle* (pp. 87–104). Leiden: CRC Press.
De Terán, M. (1949). La genialidad geográfica de la Península Ibérica. In P. Vidal de la Blache & L. Gallois (Eds.), *Geografía Universal* (pp. 3–13). Barcelona: Montaner y Simón.
Del Moral, L. (2011). El nuevo Plan Hidrológico de la Demarcación del Guadalquivir: avances y dificultades en el proceso de cambio de modelo de gestión del agua. *Hábitat y Sociedad, 2*, 187–196.
Dumont, A., Mayor, B., & López-Gunn, E. (2014). Is the rebound effect or Jevons paradox a useful concept for better management of water resources? Insights from the irrigation modernisation process in Spain. *Aquatic Procedia, 1*, 64–76.
Escribano, M. J. (2006). *Análisis de la adopción de tecnologías de riego en contextos de incertidumbre: aplicación a la horticultura del Valle de Guadalhorce*. PhD thesis, Universidad de Córdoba.
Esteban, E., & Albiac, J. (2012). The problem of sustainable groundwater management: The case of La Mancha aquifers. *Spain Hydrogeology Journal, 20*(5), 851–863.
Ferrer, J., & Garijo, L. (2013). *Mercados del agua y flexibilización del marco concesional XI. Seminario Nacional Transparencia y Concesiones*. Madrid: Observatorio del Agua de la Fundación Botín.

Gallego Bono, J. R. (1996). Instituciones aprendizaje y liderazgo en la difusión de innovaciones: una interpretación de la desigualdad implantación del riego por goteo en la citricultura valenciana. *Revista Española de Economía Agraria, 175*, 199–228.

García Alvarez-Coque, J. M., & Martínez Gómez, V. D. (2014). La regulación de los mercados en la nueva PAC y el equilibrio de la cadena de valor. In *Reflexiones en torno a la PAC* (pp. 59–82). Almería: Cajamar Caja Rural.

García-Mollá, M., Sanchis-Ibor, C., Ortega-Reig, M. V., & Avellà, L. (2013). Irrigation associations coping with drought: The case of four irrigation districts in Eastern Spain. In K. Schwabe, J. Albiac, J. D. Connor, R. M. Hassan, & L. Meza (Eds.), *Drought in arid and semi-arid regions: A multi-disciplinary and cross-country perspective* (pp. 101–121). New York: Springer.

García-Mollá, M., Ortega-Reig, M., Sanchis-Ibor, C., & Avellà, L. (2014). The effects of irrigation modernization on the cost recovery of water in the Valencia Region (Spain). *Water Science & Technology: Water Supply, 14*(3), 414–420.

Gil Meseguer, E. (2010). La Región de Murcia, un laboratorio de experiencias de ahorro y eficiencia en el uso del agua. *Papeles de Geografía, 51-52*, 131–145.

Gil Olcina, A. (2002). De los planes hidráulicos a la planificación hidrológica. In A. Gil Olcina & A. Morales Gil (Eds.), *Insuficiencias hídricas y Plan Hidrológico Nacional* (pp. 11–44). Alacant: Universitat d'Alacant.

Gómez Espín, J. M., Gil Meseguer, E., & García Marín, R. (2007a). *El antes y después de la modernización de regadíos. La experiencia de Mula*. Murcia: Universidad de Murcia and Consejería de Agricultura y Agua de la CARM.

Gómez Espín, J. M., García Marín, R., & Gil Meseguer, E. (2007b). Modernización de regadíos con aguas subterráneas en ámbitos semiáridos. *M+A Revista Electrónica de Medioambiente, 3*, 1–22.

Gómez Gómez, J. D., López Geta, J. A., & Garrido Schneider, E. (2003). *The state of sea water intrusion in Spain in Tecnología de la Intrusión de Agua de Mar en Acuíferos Costeros: Países Mediterráneos* (pp. 169–185). Madrid: IGME.

Gómez Mendoza, J. (1992). Regeneracionismo y regadíos. In A. Gil Olcina & A. Morales Gil (Eds.), *Hitos históricos de los regadíos españoles* (pp. 231–262). Madrid: MAPA.

Gómez-Limón, J. A. (2014). La agricultura de regadío en el contexto de la nueva PAC en. In I. Bardají (Ed.), *Reflexiones en torno a la PAC* (pp. 109–140). Almería: Cajamar Caja Rural.

Gómez-Limón, J. A., Berbel, J., & Gutiérrez, C. (2007). Multifuncionalidad del regadío: una aproximación empírica. La multifuncionalidad de la agricultura española. Conceptos, aspectos horizontales, cuantificación y casos prácticos. Eumedia y Ministerio de Agricultura, Pesca y Alimentación, Madrid, 207–224.

Gómez-Limón, J. A., & Martin-Ortega, J. (2013). The economic analysis in the implementation of the water-framework directive in Spain. *International Journal of River Basin Management, 11*(3), 301–310.

Gómez-Limón, J., & Calatrava, J. (Eds.). (2016). *Los mercados de agua en España. Presente y perspectivas*. Almería: Fundación Cajamar.

Hardy, L., & Garrido, A. (2010). *Análisis y evaluación de las relaciones entre el agua y la energía. Papeles de Agua Virtual 6*. Madrid: Fundación Botín.

Hernández-Mora, N., & De Stefano, L. (2013). *Los mercados informales de aguas en España: una primera aproximación*. Zaragoza: XVIII Jornadas de Derecho de Aguas.

Hernández-Mora, N., & Del Moral, L. (2015). Developing markets for water reallocation: Revisiting the experience of Spanish water mercantilización. *Geoforum, 62*, 143–155.

Hernández-Mora, N., & Del Moral, L. (2016). Disfuncionalidades de los mercados de agua en España. In J. A. Gómez-Limón & J. Calatrava (Eds.), *(Coord.) Los mercados de agua en España: Presente y perspectivas* (pp. 429–460). Almería: Fundación Cajamar.

Hernández-Mora, N., & López-Gunn, E. (2001). La gestión colectiva de las aguas subterráneas: análisis comparativo. In A. Garrido et al. (Eds.), *La economía y la gestión colectiva de las aguas subterráneas en España*. Madrid: Mundiprensa.

Hernández-Mora, N., & López-Gunn, E. (2003). La gestión colectiva de las aguas subterráneas en La Mancha. In C. Coleto, L. Martínez Cortina, & R. Llamas (Eds.), *Conflictos entre el desarrollo de las aguas subterráneas y la conservación de los humedales: la cuenca alta del Guadiana*. Madrid: Fundación Marcelino Botín and Mundi-Prensa.

INE. (2012). *Encuesta sobre el suministro y saneamento del agua*. Available at: https://www.ine.es/CDINEbase/consultar.do?mes=&operacion=Estad%EDstica+de+suministro+y+saneamiento+del+agua&id_oper=Ir

Isidoro, D., Quílez, D., & Aragüés, R. (2006a). Environmental impact of irrigation in La Violada district (Spain) I: Salt export patterns. *Journal of Environmental Quality, 35*(3), 766–775.

Isidoro, D., Quílez, D., & Aragüés, R. (2006b). Environmental impact of irrigation in La Violada district (Spain) II: Nitrogen fertilization and nitrate export patterns in drainage waters. *Journal of Environmental Quality, 35*(3), 776–785.

Jiménez-Aguirre, M. T., Isidoro, D., & Barros, R. (2014). *Effect of irrigation modernization on water and nitrogen use efficiency*. EGU General Assembly 2014 Geophysical Research Abstracts, 16 EGU 2014–775 27 April–2 May 2014.

Lapuente, E. (2012). *El full cost de la desalación de agua de Mar. Aplicación al sudeste peninsular*. IX Congreso Internacional de la Asociación Española de Desalación y Reutilización. Accessed in 11/11/2015.

Lechuga, J. (2007). Análisis de los procesos para desalinización de agua de mar aplicando la inteligencia competitiva y tecnológica. *Ingeniería. Revista Académica de la FIUADY, 11*(3), 5–14.

Lecina, S., Isidoro, D., Playán, E., & Aragüés, R. (2010a). Irrigation modernization and water conservation in Spain: The case of Riegos del Alto Aragon. *Agricultural Water Management, 97*(10), 1663–1675.

Lecina, S., Isidoro, D., Playán, E., & Aragüés, R. (2010b). Irrigation modernization in Spain: Effects on water quantity and quality-a conceptual approach. *International Journal of Water Resources Development, 26*(2), 265–282.

Llamas, M. R. (1975). Non–economic motivations in ground water use: Hydro-schizophrenia. *Ground Water, 13*(3), 296–300.

Llamas, M. R., & Garrido, A. (2007). Lessons from intensive use in Spain: Economic and social benefits and conflicts. In M. Giordano & K. G. Villholth (Eds.), *The agricultural groundwater revolution: Opportunities and threats to development* (pp. 266–295). Wallingford: CAB International.

Llamas, M. R., Custodio, E., de la Hera, A., & Fornés, J. M. (2015). Groundwater in Spain: Increasing role evolution present and future. *Environmental Earth Sciences, 73*(6), 2567–2578. https://doi.org/10.1007/s12665-014-4004-0.

López-Gunn, E. (2009). Agua para todos: A new regionalist hydraulic paradigm in Spain. *Water Alternatives, 2*(3), 370–394.

López-Gunn, E., & Rica, M. (2012). La participación activa de los usuarios: la co-gestión como forma de gobernanza del agua subterránea. In Lamban et al. (Eds.), *Las aguas subterráneas. Desafíos para la gestión en el siglo. XXI* (pp. 15–35). Zaragoza-Barcelona: AIH-GE.

López-Gunn, E., Dumont, A., & Villarroya, F. (2012a). Tablas de Daimiel National Park and groundwater conflicts. In L. De Stefano & R. Llamas (Eds.), *Water agriculture and the environment in Spain: Can we square the circle?* (pp. 259–218). Madrid: Fundación Botín.

López-Gunn, E., Zorrilla, P., Prieto, F., & Llamas, R. (2012b). Lost in translation? Water efficiency in Spanish agriculture. *Agricultural Water Management, 108*, 83–95.

López-Gunn, E., Mayor, B., & Dumont, A. (2012c). Implications of the modernization of irrigation systems. In L. De Stefano & R. Llamas (Eds.), *Water agriculture and the environment in Spain: Can we square the circle?* (pp. 241–255). Madrid: Fundación Botín.

MAGRAMA. (2002). *Plan Nacional de Regadíos*. Madrid: Ministerio de Agricultura Alimentación y Medio Ambiente Available at: https://www.mapa.gob.es/es/desarrollo-rural/temas/gestion-sostenible-regadios/plan-nacional-regadios/. Accessed November 2018.

MAGRAMA. (2013). *Encuesta sobre superficies y rendimientos de cultivos. Informe sobre regadíos en España*. Madrid: Ministerio de Agricultura Alimentación y Medio Ambiente.

Martínez Vicente, D. (2009). Coste del agua del mar desalada por ósmosis inversa. In J. A. Ibáñez (Ed.), *Desalación de aguas. Aspectos tecnológicos medioambientales jurídicos y económicos* (pp. 521–550). Murcia: Fundación Instituto Euromediterráneo del Agua.

Mateu Bellés, J. F. (2011). *La primera Confederación Hidrográfica del Júcar (1934–1942)*. València: Confederación Hidrográfica del Júcar.

Medina, J. A. (1988). *Riegos localizados: teoría y práctica*. Madrid: Mundiprensa.

Melgarejo, J. (1995). *La intervención del Estado en la Cuenca del Segura 1926–1986*. Alicante: Generalitat Valenciana-Institut Juan Gil-Albert.

Melgarejo, J. (2009). Efectos ambientales y económicos de la reutilización del agua en España. *CLM Economía, 15*, 245–270.

MIMAM Ministerio de Medio Ambiente. (2007). *La planificación hidrológica. Síntesis de los estudios generales de las demarcaciones hidrográficas en España*. Madrid: MIMAM.

Molinero, J., Custodio, E., Sahuquillo, A., & Llamas, M. R. (2013). Groundwater in Spain: Legal framework and management issues. In A. Findikakis & K. Sato (Eds.), *Groundwater management practices* (pp. 123–137). Leiden: CRC Press.

Molle, F. (2009). River-basin planning and management: The social life of a concept. *Geoforum, 40*(3), 484–494.

Molle, F., Wester, P., & Hirsch, P. (2010). River basin closure: Processes implications and responses. *Agricultural Water Management, 97*, 569–577.

Moreno, J. (coord). (2005). *Evaluación Preliminar de los Impactos en España por Efecto del Cambio Climático*. Madrid: Ministerio del Medio Ambiente. http://digital.csic.es/handle/10261/79951.

Moreno, M. A., Ortega, J. F., Córcoles, J. I., Martínez, A., & Tarjuelo, J. M. (2010). Energy analysis of irrigation delivery systems: Monitoring and evaluation of proposed measures for improving energy efficiency. *Irrigation Science, 28*(5), 445–460.

Morote, A. F., Rico Amorós, A. M., & Moltó, E. (2017). La producción de agua desalinizada en las regiones de Murcia y Valencia. Balance de un recurso alternativo con luces y sombras. *Documents d'Anàlisi Geogràfica, 63*(2), 473–502.

Naranjo, J. E. (2010). Problemática de la modernización de regadíos. In *XII Congreso Nacional de Comunidades de Regantes*, Tarragona. http://www.fenacore.org/congresotarragona/wp-content/uploads/2009/10/ponencia-3-J.E.Naranjo.pdf

OECC. (2006). *Plan Nacional de Adaptación al Cambio Climático Oficina Española del Cambio Climático*. Madrid: Ministerio de Agricultura, Alimentación y Medio Ambiente.

Ortega Cantero, N. (1979). Política hidráulica y política colonizadora durante la Dictadura de Primo de Rivera. *Cuadernos Económicos del ICE, 10*, 353–381.

Ortega Cantero, N. (1995). El Plan General de Canales de Riego y Pantanos de 1902. In A. Gil Olcina & A. Morales Gil (Eds.), *Planificación hidráulica en España* (pp. 107–136). Murcia: Fundación Caja del Mediterráneo.

Ortega Cantero, N. (1999). La política hidráulica española hasta 1936. In R. Garrabou & J. M. Naredo (Eds.), *El agua en los sistemas agrarios. Una perspectiva histórica* (pp. 159–180). Madrid: Fundación Argentaria.

Palomo-Hierro, S., Gómez-Limón, J. A., & Riesgo, L. (2015). Water markets in Spain: Performance and challenges. *Water, 7*, 652–678.

Perry, C., Steduto, P., Allen, R. G., & Burt, C. M. (2009). Increasing productivity in irrigated agriculture: Agronomic constraints and hydrological realities. *Agricultural Water Management, 96*(11), 1517–1524.

Playán, E. (2002). *El potencial de la rehabilitación modernización y conservación de sistemas de riego para un mejor uso del agua Seminario Internacional. El agua de riego a debate*. Córdoba: Fundación para el Fomento de la Ingeniería del Agua http://digital.csic.es/handle/10261/8987.

Playán, E., & Mateos, L. (2006). Modernization and optimization of irrigation systems to increase water productivity. *Agricultural Water Management, 80*, 100–116.

Playán, E., Cavero, J., Mantero, I., Salvador, R., Lecina, S., Faci, J. M., Andrés, J., Salvador, V., Cardena, G., Ramon, S., Lacueva, J. L., Tejero, M., Ferri, J., & Martínez-Cobo, A. (2007). A database program for enhancing irrigation district management in the Ebro Valley (Spain). *Agricultural Water Management, 87*(2), 209–216.

Rocamora, C., Vera, J., & Abadía, R. (2013). Strategy for efficient energy management to solve energy problems in modernized irrigation: Analysis of the Spanish case. *Irrigation Science, 31*, 1139–1158.

Rodríguez Cabellos, J. A. (2014). Towards IWRM in the upper Guadiana basin Spain. In P. Martínez-Santos, M. Aldaya, & R. Llamas (Eds.), *Integrated water resources management in the 21st century: Revisiting the paradigm*. Boca Raton: CRC Press.

Rodríguez-Díaz, J. A., Perez-Urrestarazu, L., Camacho-Poyato, E., & Montesinos, P. (2012a). Modernizing water distribution networks lessons from the Bembézar MD irrigation district Spain. *Outlook on Agriculture, 41*, 229–236.

Rodríguez-Díaz, J. A., Montesinos, P., & Camacho-Poyato, E. (2012b). Detecting critical points in on-demand irrigation pressurized networks – a new methodology. *Water Resources Management, 26*, 1693–1713.

Romero, J. (1995). El Plan Nacional de Obras Hidráulicas Precedentes y condicionantes. In A. Gil Olcina & A. Morales Gil (Eds.), *Planificación hidráulica en España* (pp. 257–282). Alicante: Fundación Caja del Mediterráneo.

Rosell, J. (2001). Aspectos económicos de la utilización de las aguas subterráneas en La Mancha. In N. Hernández-Mora & R. Llamas (Eds.), *Economía de las aguas subterráneas y su gestión colectiva*. Santander: Fundación Marcelino Botín.

Sahuquillo, A. (2009). La importancia de las aguas subterráneas. *Rev R Acad Cienc Exact Fís Nat, 103*, 97–114.

Salvador, R., Latorre, B., Paniagua, P., & Playán, E. (2011). Farmers' scheduling patterns in on-demand pressurized irrigation. *Agricultural Water Management, 102*, 86–96.

Sanchis-Ibor, C. (2012). La confederación fallida. Administración y usuarios en la cuenca del Turia (1928–1936). *Cuadernos de Geografía, 91/92*, 19–42.

Sanchis-Ibor, C., García-Mollá, M., Avellà, L., & Carles-Genovés, J. (2011). Reaching the limits of water resources mobilization: Irrigation development in the Segura river basin Spain. *Water Alternatives, 4*(3), 259–278.

Sanchis-Ibor, C., García-Mollá, M., & Avellà, L. (2016). Las políticas de implantación del riego localizado. Efectos en las entidades de riego de la Comunidad Valenciana. *Boletín de la AGE, 72*, 9–35.

Sanchis-Ibor, C., García-Mollá, M., & Avellà, L. (2017). Effects of drip irrigation promotion policies on water use and irrigation costs in Valencia Spain. *Water Policy, 91*, 165–180. https://doi.org/10.2166/wp.2016.025.

Sanchis-Ibor, C., Pulido-Velazquez, M., Macián-Sorribes, H., Marcos-García, P., García-Mollá, M., & Ruiz-Rodríguez, M. (2018). Uso intensivo del agua en la agricultura en cuencas mediterráneas: aspectos económicos e hidrológicos. In A. Garrido & A. Pérez-Pastor (Eds.), *El uso del agua en la agricultura*. Almería: Cajamar Caja Rural.

Sanz, D., Calera, A., Castaño, S., & Gómez-Alday, J. (2015). Knowledge participation and transparency in groundwater management. *Water Policy, 18*(1), 111–125. https://doi.org/10.2166/wp.2015.024.

Serrano, A. (2008). *El papel de los mercados y de la gestión del agua en España*. Thematic Week: Water Economics and Financing, Expo Zaragoza 2008.

Sese-Minguez, S., Boesveld, H., Asins-Velis, S., Van Der Kooij, S., & Maroulis, J. (2017). Transformations accompanying a shift from surface to drip irrigation in the Cànyoles Watershed Valencia Spain. *Water Alternatives, 10*(1), 81–99.

Sevilla, M., Torregrosa, T., & Moreno, L. (2010). Las aguas subterráneas y la Tragedia de los Comunes en el Vinalopó (Alicante España). *Estudios de Economía Aplicada, 28*, 305–331.

Stambouli, T., Faci, J. M., & Zapata, N. (2014). Water and energy management in an automated irrigation district. *Agricultural Water Management, 142*, 66–76.

Swyngedouw, E. (1999). Modernity and hybridity: Nature Regeneracionismo and the production of the Spanish waterscape 1890–1930. *Annals of the Association of American Geographers, 89*(3), 443–465.

Swyngedouw, E. (2013). Into the sea: Desalinization as hydro-social fix in Spain. *Annals of the Association of American Geographers, 103*, 261–270.

Varela-Ortega, C. (2007). *Policy-driven determinants of irrigation development and environmental sustainability: A case study in Spain irrigation water pricing policy in context: Exploring the gap between theory and practice. Comprehensive assessment of water management in agriculture* (pp. 328–346). Wallingford/Cambridge, MA: IWMI/CABI.

Verdú, A., Sanchis-Ibor, C., & Marco, J. B. (1999). Regadío y saneamiento urbano en l'Albufera de València: Análisis cartográfico. *Cuadernos de Geografía, 65*, 61–80.

Vicente, E., & Miracle, M. R. (1992). The coastal lagoon Albufera de Valencia: An ecosystem under stress. *Limnetica, 8*, 87–100.

Zapata, N., Playán, E., Skhiri, A., & Burguete, J. (2009). Simulation of a collective solid-set sprinkler irrigation controller for optimum water productivity. *Journal of Irrigation and Drainage Engineering, 135*(1), 13–24.

Chapter 5
France

Sébastien Loubier, Thierry Ruf, and Patrice Garin

Abstract The development of irrigation in France is several centuries old. During the twentieth century the state constructed most of the irrigation networks. Individual irrigation development dates back to the 1960s and is characterized by an easier uptake of technological innovations. Today, the total equipped area is leveling off, and trends include a development of drip irrigation and the near disappearance of gravity irrigation. Public policies with a direct or indirect impact on irrigation have also changed significantly in recent decades. Agricultural policies, once very favorable to the development of irrigation, have moved to policies that are increasingly oriented toward the protection of water resources. In some basins, farmers face severe administrative reduction of their pumping authorization. Water resources management at the watershed or territory levels also changed significantly. In basins facing large quantitative deficits, special institutions are created to manage the volume of water available for agriculture. However, concerns remain regarding the effectiveness of such institutions and more generally the future of irrigation in a context of global change.

Keywords Irrigation · France · Water user association · Water management

5.1 Historical and Institutional Context of Irrigation Development

Can the origins of irrigation in France be accurately determined? The first difficulty lies in defining the word itself. Today, irrigation refers almost exclusively to the artificial supply of water used to make up for summer water deficits. This was not always the case. In the eighteenth century, most irrigation treaties dealt with

S. Loubier (✉) · P. Garin
UMR G-Eau, Irstea, Montpellier Cedex 5, France
e-mail: sebastien.loubier@irstea.fr

T. Ruf
IRD – UMR GRED 220, Montpellier, France

F. Molle et al. (eds.), *Irrigation in the Mediterranean*, Global Issues in Water Policy 22, https://doi.org/10.1007/978-3-030-03698-0_5

watering pastures during the wet season (Bertrand 1764). Tatham (1799) viewed this function of irrigation as essential. His aim was land improvement from a broader perspective, involving an entire territory. Around the same time, numerous canals were built along the islands and banks of the Seine to divert river water. Their origin may date back to Carolingian times, between the seventh and tenth centuries, with the installation of mills and later wetland drainage and their extensive watering in winter, between the tenth and twelfth centuries, to promote sedimentation and soil fertility (Rouillard 2013). Further south, in the Rhone Valley, wetland development sites dating back to Gallo-Roman times have been found, probably related to local interactions between the "peasant" economy and the development of the Roman Villa Model, associated with more structured hydraulics (Leveau 1998). The story of the Nimes aqueduct, which operated between the first and sixth centuries, illustrates the complementarity – or rivalry – between water use in the city and in lands served by the hydraulic system to the northeast of the city (Fabre et al. 2013).

The weakening of cities in favor of the countryside is thought to have provoked the decline of networks which had become too difficult to maintain. Hydraulic schemes went through several phases, from design and testing to enduring use and, later, the emergence of technical and organizational shortcomings. It is probable that some drained areas became wet again or that watershed areas became more or less dry. In France, however, the agricultural hydraulic infrastructures dating back to Antiquity are no longer in use. They are merely the subject of archaeological studies. The oldest irrigation or drainage systems still in use today date back to the Middle Ages.

5.1.1 *Medieval Watering in South-eastern France and the Emergence of Common Property Goods*

One of the oldest structures still in use today is the aqueduct of Ansignan, in the upper Agly Valley, in the Pyrénées-Orientales (Dubois and Ruf 2003), which services a mere 10 ha area (Fig. 5.1).

Fig. 5.1 The so-called "Roman" aqueduct of the Ansignan canal, a medieval aqueduct

Such examples remain quite rare and are the result of local initiatives. Between the twelfth and fourteenth centuries, water development took on greater importance in Provence, the Alps, and the Pyrenees, following a general trend in the Mediterranean world (Ruf and Riaux 2008). Often, these infrastructures connect successive villages together. An example in the Vaucluse is the Saint-Julien Canal, built in 1171. Today, it irrigates 6000 ha from the waters of the Durance River. Further upstream, many canals were dug in the fourteenth century around Briançon (Chenard et al. 2010; Dumond 2002). In general, these irrigation systems were initiated by tenant farmers acting as collective bodies, local lords, or religious institutions, especially monasteries, which reclaimed wetlands. Towns also carried out water conveyance or farmland drainage works. Therefore, rivalries were bound to arise around the sharing of water. Depending on the local context, certain social authorities placed themselves as masters of the land and waters or as judges of water use conflicts.

Thus, as early as the fourteenth century in the Pyrénées-Orientales, institutional relations surrounding river water set the scene for tenant farmers organized as "universities" – a word which at that time defined an assembly much like the Arab Jema'ah. As project leaders, tenant farmers faced the opposition of local authorities: on the one hand, religious elites, who wanted their dependents to submit to their will, and on the other, royal power representatives who claimed to be the only ones to permit the construction of canals in exchange for considerable one-off payments. These rivalries carried on for seven centuries, across all political regimes, as Ruf (2001) illustrates about Prades en Conflent.

5.1.2 Irrigation After the Revolution: Struggle for a Compromise Between Public, Private, and Association Actors

With regard to water, the French Revolution was a decisive turning point: the night of August 4, 1789, abolished privileges, ended feudalism, and resulted in the selling off of church assets. All the local institutions which managed canals therefore potentially lost the legal basis for their activity. In many places, this threat triggered a reactivation of assemblies and claims to customary rules and rights. As relayed by Ribes, a Catalan lawyer, these worries inspired Article 645 of the Civil Code, which validated local rules for water use (Assier-Andrieu 1985). Until 1898 and the first French water law, a period of indecisiveness reactivated common property management institutions (Ingold 2011). Universities reestablished their leadership of irrigation schemes and turned into landowner associations, with their own local institutional rules (Ruf 2002).

Throughout the nineteenth century, the provision of irrigation was the object of controversy:

- In 1815, the restoration of the French monarchy created conditions discouraging new collective projects.

- The Saint-Simonian engineer movement advocated a central role for the state, in order to undertake major water works. Supporters of a new kind of development, and radically opposed to the monarchy, Saint-Simonians had acquired their hydraulic expertise in Egypt, where they built the first great diversion dam on the Nile, near Cairo, between 1820 and 1830 (Regnier 1989).
- In 1846, François Jaubert de Passa, another Catalan lawyer with Republican views, compiled a 2000-page book on irrigation institutions in various countries and civilizations. He advocated a pragmatic, local approach, anchored in local and social history. Rejecting any rush to reform water management, he instead suggested adapting policies to the diversity of local practices.

These three approaches to water governance gave way to new formal institutions:

- On the public side, the government established water services in administrative centers, which oversaw the enforcement of the new 1847 irrigation law dealing with the establishment of infrastructures on private properties (De la Moskova 1847).
- On the private side, the Société Générale des Eaux was inaugurated in the 1850s, with a dual purpose: to build and manage drinking water networks and manage irrigation networks (Goubert 1986). The first objective was reached and reinforced during the second half of the nineteenth century, while the latter ran into difficulties and was not pursued.
- On the territorial collective side, Jaubert de Passa's ideas were translated in 1865 into a law which created landowner associations (ASAs) to manage common properties such as irrigation networks.

The 1865 law will deeply influence irrigation in France, since it would be the basis, until its amendment in 2004, for building and managing most of the collective irrigation networks run by farmers, with the support of the government.

5.1.3 Public Irrigation Investment, Mainly Through Landowner Associations (from the Late Nineteenth Century to Late Twentieth Century)

From an institutional point of view, until World War II, the government entrusted the water infrastructures it built or renovated to landowner associations. The 1865 law turned these associations into public institutions entrusted with public missions. Cost sharing between public authorities and landowners varied, reaching 80% of subsidies during the extension phase, between the 1960s and the 1990s.

Landowners govern these associations. The owners of the fields come together at general assemblies, electing representatives among their ranks, who are truly in charge of managing the network and who elect a president. If they wish to do so, landowners can delegate this right to their tenant, who may be the real water user.

ASAs are subject to a very strict regulatory framework. Although they are managed by users, ASAs may receive public funding. In return, a system guarantees the sustainability of the infrastructures and of their management. This system is based on five key principles which reflect the government's oft-repeated determination to use ASAs to support its land planning policies (Garin and Loubier 2002): (i) water rights and duties are almost permanent and tied to landownership but not linked to the water user; (ii) fees are levied like taxes from landowners; (iii) Government authorities can impose a maintenance program and expenses if they consider it appropriate; (iv) a yearly audit of accounts is carried out by government services; and (v) the legality of most ASAs' decisions is verified before they are implemented. In return for this relatively binding framework, ASAs have considerable decision-making autonomy in terms of the operating rules governing the relationships between members and the operational management of the area. As the bylaws specify, decision-making mechanisms, resource sharing, and management rules are defined by the assembly. Staff management, division of labor, and pricing policy are the board's (members elected within the assembly) responsibility. These principles offer an institutional framework that is compatible with the sustainability principles set out by Ostrom (1992).

It is estimated that there are 2000 irrigation ASAs for over 300,000 ha with facilities. A quarter of these manage gravity irrigation networks, while the remaining three quarters manage pressurized irrigation networks. In gravity irrigation, a handful of very large ASAs include several thousand hectares and members, as in Carpentras, in the South of France, which has 13,000 landowners for 11,000 ha. Besides these, there are several hundred mostly small ASAs with tens of hectares and members. The ASAs managing pressurized irrigation networks are less heterogeneous. Their most likely size lies between 100 and 400 ha, for 10–50 landowners. In all these cases, collective action is weakened by two factors: first, many plots are farmers renting land who do not always have the delegation of rights of the landowner to participate in decision-making; second, multifunctionality means that non-agricultural users tend to grow in number (Garin and Loubier 2002).

5.1.4 A Second Model Inspired by Major American Post-WWII Developments and Past Colonial Experience

After World War II, the government planned major water investments in several regions. Although the regions were as yet non-existent in the country's political structure, the Fourth Republic set up *Compagnies Régionales d'Aménagement* (regional water resources development companies), later called *Sociétés d'Aménagement Régionales* (SAR), such as the Bas-Rhône-Languedoc (BRL) company, founded in 1955. The *Société du Canal de Provence* (SCP) and the *Compagnie des Coteaux de Gascogne* (CACG) followed a few years later.

The SARs' relative success in developing hydraulic infrastructure is, of course, due to the post-Second World War reconstruction programs, but it was facilitated by the experience of hydraulic engineers and other water specialists from related fields. These staff, generally civil servants from the former French colonies, were partly integrated into the SARs and transferred their southern Mediterranean experience and knowledge to southern French regions (Pritchard 2012).

Their mission was to overcome local opposition to equip large rural areas with irrigation and drainage infrastructure. SARs equipped France's southern regions with raw water distribution networks, storage facilities, and interbasin transfers. The government granted 75-year concessions, after which the assets would return to the state, which would then decide how to manage the infrastructure (direct management, concession, lease, privatization, etc.). Recently, in line with decentralization laws, the SCP and BRL changed their status, becoming regional concessions, with private entities acquiring 20–30% of the capital (Fig. 5.2), depending on the SAR. To our knowledge, this is the only example of a private-public irrigation partnership in France, although the influence of the private sector is marginal compared with that of regional authorities.

Currently, SARs total 300,000 ha (Table 5.1) of irrigated land, but the actual irrigated area is lower, around 10–20% of the area potentially irrigable for the SCP and the BRL and 40–50% for the CACG. Several factors explain these low percentages for SCP and BRL: water is relatively expensive, land is progressively absorbed by urban development, and certain political expectations about the conversion from rain-fed to irrigated agriculture have not been met. Unlike for the ASAs, irrigators are contractually bound to SARs for short (1 year) to average (5 years) durations, and irrigation may be discontinued. These low rates, especially for the SCP and the BRL, have led them to diversify their operations and, consequently, their revenue. They may supply raw water for industrial use, drinking water, environment purposes, and watering public green spaces. The SCP provides the most extensive example of diversification, since water volumes delivered for agricultural use represent only

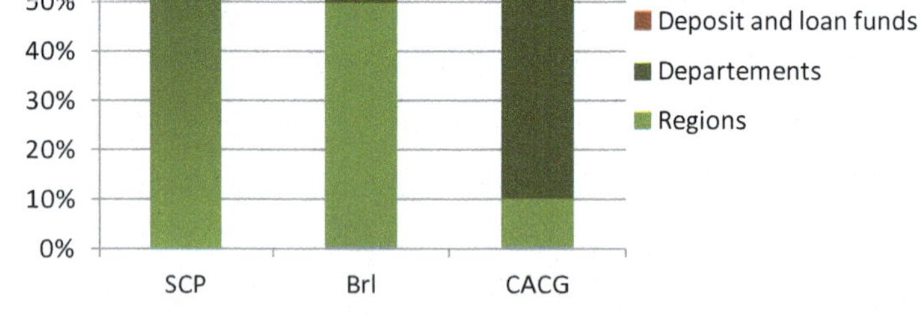

Fig. 5.2 Capital structure of the three SARs

Table 5.1 Main characteristics of SARs

	CACG	BRL	SCP
Equipped surface (ha)	130,000	100,000	80,000
Volume delivered for agricultural irrigation (Mm^3)	104	50	29
Volume delivered for agricultural irrigation as a percentage of total	84%	58%	15%
Estimated irrigated agricultural area (ha)	52,000–61,000	17,000–20,000	8000–10,000
Estimated irrigation rate (IS/ES)	40–47%	17–20%	10–12%
Average price of agricultural water (€/m^3)	0.10	0.21	0.21
Price of water for other uses	0.028–1.08	0.29–0.56	0.43–0.79
Proportion of irrigation revenue to total revenue	93%	38%	8%
Farms	5000	4300	6000
Irrigated area per farm (ha)	11	4	2

Adapted from Rollin et al. (2013)

15% of total volumes and 8% of the revenue, with the remainder serving industry and part of the metropolitan area of Marseille. Only the CACG has clearly maintained its initial objectives: irrigation represents 84% of the water sold and 93% of its revenue.

These data relate only to concession operations. The three SARs have also established lease contracts with some ASAs which faced too many management constraints and have developed engineering offices which assist the ASAs served by their networks.

As with the ASAs, pricing systems vary greatly, although they are always binomial. Only the CACG has a water tariff comparable to that of the ASAs (0.1 €/m^3). This can be explained in part by the widespread use of natural watercourses to convey water, rather than main canals which are costly to construct and maintain. Agricultural water costs at SCP and BRLs are twice as high, despite the greater share of nonagricultural uses in their water sales revenues.

5.1.5 Various Marginal Legal Status Managing Networks

SARs and ASAs make up the bulk of collectively irrigated areas. The remaining area is equipped by organizations with various legal statuses, mostly private except municipalities or inter-municipalities networks (rarely interdepartmental). The relative part of each status is unknown at national level, mainly because those with private status are not obliged to report their activities to the state. It mainly concerns simple associations and CUMAs (*Coopérative d'utilisation de matériel agricole*, cooperatives for the use of agricultural equipment), which share either a water abstraction point or irrigation equipment between several irrigators. Recently, common economic interest groups (GIE) have been created to substitute summer individual pumping by winter storage and collective water use. These GIE can include from a few to more than a hundred farmers, for larger systems.

5.2 The Evolution of Water Management Policies

5.2.1 Irrigation and Political Changes over the Last 50 Years

Official data collected since the first agricultural census in 1950 point to a continuous increase in irrigation up to recent times. From the 1960s onward, irrigation expanded to new areas through the development of individual and collective irrigation systems, with a high level of public subsidies (70–80% of the investment cost) from municipalities, *départements*, regions,[1] and the Agricultural Ministry. Two investment waves must be distinguished. The first, initiated by the state during the 1950s and 1960s, consisted of creating large regional infrastructures (managed by SARs) to secure water, to facilitate farm reconversion to face specific crises (e.g. wine production in the Languedoc-Roussillon region), to slow the rural exodus and to maintain competitive agricultural sectors. The state supported such investment in that it contributed to the overall development policy. The mid-1960s to the mid-1980s period was characterized by a concomitance of easier access (lower cost) to technological progress in irrigation and new ambitious agricultural objectives through the Common Agricultural Policy. To satisfy its production objective, the state and the EU significantly subsidized collective irrigation systems (ASAs). In 1972, an article by Martin (1972) analyzed the evolution of irrigation in France, perfectly illustrating the attitude to irrigation in this period. Irrigation development is regarded as a performance indicator, an essential feature of modern agriculture, with regions and crops ranked by the percentage of irrigated area. Martin never mentions the water resource, implicitly considered as being unlimited, nor the environment. The policies were a great success and even resulted in overproduction.

At the same time, the 1964 Water Act led to the establishment of six water agencies in metropolitan France (one for each major catchment area). However, effective water management was only implemented with the 1992 Water Act, which established water as the "nation's common heritage." Since the 1964 Water Act, water agencies have levied abstraction and pollution charges from their users as a way of facilitating, through financial support, the implementation of the measures in their SDAGE (Strategic Water Basin Masterplans). In 1992, other obligations were introduced as a means of monitoring water use and charging users for abstraction: (i) to declare all water intake points; (ii) to apply for a yearly irrigation authorization if withdrawals are higher than 80 m^3/h; and (iii) to meter abstracted volumes.

[1] Municipalities, *départements,* and regions are the three French decentralized administrative units. The 13 regions (not considering overseas territories) are composed of *départements* (96) themselves composed of municipalities (35,000).

Since 1990, individual systems have expanded more rapidly, and also benefited from public subsidies. Individual irrigation equipment, based on wells or small on-farm reservoirs, is now the dominant system in the country.

In the 1990s, irrigation continued to grow. Investments were still largely subsidized, mainly because of the land use planning policy objectives, while agricultural production became a secondary objective. But environmental restrictions (1992 Water Act) and civil society's environmental expectations signaled a turning point. Qualitative and quantitative aspects of the water balance, especially with the enactment of the Water Framework Directive (WFD) in 2000, took precedence over collective irrigation objectives. The French government began to rely on SARs and ASAs to partly regulate the agricultural uses of water (levying charges on behalf of basin agencies, participation in water planning institutions such as the Basin Committee or the Local Water Commission, etc.).

The WFD was founded on the overarching objective of achieving good environmental (surface waters) or chemical status (groundwater) in all water bodies (watercourse sections, groundwater, wetlands, and coastal regions) by 2015, 2021, or 2027, depending on the efforts deemed necessary. In 2006, the WFD's overall objectives were translated and integrated into a new French Water Act, and new irrigation projects are less common. The main objective is to substitute collective irrigation systems for individual withdrawals from water bodies facing quantitative deficits. This is motivated by a need to develop extra resources through small-scale storage as well as achieving economies of scale (and therefore reducing subsidies).

Hence, within four to five decades, irrigation first came to be seen as a lever for rural development, serving agricultural and land use planning policies, before facing environmental policy constraints.

5.2.2 Present Legal Framework and Water Use Regulation

Water management occurs at three nested levels: large hydrographic districts or river basins (six in France), watershed level, and collective irrigation networks.

5.2.2.1 River Basin Level

At the river basin level, the Water Framework Directive's (WFD) subsidiarity principle is applied in order to adapt to the specificities of local resources, under the supervision of the Ministry of Environment. This Ministry is in charge of overall water policy, particularly the implementation of the European WFD, through the French National Agency for Water and Aquatic Environment (ONEMA), created in 2007 to protect the aquatic environment by monitoring use and enforcing regulations.

Although the decentralized water agencies' involvement has increased over time, they do not replace the central government administration (particularly the ONEMA), which is responsible for the implementation of regulatory measures. The central administration still receives abstraction declarations, issues abstraction authorizations, and is responsible for law enforcement and the water police. Water agencies have an important role with regard to the WFD objectives in supporting investment that contributes to the improved ecological status of water bodies. For the first time in French history, restoring good ecological status (qualitative and quantitative) became the top priority in terms of water management. Any member state where water bodies are still classified as having a "bad ecological status" after two 6-year management plans faces EU sanctions. Derogations can be obtained according to the initial status but remain rare.

5.2.2.2 The Watershed or Aquifer Level

For each watershed or aquifer, the regional representation of the Ministry of Environment defines river flows or groundwater piezometric levels beyond which gradual restrictive measures are to be implemented. Each threshold (vigilance, alert, reinforced alert, and crisis) corresponds to identified measures designed to reduce impacts on the environment and on other uses (e.g. gradual reduction of abstractions, establishing water turns, nighttime irrigation, prohibition of abstraction).

Watershed or aquifers identified as showing considerable imbalance between resources and needs are named ZREs (*Zones de Répartition des Eaux,* water allocation areas). In ZREs, even small water users (i.e. abstracting more than 1000 m^3/year or 8 m^3/h) must obtain abstraction authorizations and thus pay abstraction charges. Due to the limited knowledge about the environment, the lack of environmental concern in the past, and the importance of water for the local economy and politics, abstraction authorizations frequently exceeded the available resource. To improve the system, however marginally, volumetric water management policies were implemented in some catchments. These policies, which are particularly suited to catchments where resource stocks are relatively well known, consist in granting each irrigator an annual water quota and setting conditions for its use in different periods (generally in weeks or decades), during the irrigation campaign. The initial expectation was that such instruments would avoid crises 4 years out of 5. However, crisis management instruments (use restrictions), intended for one exceptional year out of five, are in fact implemented almost every year in many catchments, even in wet years.

Since the 2006 Water Act, in places where considerable quantitative imbalance remains (mainly in ZREs), state authorities should define a maximal volume which may be abstracted for irrigation. Based on this volume, a single authorization, valid for several years, is granted to an *Organisme Unique de Gestion Collective* (OUGC, collective management organization) for irrigation water. This group authorization replaces the annual authorizations previously granted to individual irrigators. If the total annual requests are above the authorized withdrawal volume, the OUGC, in accordance with certain criteria, must propose a plan for distribution between each irrigator, and is then responsible for its fulfillment. The water police remains in the hands of government services. In exchange for the OUGC's involvement in management, the abstraction charge is reduced. The OUGC receives specific management subsidies for a few years after its creation and may devise a system for recovering operating costs, if possible with an incentive mechanism. In 2015, most OUGCs began to operate in earnest for the first time. In most cases, the Chambers of Agriculture were designated as OUGCs by the prefect, following a call for applications.

The intent of the 2006 law, which created OUGCs, was inspired by the relative success of SARs and ASAs in sharing water and maintenance costs between farmers depending on the same irrigation network. But the OUGCs' mission will be much harder to achieve, especially when it comes to handling many individual and dispersed groundwater abstractions.

5.2.2.3 The Collective Irrigation Systems

Within a collective irrigation network, managers (ASs, SARs, and other structures) are responsible for operating and maintaining the system, sharing water and costs among users, and resolving conflicts. They can also get policy prerogatives on users since they can control and sanction if needed. The state can help managers in different ways: subsidising specific investments, being more lenient with collective systems than with individual ones, controlling accounts and expenses, and supporting managers in the enforcement of sanctions, if needed. Regarding water policy and the implementation of water management instruments, collective systems are more or less treated as individual ones since they are considered as one abstractor. When restriction measures are taken at the watershed level, institutions managing collective systems must agree water-sharing measures among their members.

5.2.3 Private vs Collective Irrigation and Surface vs Groundwater Resource

Water that is abstracted to be distributed via collective networks is generally taken from watercourses or reservoirs, whereas individual irrigators mainly abstract groundwater. Groundwater abstraction applies to 38% of the farms with irrigation facilities and 58% of the equipped area (Fig. 5.3). Conversely, surface resources are used for 80% of farms[2] and 42% of the equipped area.

Individual irrigation is more developed in western and southwestern France. It is generally based on groundwater (Fig. 5.4) and is used for crops with relatively low

[2] "Mixed" farms have both types of access to the water (individual and collective), which explains why the total percentage is above 100%.

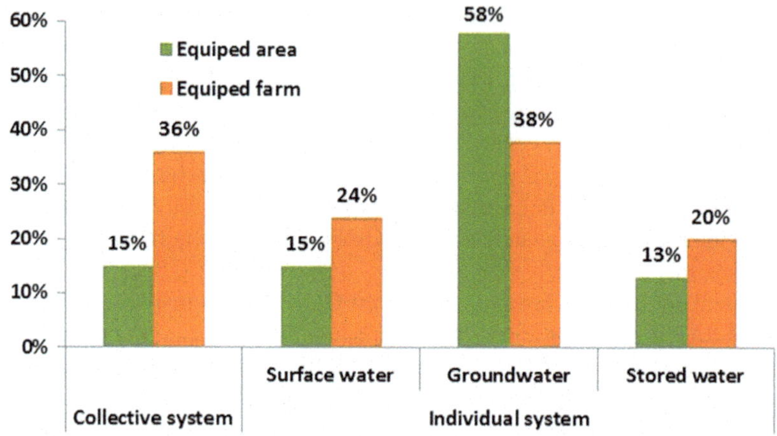

Fig. 5.3 Percentage of the equipped area and of farms equipped for irrigation in 2010, according to the type of system (Loubier et al. 2013)

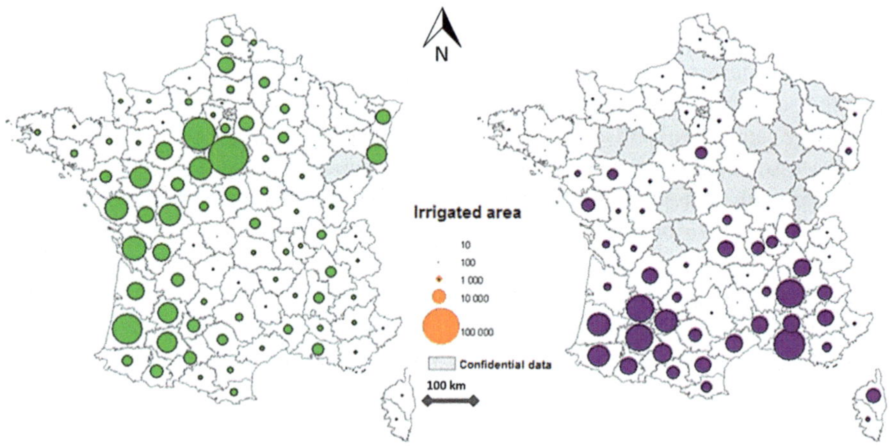

Fig. 5.4 Geographical localization of 2010 irrigated areas based on individual (green) and collective (purple) systems (confidential data correspond to departments where only three or less observations are available or one observation represents more than 80% of the observations) (Garin et al. 2013)

water requirements and/or supplement irrigation. Irrigation networks depending on surface water are more developed in the south and meet higher crop water requirements. Finally, the volume of groundwater used is believed to be roughly equivalent to that abstracted from rivers, dams, and small reservoirs.

The spread of individual irrigation poses specific management challenges to OUGCs (Loubier and Garin 2013). Individual equipment decisions are specific to individual farms and have not been thought through collectively, with neighboring farmers. They are regarded as private investments, authorized by state services in the past without taking the area's water potentialities into consideration. These abstractions are seen by farmers as "individual rights" which have been acquired. In some regions, many illegal abstraction points (without authorization) exist and make the collective management of individual pumping difficult. This is all the more true because the resources of the water police are continuously decreasing, compromising its work, and when sanctions are warranted, justice does not systematically follow.

Groundwater is no different from surface water in that the same instruments of volumetric management apply and in this case are based on piezometric (water level) thresholds. Groundwater management issues are not as significant as in other Mediterranean countries. However, some aquifers can be quantitatively overexploited or qualitatively damaged.

The Beauce aquifer, southwest of Paris, is an interesting example from a management point of view. Irrigation development based on individual groundwater pumping started at the end of the 1970s. The aquifer extends over 10,000 km^2, and 90% of the area is agricultural. Soils are somewhat superficial; rainfall tends to be low (630 mm/year); and the available infiltration for aquifer recharge is low (120 mm/year) (Lejars et al. 2012). Irrigation developed alongside the establishment of transformation industry (sugar beet refinery and vegetable industries). Farmers with irrigation facilities also used them, if needed, for spring wheat irrigation. Water demand continuously increased until the mid-1990s. This period was characterized by the lowest piezometric level ever observed, dry river beds, and endangered wetlands (Petit 2009), which triggered a protest by environmental groups and a long new governance process. Quotas and restrictions were implemented. As hydrogeological knowledge improved, measures were adjusted, and sub-management areas were established in accordance with the environmental impact of abstraction. Today, the area equipped for irrigation represents 53% of the agricultural area and nearly 60% of farms. In total, a maximum of 420 million m^3/year can be abstracted for irrigation in favorable conditions and only 200 million m^3/year in an average year. At the end of winter, depending on piezometric levels, state services define an aquifer coefficient of between 0 and 1 (1 corresponding to a potential abstraction of 420 million m^3/year). An OUGC manages farmers' water demands and proposes to the state services a water allocation plan among farmers. The large interannual variation in aquifer coefficient can lead to cropping pattern adjustments and, in extreme cases, to competition between food processing sectors which depend on specific crops (Lejars et al. 2012). Although farmers and food processing industries face new uncertainties, environmental impacts are now believed to be controlled and other aquifer uses are secured.

Several other smaller aquifers are overexploited, including the Roussillon aquifer, as discussed by Montginoul et al. (2016). Here, abstraction points are rarely declared, and farmers contest their impact on the piezometric levels of the aquifer and the consequences in terms of saline water intrusion, challenging effective management. Rinaudo and Herivaux (2016) identified other interesting groundwater management instruments adapted to local conditions in Aisne *Département* and Tarn et Garonne, for example. Based on these, the authors highlight (i) a large diversity of water allocation criteria to face overexploitation and (ii) the relative latitude of OUGCs to implement local and accepted instruments.

5.3 Present Situation and Recent Developments in Irrigation Practice

The development and adaptation of irrigation in France are the result of changes in many domains. The irrigated area and irrigation practices slowly adapt to global, sometimes conflicting, external factors: the EU Common Agricultural Policy, environmental issues, market volatility, farmers' constraints, increased energy costs, water prices change etc. The current changes also illustrate adaptation strategies deployed to face future global change (climate, political, and economic).

5.3.1 Irrigated Areas Are Leveling Off

Irrigation in France is a structural economic imperative only for vegetables and arboriculture. Major rain-fed crops (cereal, protein, and oil crops) and forage crops can be grown nearly everywhere, even if potential yields are lower, such as on the Mediterranean coast. Actual irrigated areas vary from year to year, depending on the climate, local resource restrictions (temporary irrigation bans), and international markets. In 2010, the total irrigated area amounted to 1.57 Mha. The actual irrigated area was roughly the same in 2000 and 2010. However, this stability hides the fact that, in the very dry spring of 2010, all over the country, farmers with irrigation equipment were forced to irrigate spring crops (mainly cereals), thereby artificially maintaining the national irrigated area. In reality, a more meaningful indicator of irrigation dynamics is the equipped area, and, for the first time in France, it decreased by 12%, as the sociopolitical and economic contexts evolved and shifted support away from this kind of investment (Loubier et al. 2013). This decline mainly affects southern and southwestern *départements*, where the proportion of the agricultural area with irrigation facilities is high (Figs. 5.6 and 5.7). Some *départements* in central and eastern France may show some increase, depending on water availability, local farm dynamics, and market opportunities (Fig. 5.5).

Variations in the area equipped for irrigation and actual irrigated area can be explained by several factors:

- Over a period of 10 years, maize (grain and seeds) has lost over 135,000 ha (17%) while still ranking as the main irrigated crop (Fig. 5.8). The explanation is mainly political, since, in the 1990s, targeted subsidies for irrigated crops, which had until then been the core engine of expansion, were discontinued. Gleyses (2006) and Loubier and Gleyses (2009) showed how this political reform contributed to a 13% decline in the irrigated area of maize, while the total water consumption remained stable, pointing to a more intensive use of water per hectare.

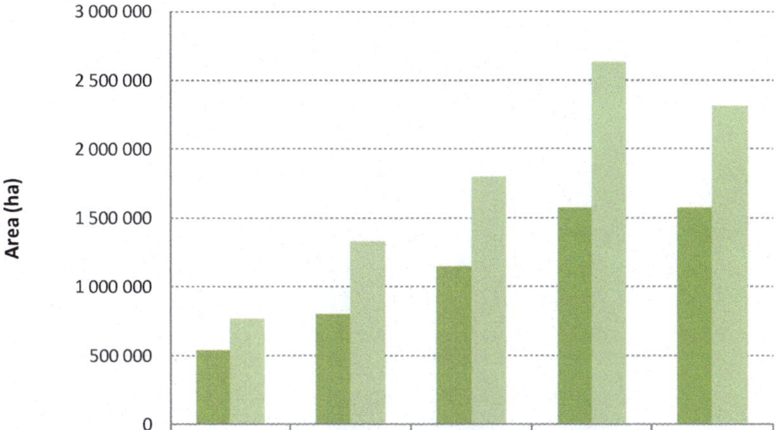

	1970	1979	1988	2000	2010
Irrigated area (ha)	538737	800533	1146988	1575625	1 574 789
Equiped area (ha)	767200	1325227	1796469	2633682	2 308 707

Fig. 5.5 Irrigated and equipped area over the last five agricultural censuses in France (Loubier et al. 2013)

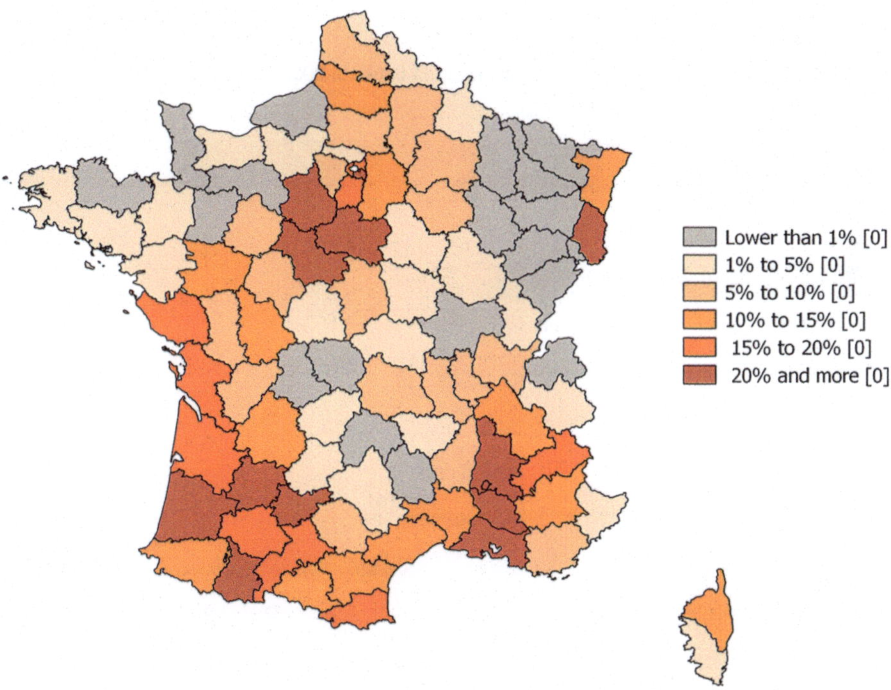

Fig. 5.6 Agricultural area equipped for irrigation in France in 2010 (Campardon et al. 2012)

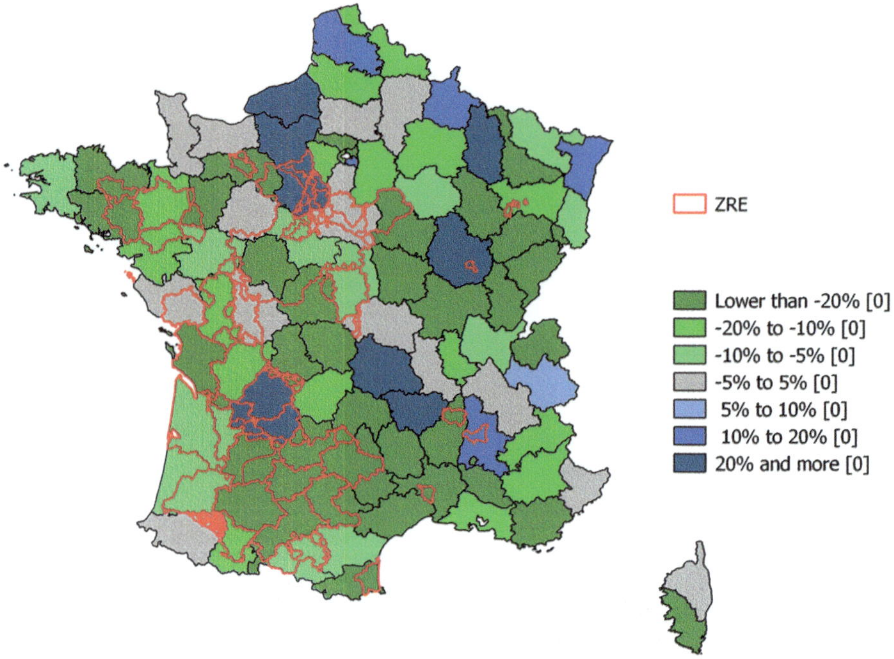

Fig. 5.7 Variation in the area equipped for irrigation between 2000 and 2010 and river basins facing structural quantitative imbalance

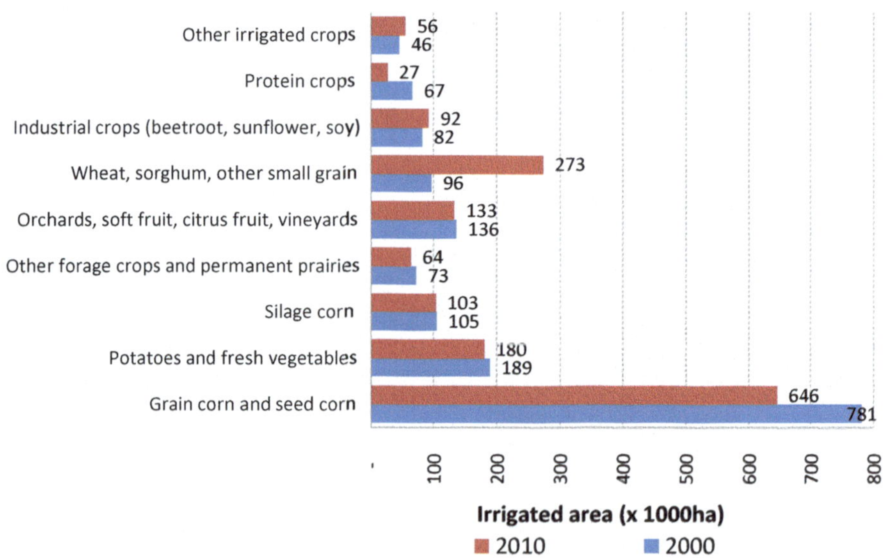

Fig. 5.8 Area of main irrigated crops in 2000 and 2010 (SSP – 2000 and 2010 agricultural census)

- Administrative measures restricting water use and reducing global abstraction authorizations were adopted. Figure 5.7 highlights that the *départements* where the equipped area decreased are highly correlated with ZRE (basins facing structural water deficits).
- In the west and southwest of France, where irrigated maize is by far the main irrigated crop, the decrease in equipped areas can also be explained by extremely volatile agricultural prices and the substantial increase in the price of wheat, to the detriment of maize. In southern France the decrease is also due to recurrent fruit and vegetable crises in a context of high competition between irrigated areas (mainly in plains, since these departments are equipped with gravity systems) and urban development.

The decline in irrigation over a 10-year period is also observed for fodder crops and permanent prairies (−12%) and silage maize (−2%). Irrigated protein crops have dropped by close to 60% in nearly all regions. Potatoes and fresh vegetables have also dropped (−5%), along with fruits and vineyards (−2%), although there are strong regional discrepancies. In Languedoc-Roussillon, for example, the drop in irrigated areas devoted to fruit has been more than compensated for by the increase in the area of irrigated vineyards (+10,000 ha). The considerable expansion of micro-irrigation in this area suggests that a major portion of recently planted vines are irrigated with drip systems.

5.3.2 Long-Term Trends and New Irrigation Practices

Successive agricultural censuses show that gravity irrigation, practiced by one out of three farms in the late 1970s, is now practiced on only 8% of the farms equipped for irrigation and corresponds to less than 3% of the equipped area. They are all in the Mediterranean region, where old canal networks have been preserved. There are several very large rice-growing farms, while the vast majority of small farms grow fruits and vegetables or irrigate meadows. This gradual disappearing of gravity irrigation is due to three factors. First, the operation of gravity systems is perceived to be too time consuming and constraining. When farmers have the choice, they prefer groundwater individual pumping or to modernize their material into drip and automated irrigation systems. Sometimes networks are gradually abandoned and, without maintenance, decay rapidly. Second, as presented earlier, there is growing competition in plains between urbanization and gravity networks. In some cases, this leads to a simple demise of irrigation systems, but, most of the time, gravity systems are turned into multiple-use systems that deliver raw water to municipalities, private gardens, and sometimes industry, and are partially modernized into pressurized systems. Given the superposition of canal systems and urbanized areas, networks are also used as rainwater evacuation systems to prevent flooding. The financial equilibrium is then found thanks to new users' greater willingness to pay, allowing farmers to maintain their level of contribution. The last reason is due to new environmental constraints. Even if gravity systems return most of their withdrawals to the environment (aquifers or downstream rivers), rivers can face local water deficits. In such cases, the state, before enforcing administrative restrictions, encourages the modernization of the system into a pressurized network. However, this is very costly, requiring considerable subsidies, and therefore cannot be carried out more broadly. In other words, the substantial decrease in the practice of gravity irrigation does not always mean a disappearance of irrigation systems but also modernization and the development of multipurpose systems.

High-pressure irrigation systems are the most widely used. They are found on 78% of farms, 90% of the area equipped for irrigation (source: agricultural census 2010), and all crops except most fruit and vegetable areas. Thus, this equipment is largely dominant except in the Mediterranean region where both fruit and vegetable farming are commonplace and gravity irrigation systems are still in use. Because of topographic constraints and farm land fragmentation, reel machines are by far the most commonly used system, before center pivot and lateral move systems (Table 5.2).

Micro-irrigation (drip irrigation and low-pressurized systems), which was adopted by only 3% of farms in 1979, is now used by one farm out of four but for relatively small areas. Its development seems to be more or less proportional to the decline in gravity irrigation. Canal networks in southern France are gradually being modernized to give way to drip irrigation of perennial crops (vineyards and fruit trees) and vegetables.

The choice of irrigation system almost never depends on water scarcity, but on labor constraints, agricultural sector specifications, investment capacity, and crop type. However, when farmers face quantitative constraints or want to extend their irrigated area, they frequently invest in automation, tensiometers, or any equipment preventing water waste rather than changing their irrigation system.

Table 5.2 Recent changes in irrigation technology (2000 and 2010 censuses)

	Gravity irrigation	Sprinkler irrigation	Micro-irrigation
No. of farms equipped for irrigation			
2000	13,753	85,051	18,385
2010	6981	63,865	20,481
Total equipped area			
2000	133,220	2,402,970	97,492
2010	108,824	2,156,691	203,660

Farms may use several types of system and be counted several times. The area of farms with more than one system is also counted twice, which explains the total area (2.47 Mha) being higher than in Fig. 5.6 (2.3 Mha)

If the adoption of micro-irrigation is largely unrelated to water scarcity, this is also partly true for irrigation in general. Irrigation is a means to secure production but also to ensure the quality of the product in terms of caliber, shape, sugar content, etc. For example, the famous Agen plums (*pruneaux d'Agen*) are irrigated to ensure their conformity with the demand of both the processing sector and supermarkets, which want uniform and "flawless" products.

5.3.3 Present Water Pricing

5.3.3.1 Water User Associations

Water tariffs vary significantly. In gravity networks, the flat fee for water ranges from tens of euro per subscribed hectare to €300, and sometimes even €400 for modern networks, to which high financial costs must be added. For pressurized networks, the price structure is most often binomial and determined by the ASA based on its specific local costs. The fixed part of the price is composed of fixed charges (loan reimbursement for the 20% of the investment cost shouldered by the landowners: electricity subscriptions, salaries, etc.). The cost of electricity makes up the bulk of the tariff's variable part. For example, the average tariff in the Loire-Brittany Water Agency basin in 2004 was €210 per hectare and €0.06 per cubic meter consumed (Montginoul et al. 2015).

The value added generated by irrigation also varies widely (Acteon et al. 2011). The gross margins for irrigated crops (estimated as price*yields + agriculture subsidies – variable costs) differ according to several parameters, even in the same region. For example, in the Garonne basin in 2011, gross margins varied between €430 and €469 for 1 ha of durum wheat or sunflower and between €550 and €650 for 1 ha of maize but were around €17,000 for 1 ha of vegetables grown in open fields. Since the size of the farms and their structural charges vary extensively, it is difficult to estimate the net income due to irrigation.

5.3.3.2 Regional Development Agencies (SARs)

SAR's water tariffs are generally higher than for ASAs (see Table 5.1) since they correspond to a different logic: infrastructure precedes water demand and includes a gamble on future developments. As a result, they are economically less optimized than smaller systems managed by water user associations. In addition, water management in public networks is partly demand-driven, which requires a degree of oversizing compared to smaller systems where water turns allow for lower investment costs. Users only pay a modest part of the investment, which received 70–90% of public subsidies.

5.3.4 The Hérault River Basin: An Example of Rapid Change

Changes in irrigation at national level are rather slow but hide rapid local change, as in the Hérault River basin in southern France (Ponchant et al. 2017). In this 2500 km^2 basin with a Mediterranean climate, vineyards account for 40% of the agricultural area. Until 2000, irrigation continuously decreased because of the demise of old gravity systems. Policies aimed at limiting the overproduction of wine and improving its quality led to a significant reduction in the vine area (−30% in 20 years). Before 2006, the irrigation of vineyards was prohibited in order to limit the overproduction of low-quantity wine, but an easing of this regulation generated rapid and significant change. In 2000, only 4% of vineyards were irrigated, and vine irrigation represented 58% of the total irrigated area, in 2013, while the total vineyard area had decreased by 20%, its irrigation had tripled, as a result of local policies supporting drip irrigation projects. Ponchant et al.'s (2017) surveys reveal that irrigation is used in order to remain competitive on wine markets rather than to adapt to drought and future climate change.

5.4 Challenges for the Future

5.4.1 Reduction in the Water Available for Agriculture

In many catchments in southern France, environmental objectives are translated into minimum discharge requirements at several specific points of river systems. These requirements are defined as a percentage of the theoretical average natural discharge (i.e. not influenced by human activity). All water in excess (if any) is then available for human use, and prioritised for domestic use. Faced with a common disregard for such minimum discharge, the state conducted new assessments of the available water and revised previous abstraction authorizations. Consequently, the available water for human use should be reduced with regard to past abstraction rates, with reductions as high as 90% in basins such as the Seudre basin, in the Charentes region (Hébert et al. 2012). In a context where the government has previously delivered more abstraction authorizations than the environment was able to sustain, and where farmers still have considerable political weight, compromises have to be negotiated:

- In catchments where this is possible, farmers advocated for the creation of winter storage infrastructure to be filled when water is abundant. These are generally collective and benefit from public subsidies.
- In other catchments, exemptions to seasonal or annual volumetric caps are being granted, and abstraction can be based on the observation of actual discharge and the respect of minimum thresholds. Therefore, during wet and "normal" years, the maximum volume which may be abstracted is much higher, while negotiations

take place during drier years, with the agricultural sector trying to avoid any irrigation ban or restriction.

Local negotiations involve the Water Agency, farmer representatives, state services, and regional authorities and may result in cost-sharing agreements for infrastructure. But their implementation is facing increasingly strong opposition from civil society which criticizes both (i) public subsidies that support a form of intensive agriculture which generates a number of negative externalities and (ii) the legitimacy and transparency of the administrative/political processes validating these projects. Consequently, many of these projects face legal challenges and even violent demonstrations and opposition (e.g. the case of the Siven dam, in the southwest). Because of such opposition, it is doubtful that the initial compromises negotiated between the state and farmers will allow excessive past agricultural abstractions to be regulated.

5.4.2 Uncertainty Concerning OUGC Effectiveness

In basins facing structural deficits, the state cancels all individual abstraction authorizations and entrusts the OUGCs with establishing a new water sharing plan among individual farmers and collective organizations. If the plan is compatible with the total available volume for agriculture, the state approves it. The logic behind the creation of OUGCs was to confer on farmer organizations the responsibility of managing "their water," where the state had partly failed. The state retains most of its past water policy remit but transferred responsibility for water sharing, pricing, monitoring individual abstraction points, and penalizing if needed. If the OUGCs fail to enforce regulations, the state will step in, but more easily than in the past, since the OUGCs prepare an annual report showing whether the total available volume is being respected overall and by each farmer. If the limit is being adhered to, the state considers it the OUGC's responsibility to take action against any farmer exceeding their individual allocation. Another way to empower OUGCs has been to require them to propose measures to be implemented in dry years (1 year out of 5 theoretically) so as to avoid administrative water restrictions. OUGCs are too recent to allow an assessment of the effectiveness of that reform. However, the analysis of the content of their status and internal rules shows a large diversity in the way they intend to fulfill their missions. Some will simply propose to the state a water-sharing plan, whereas others will seize all the opportunities offered, including monitoring and enforcement. The effectiveness of that new water management policy remains to be seen.

5.4.3 Intersectoral Coordination

Irrigation regulations and policies may also resort to indirect economic incentives. Several past policies conflicted with water management objectives. The energy policy, with its artificially low cost of electricity, lowered costs for irrigators. The Common Agricultural Policy (CAP) encouraged the expansion of irrigated areas since agricultural subsidies were higher for irrigated than rain-fed crops. Local policies (municipalities, *départements*, and regions) also encouraged irrigation because of its role in preserving a farming population and spillover effects on upstream and downstream sectors. They finance part of the investments, modernization or extensions of infrastructure.

Current sectoral policy seems better equipped to achieve the common objective of more sustainable irrigation.

- Energy costs regularly increase more than inflation, particularly after anticipating the dismantling of production infrastructure. The increase in electricity tariffs has the potential to assist in reducing wastage but can sometimes affect the profitability of energy-intensive pressurized irrigation networks. Electricity prices increased by 60% between 2008 and 2015 but remain 20% lower than the European average.
- Agricultural policies sanction behaviors which go against water management objectives (Roy 2013). Irrigators who do not comply with their abstraction authorizations will see their main European aid (Single Payment Entitlements, DPU) lowered by 3% and by an additional 1% if they do not have a metering system. Several measures have been implemented as part of the second Common Agricultural Policy. In some regions, irrigators may adopt agri-environmental measures (AEMs), whereby in exchange for closing a water abstraction point or substituting legume crops for maize in some rotations, they may benefit from a 5-year compensatory aid. Within the framework of the *Programme de Développement Rural Hexagonal* (PDRH, French rural development program), investment aid is extended for constructing "substitution storage" (filled in winter), modernizing networks, water transfers, or creating new irrigated schemes, as long as the resource is maintained in an acceptable state. These schemes, which were implemented from 2007 to 2013, have been extended to the 2014–2020 period.
- As the benefits in terms of development are potentially considerable, water agencies will make their support conditional upon irrigation projects being part of local development projects that enhance collective objectives and benefits.

With this in mind, water agencies are currently considering the future of surface irrigation. This technique results in the withdrawal of considerable amounts of water from rivers, but much of the water returns to the environment (drainage system, groundwater), with potentially positive effects (environment, pumping for drinking water, aquifer recharge); drainage networks evacuate part of the water during heavy rain periods; landscapes remain attractive in summer; engineering works become local heritage sites. These benefits sometimes explain why regional authorities are attached to maintaining old systems, even if it means modernizing them and reconsidering their governance (financing and decision-making) (Kuhfuss and Loubier 2013; Ladki and Garin 2011).

5.4.4 *Wastewater Reuse*

Some countries resort massively to treated wastewater (e.g. 90% of wastewater in Jordan is treated and reused for irrigation), but in France it is estimated that only 0.15% of the annual volume of wasted water is put to another use (Hubert 2017). The regulatory context did not encourage the reuse of treated wastewater in the past, but this is slowly changing. Many decision-makers consider reuse as a real opportunity for reducing pressure on the environment and contributing to local development. However, in many rivers in the south of France, the discharge is mainly made of wastewater in summer, helping to meet environmental needs (minimum river flow). The future of reuse in France remains uncertain.

5.4.5 *Climate Change and Adaptation Strategies*

All water agencies have started to study the effects of climate change, especially in the southern half of France, where IPCC models agree that an increase in ET is to be expected, along with a rainfall decrease in the summer, a snowpack decrease, and more frequent and severe droughts.[3]

These studies highlight an increased risk of pressure upon water resources in summer, with an increased agricultural demand, river discharges 20–80% lower than today, growing uncertainty with regard to dam filling, and higher risks of eutrophication. The demand for groundwater will also be higher.

It is already anticipated that a more systematic use of irrigation will be the preferred approach, alongside transforming cropping systems and practices or developing appropriate cultivars. But, as highlighted by participatory prospective exercises (e.g. Richard-Ferroudji et al. 2013), farmers do not see climate change as a threat – climate vagaries are part and parcel of life. They mostly feel vulnerable to medium-term unforeseen changes in the economic (market, international competition) and sociopolitical (regarding social support to farming) environments. This latter issue will be crucial to the evolution of how a rarer, and more

[3] For example, http://oai.eau-adour-garonne.fr/oai-documents/60718/GED_00000000.pdf; www.eaurmc.fr/fileadmin/grands-dossiers/documents/Changement_climatique/bilan_connaissances_Chgt__Clim_AERMC_couv_def.pdf, respectively, for the Adour Garonne basin in the southwest and the Rhone basin in the southeast.

vulnerable water supply will be shared between competing sectors. A further decisive factor will be whether there is sufficient social consensus for new water infrastructure (dams, interbasin transfers), which farmers see as the main response to climate change.

5.5 Conclusions

Irrigation developed in France over several centuries and totals about 2 million hectares. Early development consisted of collective irrigation systems abstracting water directly from rivers, while, more recently, individual pumping systems in aquifers, rivers, and reservoirs have come to prevail. The two main collective irrigation management systems are water landowners' associations (ASAs) and regional water management companies (SARs). ASAs manage their system according to common pool management principles identified by Ostrom (1992), defined in an 1865 law and modernized in 2004. SARs are an example of public-private partnership. Both systems have shown their capacity to manage irrigation schemes for over 60 years, covering operational and maintenance expenditures but only very partially investment and long-term replacement costs.

Individual irrigation has been developing for 30 years in the southwest and gradually reached large plains in the center of France. This development took place with little concern for actual water availability, and as a result many river basins face water use conflicts between irrigation, environment, and other users.

Collective water management has thus become problematic. The state, which is partly responsible for this situation, lacks the legal framework and human resources to enforce drastic irrigation restrictions, except in collective irrigation systems run by ASAs. Furthermore, the legitimacy and the financial means to develop irrigation water storage infrastructure are not ensured since there is no social or political consensus regarding the necessity of irrigation to maintain productive agriculture. Therefore, a reform of the water management policy took place to transfer to OUGCs part of the water management burden in watersheds facing structural deficits. Much hope has been pinned on the effectiveness of this measure, but, at the same time, agricultural organizations expect subsidies to build new controversial infrastructure to limit current environmental impacts. Among alternative solutions, reuse, poorly developed in France, seems to be a marginal solution to the reduction of deficits.

The climate change argument, advanced by farmers, fails to convince the opponents to water transfer or dams projects, who reject intensive irrigated agriculture models. These opponents challenge the economic interest of irrigation on account of its environmental consequences, such as wetland destruction, pollution risks, and competition among uses during dry periods.

The future of irrigation is thus very uncertain in France. In regions which are being equipped with on-farm or local water storage infrastructure and are most likely to face climate change (e.g. the southeast), land competition with urbanization and the economic weakness of the farming sectors are problematic. In other regions (southwest, center, and centre west), water resource vulnerability, water use conflict, and the lack of consensus on what agriculture should be are all central issues. However, irrigation will remain a key option for reducing production risk and responding to demand in the quality of products in highly competitive European sectors.

The future of water management is also uncertain. In a context characterized by the shared responsibility of farmers and the state in the current overabstraction of water and the poor enforcement of laws, the effectiveness of decentralizing water management for irrigation is in question, as is the development of water storage capacities to secure water, develop irrigation, and anticipate climate change impacts.

References

Acteon, BRGM, Cemagref. (2011). Révision des autorisations de prélèvement d'eau pour l'irrigation sur le bassin Adour-Garonne. Rapport de phase I: Évaluation territorialisée de l'impact sur l'économie agricole. Agence de l'eau Adour-Garonne (p. 237).

Assier-Andrieu, L. (1985). Le destin de la loi Stratae: droit des puissances et droits du peuple en Catalogne. Thèse en Droit, Université des Sciences Sociales de Toulouse.

Bertrand, J. (1764). *De l'eau relativement à l'économie touristique, ou traité de l'irrigation des prés*. Avignon: Regnault.

Campardon, M., Loubier, S., Morardet, S. (2012). *L'irrigation en France: Etat des lieux 2010 et évolution. 2011/2012*. RFCBI.

Chenard, G., Kabashi, S., & Manganelli, L. (2010). *Eaux courantes, du torrent au canal du Moyen Âge au milieu du XXe siècle*. Gap: Archives départementales des Hautes Alpes.

De la Moskova, P. (1847). Rapport fait à la Chambre au nom d'une commission spéciale chargée de l'examen du projet de loi sur les Irrigations.

Dubois, A., & Ruf, T. (2003). Ansignan, A Pyrenean valley with many hydraulic historical enigmas. IWHA 2003 Conference Abstracts.

Dumond, C. (2002). Les canaux d'arrosage du Briançonnais. Modalités de gestion et droits d'eau. In O. Aubriot & G. Jolly (Eds.), *Histoire d'une eau partagée: Provence-Alpes-Pyrénées* (pp. 101–125). Aix-en-Provence: Publications de l'Université de Provence.

Fabre, G., Fiches, J. L., & Paillet, J. L. (2013). Vie et mort de l'aqueduc de Nîmes. *Pour la Science, 322*, 52–59.

Garin, P., & Loubier, S. (2002). Durabilité et fragilité des Associations Syndicales Autorisées (ASA) d'irrigation en France. *Comptes Rendus de l'Académie d'Agriculture de France, 88*(3), 61–71.

Garin, P., Loubier, S., & Campardon, M. (2013). Irrigation individuelle – irrigation collective: état des lieux et contraintes. *Sciences Eaux & Territoires, 11*(2), 86–89.

Gleyses, G. (2006). Mise en oeuvre de la PAC: impact de la réforme de juin 2003 sur la demande en eau d'irrigation. Rapport final, Série G-EAU-Cemagref «Rapports» 2006-Convention EAHER DGFAR–Cemagref.

Goubert, J. P. (1986). *La conquête de l'eau*. Paris: Robert Laffont.

Hébert, N., Grandmougin, B., Loubier, S., Graveline, N., Marsac, S., Amen, J. F., & Brunel, L. (2012). Réforme des autorisations de prélèvement dans le bassin Adour-Garonne: impacts sur l'économie agricole. *Revue AE&S, 2*(2), 113–126.

Hubert, F. (2017). La réutilisation des eaux usées traitées: Développement d'une méthodologie d'analyse spatiale des potentialités. Mémoire de fin d'étude Irstea – Ecole Nationale Supérieure des Sciences Agronomiques de Bordeaux. 124p.
Ingold, A. (2011). Gouverner les eaux courantes en France au XIXe siècle Administration, droits et savoirs. *Annales. Histoire, Sciences Sociales, 66*(1), 69–104.
Kuhfuss, L., & Loubier, S. (2013). Comment évaluer les services rendus par les réseaux d'irrigation? Application au cas de la recharge de nappe. *Sciences Eaux & Territoires, 11*(2), 70–75.
Ladki, M., & Garin, P. (2011). Des tuyaux et des hommes. In G. Bouleau & L. Guérin-Schneider (Eds.), *Les réseaux d'eau en France*. Paris: Quae NSS Dialogues.
Lejars, C., Fusillier, J.-L., Bouarfa, S., Coutant, C., Brunel, & Rucheton, G. (2012). Limitation of agricultural groundwater uses in Beauce (France): What are the impacts on farms and on food-processing sector? *Irrigation and Drainage, 61*(suppl. 1), 54–64.
Leveau, P. (1998). Échelles d'anthropisation et archéologie des campagnes de Gaule du Sud à l'époque romaine. *Méditerranée, 90*(4), 17–26.
Loubier, S., & Garin, P. (2013). Un avenir incertain pour les associations syndicales autorisées d'irrigation. *Sciences Eaux & Territoires, 11*(2), 90–95.
Loubier, S., & Gleyses, G. (2009). Evaluation de l'impact de mesures de réduction des prélèvements agricoles proposées par l'IGE et le CGAAER (2007), Rapport: Action N°30 – Convention Cemagref – ONEMA. 41p.
Loubier, S., Campardon, M., & Morardet, S. (2013). L'irrigation diminue-t-elle en France? Premiers enseignements du recensement agricole de 2010. *Sciences Eaux & Territoires, 11*(2), 12–19.
Martin, J. (1972). L'évolution de l'irrigation en France. *Revue de Géographie Alpine, 60*(3), 419–443.
Montginoul, M., Loubier, S., Barraqué, B., & Agenais, A. L. (2015). In A. Dinar, V. Pochat, & J. Albiac (Eds.), *Water pricing experiences and innovations* (pp. 139–160). Switzerland: Springer International Publishing.
Montginoul, M., Rinaudo, J. D., Brozović, N., & Donoso, G. (2016). Controlling groundwater exploitation through economic instruments: Current practices, challenges and innovative approaches. In A. J. Jakeman, O. Barreteau, R. J. Hunt, J.-D. Rinaudo, & A. Ross (Eds.), *Integrated groundwater management* (pp. 551–581). Cham: Springer Open.
Ostrom, E. (1992). *Crafting institution for self-governing irrigation systems*. San Francisco: Institute for Contemporary Studies Press.
Petit, O. (2009). Un regard rétrospectif sur l'évolution de la gouvernance de l'irrigation en Beauce (1993–2008). *Sécheresse, 20*(3), 262–270.
Ponchant, L., Gary, C., Molenat, J., Prévot, J., & Rollin, D. (2017). *Ressources et Usages de l'eau sur le Bassin Versant de l'Hérault: Enjeux de gestion des ressources et d'adaptation des systèmes viticoles aux changements globaux, G-Eau Working Paper No.2*. France: Montpellier.
Pritchard, S. (2012). From hydroimperialism to hydrocapitalism: 'French' hydraulics in France, North Africa, and beyond. *Social Studies of Science, 42*(4), 591–615.
Regnier, P. (1989). *Les saint-simoniens en Égypte (1833–1851)*. Le Caire: BUE-AF Abdelnour.
Richard-Ferroudji, A., Garin, P., Matignon, M., Maton, L., Montginoul, M., Rinaudo, J. D., et al. (2013). Engager des agriculteurs à répondre à l'injonction d'adapter la gestion de l'eau au changement climatique. Une démarche de prospective participative avec des agriculteurs usagers des nappes du Roussillon (France). In V. Ancey, B. Dedieu, M. Antona, et al. (Eds.), *Agir en situation d'incertitude en agriculture. Regards pluridisciplinaires au Nord et au Sud* (Vol. 17, p. 419). Switzerland: Peter Lang, Editions scientifiques internationales. isbn:978-2-87574-037-3.
Rinaudo, J. D., & Herivaux, C. (2016). Quels instruments pour une gestion collective des prélèvements individuels en eau pour l'irrigation? Les Rencontres de l'Onema – Synthèse, 52.
Rollin, D., Brelle, F., Citeau, J. M., & Villocel, A. (2013). Avec leur statut original, quels rôles pour les sociétés d'aménagement régional dans les politiques de l'eau et de l'irrigation? *Sciences Eaux & Territoires, 11*(2), 96–98.

Rouillard, J. (2013). The contribution of iconographic, cartographic and written sources, prior to the French Revolution, to the knowledge of the developments of the valleys of the Seine basin. 9th IWHA Conference Montpellier, France – June 2013.

Roy, L. (2013). Gestion quantitative de l'eau et irrigation en France. *Sciences Eaux & Territoires, 11*(2), 4–5.

Ruf, T. (2001). Droits d'eau et institutions communautaires dans les Pyrénées-Orientales: les tenanciers des canaux de Prades (XIVe-XXe siècle). *Histoire et Sociétés Rurales, 16*(2), 11–44.

Ruf, T. (2002). *Les cycles institutionnels de l'irrigation paysanne, approches théoriques et exemple dans les Pyrénées-orientales et dans les Andes équatoriennes, Travaux et recherches, n° 22, mai 2002: L'eau dans tous ses états* (pp. 47–60). Lausanne: Institut de géographie de l'Université de Lausanne.

Ruf T., & Riaux J. (2008). Institutional and social innovations in irrigation Mediterranean management. ISIIMM. General synthesis. Euro Mediterranean regional program for local water management ME8/AIDCO/2001/0515/59763-P 016. Montpellier, Agropolis International, 108 p.

Tatham, W. (1799). *The political economy of inland navigation, irrigation, and drainage*. London: Faulder.

Chapter 6
Italy

Gabriele Dono, Simone Severini, Davide Dell'Unto, and Raffaele Cortignani

Abstract This chapter analyzes the recent evolution and current situation of the Italian irrigation sector, which presents a remarkable dualism between the north and south of the country. After a brief review of the historical origins of irrigation, the chapter examines the role of the state in the regulation and promotion of irrigation in modern Italy since unification. The chapter describes collective irrigation institutions and analyzes the role of the public sector, highlighting the lack of an authority capable of mediating conflicts between the regions involved in water transfers. It also explores the turn toward demand-side management, through an analysis of legislative changes. The last section identifies the main actual and future challenges: the incapacity of current institutions to control groundwater overexploitation, increasing salinization, and the expected impacts of climatic change.

Keywords Water policy · Legal framework for water · Water transfers · Institutions · Italy

6.1 Introduction

Compared with most other Mediterranean countries, Italy has abundant water resources. It has an average annual rainfall of 296 billion cubic meters (Bm^3), of which 168 Bm^3 are consumed by evapotranspiration, while 155 Bm^3 end up as surface water (runoff) and 13 Bm^3 as groundwater (IRSA-CNR 1999). This provides 2800 m^3 per inhabitant, greatly exceeding the values observed in most Mediterranean countries, and even Great Britain and Germany. However, much of the runoff cannot be used because rainfall occurs mainly from October to March, which requires a storage capacity for using water in summer, when irrigation takes place. According to the Istituto di Ricerca sulle Acque (IRSA), the capacity of the reservoirs at the beginning of 2000 provided an available regulated resource of approximately 40 Bm^3 from waterways and 12 Bm^3 from groundwater (Conte 2010). Table 6.1

G. Dono (✉) · S. Severini · D. Dell'Unto · R. Cortignani
Dipartimento di Scienze Agrarie e Forestali, Università della Tuscia, Viterbo, Italy
e-mail: dono@unitus.it

F. Molle et al. (eds.), *Irrigation in the Mediterranean*, Global Issues in Water Policy 22, https://doi.org/10.1007/978-3-030-03698-0_6

Table 6.1 Available water resources for hydrographic compartment

Compartment[a]	Precipitation			Usable resources (Mm^3/year)		Existing reservoirs (n.)	Underground resources (Mm^3/year)	Available resources (total and per capita)	
	10^6 m^3/year	%	m^3 per capita/year	Without regulation	With regulation			Mm^3/year	m^3 per capita/year
Bacino del Po	71.8	24	4653	11,374	16,118	2194	4468	20,586	1334
Nord Est	42.8	14	6677	4425	10,939	1069	1721	12,660	1975
Liguria	6.4	2	3553	235	372	29	307	679	377
North	121.0	41	5118	16,034	27,429	3292	6496	33,925	1435
Romagna Marche	20.7	7	6127	299	995	212	620	1615	478
Toscana	20.9	7	5847	199	543	141	440	983	275
Lazio-Umbria	24.1	8	4171	321	1399	452	1126	2525	437
Abruzzo-Molise	11.9	4	7729	621	2454	603	248	2702	1755
Center	77.6	26	5434	1440	5391	1408	2434	7825	548
Puglia	13.2	4	3425	13	523	397	325	848	220
Campania	23.2	8	4284	152	1237	77	929	2166	400
Calabria-Lucania	24.0	8	9109	650	2514	1131	595	3109	1180
South	60.4	20	5070	815	4274	1605	1849	6123	514
Sicilia	18.8	6	3862	29	738	718	1151	1889	388
Sardegna	18.3	6	11,542	29	1841	1403	217	2058	1298
Italia	296.0	100	5261	18,347	39,673	8426	12,146	51,819	921

Source: Irsa-CNR (1999)

[a]Hydrographic compartments are a subdivision of the Italian territory made in 1917 as part of the Hydrographic and Mareographic Service, formed to standardize, organize, and make available rainfall, hydrometric, and mareographic measurements. The compartmental division, amended in several cases, considers the major Italian river basins and the administrative organization of their territories

shows the regional distribution of available water in absolute and per capita values, evidencing a concentration in the Po basin and the northeast, but also in some southern regions. The combination of reservoirs and low population gives a high per capita availability in Abruzzo-Molise, Calabria-Lucania, and Sardinia. Puglia, Campania, and Sicily have a lower availability, as well as Tuscany and the Tiber basin. However, the actual availability of water is larger due to transfers between Italian basins. The transfers and reservoirs allow even the lowest-endowed regions to enjoy a per capita water availability of 400 m^3 per year, more than ten times that found in the countries of the Southern Mediterranean (Margat and Vallée 2000).

The National Institute for Environmental Protection and Research (ISPRA) recently used the Bigbang 1.0 model to estimate the Italian hydrological balance (Braca 2017). While the ISPRA methodology was different, it did however indicate a slight reduction in total precipitation, with a sharper drop in the Tyrrhenian compartments (−5/8%) and a growth in the Adriatic (4%). It also pointed to a considerable increase in evapotranspiration as a result of the increase in temperatures and a stable annual rainfall, entailing a reduction in the total availability of water.

Table 6.2 shows the evolution of water withdrawals and uses. Although caution is required when comparing results obtained with different estimation methods, we can note the reduction in agricultural withdrawals from 25.6 Bm^3 in 1970 to 17.0 Bm^3 (agriculture and livestock) in 2012. This is mainly due to the decrease in irrigated areas and irrigation needs, especially accentuated after 1999, as indicated by the last census of agriculture.

Industrial use and energy production have also fallen due to technological improvements and the transfer of more water-intensive activities abroad (Conte 2010). Instead, water withdrawals from the domestic sector continue to grow. The most significant increase occurred in the regions of the northeast and center: Lombardy withdraws the most water, with 1.4 Bm^3 out of a total of 9.1 Bm^3. The last two columns show the actual uses and losses compared to withdrawals: note the large percentage of losses in domestic uses and the much lower values for other uses.

Table 6.2 Water withdrawals and uses by main sectors (different years and estimates)

	National Water Conference (1971) withdrawals		IRSA, CNR (1999) withdrawals		ISTAT 2012			
					Withdrawals		Uses	
	Bm^3	% of total	Bm^3	% of total	Bm^3	% of total	Bm^3	Losses % on withdrawals
Irrigation	25.6	*61.1*	20.1	*49.6*	16.0	*46.8*	13.6	*15.0*
Domestic	7.0	*16.7*	7.9	*19.5*	9.5	*27.8*	5.2	*45.3*
Industry	9.0	*21.5*	8.0	*19.8*	6.1	*17.8*	5.5	*9.8*
Power	–	–	4.5	*11.1*	1.6	*4.7*	1.4	*12.5*
Livestock	–	–	–	–	1.0	*2.9*	0.9	*10.0*
Navigation	0.3	*0.7*	–	–	–	–		
Total	41.9	*100.0*	40.5	*100.0*	34.2	*100.0*	26.6	*22.2*

Source: Irsa-CNR (1999) and ISTAT (2017)

6.2 Historical Development of Irrigation in Italy

Italian irrigation is characterized by a large diversity of irrigation practices and systems resulting from a long history, with works made over the centuries that still shape the structure of the management and use of water.

6.2.1 Historical Benchmarks of Italian Irrigation

Irrigation and land drainage were practiced in Italy before the Romans conquered the peninsula, with documented evidence of Etruscan systems in central-northern Italy and Greek ones in the south. The greatest Roman hydraulic works were aqueducts providing drinking water to urban areas, including the 11 aqueducts of Rome. Suburban areas and irrigation fields were equipped with smaller water systems, of which, unfortunately, there are very few archaeological remains. Literary sources mention that the owners of villas, including Cicero, paid annual fees for using the water of those systems that was distributed in rounds of hours and days. The villas were equipped with tanks to ensure supply between water delivery shifts. Only a few instances of Roman dams to collect water for irrigation were found in Italy, while there are many examples in Anatolia, Spain, and Portugal. Catone, Varrone Reatino, Columella, and Virgil mentioned regulations governing irrigation (Fassò 2003).

After the collapse of the Roman Empire, irrigation developed differently in northern Italy and in the south and the islands. In the Po Valley, especially Lombardy, irrigation expanded in the eighth century with the founding of the Benedictine monasteries. Large-scale facilities were developed because of the extension of the plains, the abundant water of the large rivers fed by subalpine lakes, and groundwater, with spontaneous sources and springs. Irrigation canals were built from the fifth century, and irrigation was practiced from the seventh century with sewage drains in many Lombard cities. Changes in the legal framework encouraged investment in this system of channels.

The peace of Costanza (in the year 1183) constituted a juridical milestone because Emperor Frederick I Hohenstaufen recognized the independence of the municipalities of northern Italy, also granting them the ownership of public water and the considerable proceeds from its sale. This favored the financial commitment of local authorities to build channels for irrigation and navigation: as a result, in the late fifteenth century, the network of those channels was already similar to what it is nowadays. The construction of channels also benefited from another legal institution of the same period, the *servitude of aqueduct,* which prevented landowners from opposing the passage of the waters of others through their land, if damage was compensated.[1]

[1] Water-dependent business was also favored by technological innovations, notably in the field of irrigation. A key invention was the "Edificio Magistrale Milanese" by the Engineer Giacomo Soldati in 1571, to measure the water diverted and define diversion rights and related payments. Another invention was the Cipolletti weir in 1886, for measuring and controlling the water flow of

Irrigation canals and mills were also built in Piedmont from the eighth century, and some are still operating. Large facilities were built from 1448 in Vercellese, Novarese, and Lomellina, while the Venetian Republic constructed the Brentella Channel on Piave in 1443 to irrigate the plains of Treviso, a canal that is still in use. Another work was the diversion of the Po River course in its final stretch, which took place at the beginning of the sixteenth century to avoid the silting of the Venetian lagoon, and created a network of watercourses that still characterizes that part of the Po River basin (Fassò 2003; Quaglia et al. 2011; INEA 2014). Also in central Italy, from the thirteenth century, there was a resumption of the construction of aqueducts, some even able to move water uphill, as in the cases of Perugia (1276–1278), Orvieto (late twelfth century), and Spoleto (thirteenth to fourteenth century).

Since the second half of the seventeenth century, private landowners came together in various parts of the north to manage the distribution of water and solve hydraulic and reclamation problems. Many of these spontaneous associations were small, especially in the Alpine areas of Valle d'Aosta, Trentino, and Bolzano Province. Other landowner associations were larger and began to manage water resource exchanges at the interregional level in Piedmont, Lombardy, Emilia, and Veneto (INEA 2014).

The situation was different in southern Italy and the islands, where, after the collapse of the Roman Empire, small-scale irrigation developed during the Arab and Norman periods. Many hydraulic techniques were introduced in Sicily under Arab rule and were still in use in the 1960s; indeed modern Italian citrus-growing expressions are derived from Arabic. Despite these local experiences, major irrigation projects were not developed in the south for many centuries because of the narrow plains and the torrential regime even of the main rivers. Socioeconomic factors, such as the predominance of large estates, discouraged the municipalities and user associations from building large, collective irrigation facilities.

6.2.2 *Water Management and Irrigation from the Unification of Italy to World War II*

After the unification of Italy in 1861, water policies extended the efforts made since the fifteenth century to develop irrigation systems (Rainaldi 2010). The magnificent Cavour Canal, built in 1863–1866 to irrigate the Vercellese region with water from the Po River, became the hydraulic symbol of this new national era.

---

the channels, disseminated and still used in Italy and abroad, particularly in Latin America. The theoretical and experimental studies by De Marchi in the 1930s–1940s on the instruments for the determination of flow rates in channels, simple to construct and maintain, were widely used in the Italian networks of irrigation in the past 60 years.

The Italian Civil Code (1865) recognized rivers as public property and subordinated their use to public concessions whose timing and conditions, and the price payable to the State, were ruled by the Act on public works of 1865. Landowners maintained their previous rights to water use and made them official by the system of concessions. Irrigation was not subject to regulatory restrictions except for the obligation to return the excess flow to water bodies, so as not to affect the availability and rights of downstream users. No criteria were established to identify the rational use of water resources. Subsequent legislation simplified the procedures for obtaining concessions in order to promote the use of water and satisfy the needs of economic growth.

Water policies began to assign key tasks to landowner associations that had been set up to manage local water resources. Yet, the limited presence of such organizations in central-southern Italy accentuated the contrast with the north with regard to developing irrigation and collective water management. In fact, many subsidies for reclamation and irrigation projects were granted by the Cavour government to these bodies, whose absence in the southern regions limited their access to such resources. Only later were organizations established in the center-south as part of public projects to build and manage dams and water distribution networks and to implement reclamation projects against malaria.[2] The Royal Decree 368, May 8, 1904, spelled out the institutional roles of landowner associations, calling them Reclamation Consortia (*Consorzi di Bonifica*). The decree approved the regulation on the reclamation of swamps and wetlands (reclamation police).[3] This public foundation made the landowner associations in the center-south larger than the homegrown organizations that operated in the north. Furthermore, their irrigation and reclamation activities were more integrated.

The situation changed in the first half of the twentieth century, when the commitment to hydroelectric production emerged as an ideological pillar supporting industrial growth as well as reducing dependence on imported coal (Isenburg 1986; Rainaldi 2010). Public concessions in perpetuity were abolished, as was the chronological priority for allocating water rights. Only surface water was defined as public, mainly to ensure its use for military purposes. Mountain waters, springs, and groundwater remained private and excluded from control. Project evaluation began to be based on their impact on collective interests and on their economic potential. Diversion rules were reformed to promote hydropower production, even at the expense of irrigation, as dam water releases came to be governed by the need for

---

[2] Large reclamation projects in the Tyrrhenian coastal of Tuscany and Lazio were also prosecuted to enabling stable irrigation activity. The Act for the *Agro Romano* reclamation, 4642/1878, approved after proclaiming Rome as Capital and applied remediation to eradicate malaria and make marshes and wetlands suitable for agricultural use, becoming a model for remediation in first postwar period (Novello 2003).

[3] The reclamation police is the surveillance activity on reclamation works, aimed at soil protection and hydraulic safety: it is an activity of prevention, paralleled with maintenance, to ensure the full functionality of the consortium works.

hydroelectric production. Industries exploited the potential offered by the transport of energy over long distances and invested in the hydropower sector.[4]

The organization of the reclamation activities was also modified. The role of the state, defined by the *Bonifica Integrale* principle, was extended from hydraulic works in the marshes and wetlands to road works, flood protection, reforestation, and the consolidation of slopes. This intervention was mainly intended to improve the sanitary conditions of the territory, especially in the marshy and malarial areas. This established the principle that works of general interest may be carried out throughout the country, and not only in swampy areas, to increase the productivity of natural resources.

This approach was reconsidered after World War I, and some initial decisions of the Fascist regime slowed down the commitment to reclamation and land consolidation, including the abolishing of the Ministry of Agriculture in 1923. This action had a counter-effect in that, in 1928, the 3134 Act on *Bonifica Integrale* (*Mussolini Act*) reinforced reclamation, correcting previous choices. A key role was entrusted to private initiative that proved ineffective because the entrepreneurs could only mobilize limited financial resources. The law financed both the reclamation work already under way and new works, even when these included irrigation systems. In 1929, the Ministry of Agriculture was reinstated, taking over the remit of the Ministry of Public Works regarding irrigation, land consolidation, and drainage.

Despite these acts, hydropower and reclamation prevailed over irrigation. The state used the Consolidated Law on water and electrical systems [Royal decree (Rd) 11 December 1933, n.1775] and the Remediation Law (Rd 13 February 1933, n.215), also known as the *Serpieri Law*, to establish major agreements with corporate interests of hydroelectric production and main landholders (Borelli 2008; Rainaldi 2010). Based on these laws, major facilities for the artificial regulation of subalpine lakes were built in the period 1933–1947. These still represent an annual reserve of over 1 Bm^3 for irrigation in the Po Valley and contribute to hydropower generation and, to a lesser extent, flood control in the Po River basin (RBD of the Po River 2010; Barbero and Bertoli 1998; Regione Lombardia 1996).

In this context, the landowner associations were given public legal personality, as self-governed bodies responsible for managing and maintaining public reclamation and irrigation works. In particular, Article 54 of the *Serpieri Law* spelled out the role of private owners of properties benefiting from reclamation and organized in mandatory Consortia that provide for the maintenance and operation of reclamation works, and sometimes even their initiation (Galvani 2009). This institutional role

[4]Yet, the Ministry of Public Works had to monitor who appropriated water, and in the Triveneto area, its competences were reorganized with the establishing of a decentralized ministerial body (Magistrato alle acque – Water Magistrate, Law No. 257 of 5 May 1907) responsible for the protection, safety, and management of the hydraulic system (Isenburg 1986; Massarutto 2008; Rainaldi 2010). The Water Magistrate was also responsible for the reclamation and decontamination of the Venice lagoon and for the port and lighthouses regime. In 1955 the Ministry of Public Works set up a similar body (Magistrato per il Po), after the catastrophic flood of 1954, to coordinate and unify the competences of the various local authorities.

allowed the development of a network of landowner associations (*Consorzi di Bonifica*) when public works for reclamation and irrigation were realized in Lazio, Sardinia, and Puglia and, between 1938 and 1942, in Campania and Sicily.

The results of reclamation and irrigation projects were controversial. The intention enshrined in the *Bonifica Integrale Act* was to intervene on 8 million hectares (ha) and in 1933 the regime declared to have reclaimed 4.7 million ha. Yet, critics such as Bandini suggest that the outcome was less impressive: of 900,000 ha of completed public and private land reclamation, only 220,000–250,000 were satisfactorily transformed, with significant productive outcomes, denser settlements, and redistribution of private property on 70–80,000 ha, largely in the Veneto and Emilia regions (Novello 2003). Over another area of 100,000 ha, irrigation systems were completed, increasing production but without creating new settlements. In the remaining 600,000 ha, there were no significant results regarding production or settlement.

6.2.3 Water Management and Irrigation from World War II to Present

After World War II, industrial development and new energy policies, based on the use of fossil fuel, reduced the interest in hydropower, and, to a lesser extent, irrigation and land issues. As a result, less attention was paid to water management. Yet, significant irrigation investment continued. The construction of reservoirs and networks for long-distance water transfers, initiated as state projects in the early twentieth century, was completed by the Agency for the Development of Southern Italy (*Cassa per il Mezzogiorno*, or CASMEZ) (Act n. 646, 10/08/1950).[5] CASMEZ carried out the main projects for land reclamation and the restoration of the water balance, with environmental interventions on municipal discharge and industrial waste. Major schemes were also funded to rationalize the use of water, promote and market agricultural products, provide infrastructure for developing the hinterland, and for scientific research on agriculture and water. The CASMEZ investments thus completed a water distribution network extending throughout the Italian territory.

As for the north of Italy, the Po River and its tributaries feed a set of irrigation and reclamation channels that provide the main water supply for Italian agriculture, with an overall length of 55,700 km. This system is linked to large alpine lakes, with a total annual storage of over 1 Bm^3 that, in 1933, began to support the needs of irrigation and hydroelectric utilities, and at least partially controlled the floods of the Po River (RBD of the Po River 2010; Barbero and Bertoli 1998; Regione Lombardia 1996).

[5] CASMEZ was instituted by the sixth "De Gasperi" Government, with the Act n. 646, 10 August 1950.

CASMEZ constructed many dams and reservoirs to support the use of water and the development of irrigation in southern regions. Major transfer systems were built to supply Puglia, historically suffering from water shortages, and the large urban area of Naples and Caserta. Artificial reservoirs were built to store water. Likewise, particularly in the 1970s, special projects were realized to improve the functionality of the aqueduct system. Climatic vagaries in the 1980s accelerated the completion of unfinished works and the planning of new works, mainly for tapping groundwater in areas with large water deficits. Other aqueducts in Lazio and Molise were interconnected. The result is a complex system of aqueducts for interregional transfers of huge volumes of drinking water moved from the northern zones of the district toward the south, and between western and eastern territories: that is, from Lazio and Molise to Campania and Puglia and from Campania to Puglia, also involving Basilicata, and only minimally Calabria. These transfers are indicated in Fig. 6.1, with average flows expressed in million cubic meters.

Despite these efforts, Italy lacked a wider vision of the challenges posed by growing industrialization and urbanization. The interventions were oriented to respond to specific environmental emergencies and not to manage the resource, the environment, or the territory (Massarutto 2008). Introducing coordinated management on the Italian territory needed time and was not helped by the first phase of the establishment of the regions in the 1970s that generated conflicts with the state regarding the allocation of resources and responsibilities (Borelli 2008; Rainaldi 2010). Only at the end of the 1980s were new issues addressed, such as the worsening of the water balance and water quality, especially in densely industrialized and populated areas. Attention was paid to new water demands, such as those arising from mass tourism, which, in a few months, generated hugely increased demand in tourist areas.

6.3 Recent Changes in Irrigation Policy and the Institutional Environment

Major change has occurred in recent decades in Italian water policies and management. A new legal framework for water was developed, with a shift toward demand management. The first milestone was the **319/76 Act (Merli Act)**, which aimed to prevent the pollution of water. It regulated wastewater discharges into surface waters and the sea, and groundwater, establishing the *allowed maximum concentration* of various substances, based also on the load capacity of water bodies. It instructed regional authorities to systematically monitor the quality and volume of water bodies, as well as prevent their degradation and intervene in environmental disasters. Legal and administrative sanctions were provided when the concentration of pollutants exceeded acceptable limits: herbicides and pesticides were included, directly affecting agriculture.

Fig. 6.1 Water transfers among the regions of the Southern Apennines district. In red: mean annual volumes transferred (Mm^3). (Source: our analysis of data from NBA-AM 2010)

The **183/89 Act** brought a major change by identifying river basins as the governing scale for protecting soil and improving water quality and management. Italy was divided into national basins, interregional and regional, and related Autorità di Bacino (River Basin Authorities) were tasked with planning and coordinating the administrations involved in these territories. For the first time, the jurisdiction of a Public Authority was not bounded to administrative bases but to hydro-geomorphological and environmental criteria. The River Basin Authorities prepared watershed plans with guidelines for soil conservation and coordinated the plans of other administrations, including environmental protection and economic development. The 183/89 Act also impacted irrigation by emphasizing the concept of in-stream flow as a limit to withdrawals from water bodies.[6] Yet, the generic

[6]The in-stream flow is the capacity of a water body that is able to ensure its natural ecological integrity, albeit with small populations, with particular reference to the protection of aquatic life.

description of this limit did not effectively regulate water use, and additional clarifications were required in subsequent legislation.

The following **36/94 Galli Act** confirmed public ownership of surface and groundwater and determined that, when scarce, human and agricultural uses, in that order, are given priority, while the rights and expectations of future generations must also be considered. In-stream flow remained the limiting factor for protecting aquatic ecosystems, geomorphological processes, water balance, and agriculture itself. The administrative authorities of the river basin were responsible for identifying this minimum flow and ensuring the general compatibility of withdrawals, including those for irrigation. Governance was reformed to reduce fragmentation and held responsible for the lack of specialization, low-tech nature, and inefficiency of the sector. The management, government, and regulation of water use were separated. The integrated management of water for urban and industrial use was entrusted to independent supply providers (SII – Servizio Idrico Integrato) within Optimal Territorial Areas (ATO). The management of the SII could be entrusted to private companies (individual or associated), to mixed public-private corporate forms, or to self-production public companies (Santuari 2013). Regions defined the territorial extent of the ATOs, whose authorities (AATO) address and control the SII providers (HERA 2012). This system did not directly concern irrigation. Yet, its establishment introduced into the normative framework and the debate the principle that water management has to be of an entrepreneurial nature, based on efficiency, effectiveness, and economy, with the tariff policy designed to ensure full coverage of investment and operating costs. In subsequent years, these principles were taken over and applied to irrigation by the EU Water Framework Directive.

Legislative Decree 152/99 was the first major integration of European regulation on water management and protection into Italian legislation. It repealed the Merli Act to implement EU Directives 91/271 and 91/676 on urban wastewater treatment and the protection of water against pollution by nitrates from agriculture. The decree aimed at the sustainable and durable use of water bodies by preserving their self-purification capacity and environmental status. Their protection was hence based on a *maximum allowable load* for receiving water bodies. Moreover, in case of multiple uses, and in conditions of scarcity, the priority goes to the use with lower requirement and also better quality of outflow. These elements have also conditioned irrigation water withdrawals that have been limited in various territories, even water rich like central-northern Italy, to enforce admissible load conditions (INEA 2009a, b).

The **European Water Directive 60/2000/EC (WFD)** was incorporated into Italian law by the **Legislative Decree 152/06 (Environmental Code)**. According to the WFD, water is a common asset to be protected and used according to Basin Authority Plans formulated to respect the minimum vital flow and, in emergency situations, to allocate the scarce resource first to drinking use and then to irrigation. WFD affirms the *polluter pays principle* and requests member states to adopt pricing systems that encourage efficient use of water and make it available in the long term. This has triggered a review of the Italian pricing systems for a full coverage of the costs of irrigation water supply. Furthermore, the reference to in-stream flow as a limit to water withdrawals has been strengthened, which has often constrained irrigation activities. Finally, the 152 Decree attributed legal personality to the AATOs, to which the local authorities' remits concerning planning water infrastructure and management were transferred. The AATOs had to assign the SII to a *Utility Company* and verify that its activity was consistent with the objectives of the action plan. The AATOs must also protect consumers from the monopoly on the SII, yet facilitating the application of tariffs that guarantee cost recovery, including return on investment, and encourage the efficient use of water. **Law 42/10** modified this system by abolishing local AATOs in order to contain the costs of local administrations, rationalize management, and simplify the system. Political arguments delayed the institutional change by 4 years.[7] In the end, the D.l. n. 133/2014, art. 7, organized the AATO on a large scale, merging local bodies into provincial, regional, or supra-regional organizations.

The integration of environmental objectives in the CAP contributed to making water conservation a basic principle. The expansion of irrigated land was strongly discouraged unless it introduced more efficient irrigation systems, reducing the use of water per hectare. **Council Regulation (EC) 1698/2005** limited the expansion of irrigated areas in the Rural Development Program (RDP), only supporting investments that did not increase the irrigated area or the use of surface or groundwater. **Regulation (EU) 1305/2013** later allowed the expansion of the irrigated area only if the investment reduced the overall use of water and only in watersheds with a *good water status,* as defined by the River Basin plan. Elsewhere, projects must prove that they generate no major negative environmental impact. Investments that increase the net irrigated area are only eligible if the investment is combined with an investment in an existing installation where the ex ante evaluation demonstrates a water-saving potential of at least 5%–25%, or if the use of water is reduced to at least 50% of its water-saving potential. This orientation is motivated by the expectation that Climate Change will further increase water needs, as well as competition among uses for energy production, agriculture, and tourism.

This set of constraints slowed down the development of Italian irrigation in the first decade of the new century. After 30 years of expansion of the irrigation distribution network, the area that can be irrigated decreased from 3.88 million ha in 1990 to 3.75 million ha in 2010, while the areas effectively irrigated decreased from 2.72 to 2.42 million ha during the same period (ISTAT 2010).

[7] In 2011 two referendums abrogated the possibility of assigning water management to a private Utility Company, to include in the water rates a return on capital invested by this company. In much of Italian territory, the effects of the referendums are not yet applied: in March 2014 only Naples had transferred the management of the SII to the municipality, and only the Lazio Region had issued an Act (Corriere della Sera, 20/03/2014).

6.4 Collective Irrigation: Institutions and Schemes

6.4.1 The Reclamation and Irrigation Consortia

According to INEA (Istituto Nazionale di Economia Agraria), nearly 500 *Consorzi di Bonifica e Irrigazione*, i.e. Reclamation and Irrigation Consortia (RIC),[8] or landowner associations involved in collective water and land management, manage most of Italy's irrigation (Zucaro 2008; INEA 2014). The RICs of northern Italy mainly derive from spontaneous associations of landowners. Some are quite small, as in the alpine areas of Valle d'Aosta, Trentino, and the Bolzano province; others are large, and some also associate farms from different regions, such as Piedmont, Lombardy, Emilia, and Veneto. The RICs of the center, south and the islands are generally quite large because many were established in the twentieth century to manage large public hydraulic works.

The *Serpieri Law* (1933) entrusted the RICS with the management and maintenance of public reclamation and irrigation works, giving them even public juridical personality to share management costs among landowners and collect the related payments. Moreover, the *Serpieri Law* stated that, in the reclamation areas, the RICs could be established only if they involved the owners of the majority of the landholdings. This resulted in a large number of farmers being associated, especially in southern Italy. The law n.183, 12/02/1942 (Supplementary Provisions of the Act on land reclamation) strengthened the ability of the RICs to impose the payment of contributions on associated properties to build, manage, and maintain collective hydraulic structures. Because of these acts, reclamation and irrigation works were carried out in Lazio, Sardinia, and Puglia, and, between 1938 and 1942, in Sicily and Campania, which led to the creation of a network of RICs in those regions. The RICs were reorganized and merged several times to reduce and rationalize the costs of water supply, land reclamation, and consolidation and to adequately allocate these costs to the consortium members. The process is ongoing. For example, the Lazio Region law n.12/2016 unified its 11 RICs in 4 macro-RICs by coastal and internal areas, north and south.

Table 6.3 refers to the river basins defined by the 152/06 decree to implement the WFD. It provides the number of RICs per River Basin District (C); administrated hectares (A), equipped with distribution networks (E), actually irrigated (I); average administrative area (A/C), equipped (E/C), and irrigated (I/C), and equipped on administered area (E/A), and actually irrigated area on equipped (I/E).[9] Out of 489

[8] Many Consortia have added the irrigation activity to the reclamation activity for which they were initially established; many others have been set up directly to carry out reclamation and irrigation activities together. For this reason, from here on we will refer to *Consorzi di Bonifica e Irrigazione* (Reclamation and Irrigation Consortia), instead of *Consorzi di Bonifica* (Reclamation Consortia), as previously done.

[9] Section 3.4 and the following section will provide the extent of the Italian irrigated area as surveyed by ISTAT in the various Censuses of Agriculture. The latter includes the irrigation that is not managed by RIBs or other collective Agencies and is, therefore, larger than that reported in this table, albeit slightly.

Table 6.3 Areas administered, equipped with irrigation networks and irrigated in RICs in River Basin Districts

River Basin Districts	RICs (C)	Area in 000 ha Administered (A)	Equipped (E)	Irrigated (I)	Ratios A/C	E/C	I/C	E/A	I/E
Padano	240	4720.4	1325.9	983.9	19.7	5.5	4.1	0.28	0.74
Alpi Orientali	157	1371.4	598.7	586.7	8.7	3.8	3.7	0.44	0.98
Alpi Orientali Padano	5	278.8	170.0	148.2	55.8	34.0	29.6	0.61	0.87
Appennino Settentrionale	12	2082.2	135.7	49.2	173.5	11.3	4.1	0.07	0.36
Serchio	2	95.5	1.1	0.0	47.8	0.5	0.0	0.01	0.00
Appennino Centro-Nord	5	619.4	24.4	14.1	123.9	4.9	2.8	0.04	0.58
Appennino Centro	9	1881.2	92.9	74.5	209.0	10.3	8.3	0.05	0.80
Appennino Centro-Sud	2	337.9	25.2	1.0	168.9	12.6	0.5	0.07	0.04
Appennino Meridionale	37	4951.1	413.1	207.5	133.8	11.2	5.6	0.08	0.50
Sicilia	10	2382.3	143.0	74.2	238.2	14.3	7.4	0.06	0.52
Sardegna	10	937.4	161.5	59.3	93.7	16.2	5.9	0.17	0.37
Italia	489	19,657.5	3091.4	2198.7	40.2	6.3	4.5	0.16	0.71

Source: Own elaborations on Istituto Nazionale di Economia Agraria data (INEA 2014)

RICs, 414 are located in northern Italy with an average administrative area smaller than elsewhere. Conversely, the presence of water supply networks (E/A) is larger than in central and southern Italy, where many RICs are mainly committed to reclamation. In these latter areas, substantial public funding greatly expanded the area equipped with irrigation networks (E/C), yet the actual use of these networks (I/E) remains below 50%. According to INEA, the partial use of irrigation facilities is due to farmers' production choices, profitability, in some cases oversized facilities, and, above all, water availability. The abundance of water in the north favors irrigation, also through the reclamation drainage channels. The south and the islands, despite huge infrastructure, cannot fully use the network because of water scarcity (INEA 2014).

Water is mainly withdrawn from surface bodies (78%); groundwater supplies are notable in many areas of the south and the islands.[10] The total length of main and secondary canals is around 23,000 km, 11,000 in Po Valley, and 4000 in Southern Apennines. The most recent structures are in the center-south. Open channels prevail in north, where half of the Padano network serves reclamation and irrigation. This system also produces positive externalities, conserving migratory protected species and protecting biodiversity, as well as serving cultural and recreational purposes. Large channel networks in the north have key roles in recharging groundwater, as well as architectural, cultural, and historical heritage. Many programs aim to enhance these systems for multifunctional purposes.

An accurate assessment of water diverted to these schemes is provided by SIGRIAN, an information system based on GIS technology, for irrigation programming, detailed at the basin, region, and RIC levels (CREA 2018). According to SIGRIAN, national withdrawals are 15–20 Bm^3 per year, mainly from rivers in northern Italy. In the south and islands, major diversions are from reservoirs. Fifty to sixty percent of the analyzed *withdrawal licenses* are temporary; many expired licenses are being reconsidered with respect to the instream flow constraint. Some are old (even dating back to the nineteenth century) and neither show the volumes or discharges granted, nor sometimes the period of validity. In these cases, water withdrawals operate under a temporary regime that has made it difficult to plan the use of water by the basin authorities (INEA 2014).

These structural differences generate diversity in water supply. A first system is *rotational supply*, where water comes to a group of users at scheduled time intervals, constant or variable, during the irrigation season. Distribution is organized at the beginning of the irrigation season based on bookings made by farmers; water supply programs define in detail the areas, volumes, turns, and schedules of distribution, also providing for possible changes during the season. In the *withdrawal on demand* system, each user can withdraw water whenever appropriate, based on the needs of the crops and without schedules or turns. This is only possible with a large water supply and/or networks of pressurized pipes. Even in the same RIC, different types of distribution can coexist. INEA shows a certain prevalence of *rotational supply*

[10] Of 881 sources, 580 withdrawn from groundwater, 72% of wells in the South are in the Apulia region.

(24% of cases) followed by *on demand* (20%). The *reservation of irrigation water,* whereby water use is planned at the beginning of the season, is widespread in southern Italy and part of the center (37% in Sardinia). In Veneto, Lombardy, Emilia Romagna, and Friuli, distribution systems are unstructured, with free withdrawals from channels: this would become problematic if Climate Change were to permanently reduce the current abundant water availability. In fact, this practice prevents the organization of management and control of uses (INEA 2008; Zucaro and Pontrandolfi 2007). To describe the complexity of this situation, the next section presents the case of an RIC that manages multiple water schemes, faces main competition problems with other water uses, and is in need of new investment.

We can note that the use of multiple resources in the southern RICs is often from different river basins or regions, while historical supra-regional water management across the Po Basin and Valley also makes it crucial to have a supra-regional water governance system. The tension and conflict between uses and territories show the limits of governmental authorities defined at the scale of the river basin. The same applies to the negotiation between regional authorities for the transfer of water and the use of common resources. The increased competition between water uses will increase the need for supra-regional governance. But Italy does not have the supra-regional political and administrative authority to which these bodies can refer in order to settle disputes between users and areas of different regions, mediating the effects of adverse decisions to some of them. In fact, the Italian hierarchy of powers, and related governance, implies that the authority of the national ministries (agriculture, environment, and infrastructure) will not always be sufficient to settle or manage disputes in due time between administrations or technical bodies at the regional level.

6.4.2 Irrigation Management by RICs: Financial and Economic Aspects

As explained earlier, the *Serpieri Law* establishes that the RICs can recover the costs of the reclamation and irrigation activities through a specific tax power, making the owners pay, public and private, in their district area. The landowners are therefore members of RICs rather than customers of their services, and their payments for irrigation and reclamation are a specific type of tax and not a commercial rate. Therefore, these payments can only cover the costs of the RICs and cannot include or reflect, for example, the conditions of water scarcity or the willingness to pay.

Farmers' *irrigation contributions* must cover a set of costs, both fixed and variable. These include fees to be paid to the regions for the use of the public resource (based on the average flow rate granted or the area irrigated), the operating and maintenance costs of facilities managed by regional authorities, and the costs of the consortium for maintenance (e.g. reservoirs, water networks and pumps) and water distribution (in particular, energy costs for lifting and pumping, and labor costs).

Finally, the fees have to cover the administrative costs of the RICs, as well as part of their general costs. Some regional administrations contribute financially in order to offset the structural handicaps involving higher energy costs for lifting and distributing or to subsidize specific users like agriculture (Cortignani et al. 2018). However, once public contributions are subtracted and the discounted rates applied, the residual part is paid by the farmers associated with the RIC in proportion to the benefits received from irrigation.

According to INEA, RICs distribute these costs among farmers in different ways, with about 20 systems for calculating water payments, some of which are simultaneously applied in different districts of the same RIC (Dono et al. 2003; Dono 2008; Zucaro 2008; INEA 2014). These payment systems are either *binomial* or *monomial*.[11] The first type has two parts: the first is to cover fixed costs, in particular maintenance, usually proportional to the potentially irrigable area, while the second considers the water actually received by each user (or a proxy, such as the type of crop or the quality and location of the irrigated land). When this assessment is difficult, or marginal differences exist between irrigated and irrigable areas, the *monomial* system is applied. This can vary between areas of the same RIC, e.g. according to the nature of the soil, the distance from the water delivery points, the elevation, or the crop water requirements.

The *monomial* system prevails in northern and central Italy where the benefits of irrigation and drainage are associated with the same channel networks and are therefore difficult to separate (INEA 2014). The *binomial* is common in the south and the islands and where irrigation is limited to specialized areas in the center and the north. The monomial system, or the variable portion of the binomial contribution, is generally paid per irrigated hectare or per hectare equipped for irrigation. It varies markedly, from 0.62 €/ha per year in the Aosta Valley to 787 €/ha in Pontine reclamation (Lazio), and up to 2000 €/irrigated ha in the Trento Province. In general, higher payments correspond to higher energy costs for lifting or pressurizing water. Payments based on the irrigated crops also vary: 420 €/ha for rice in East Sesia, 467 €/ha for permanent grasslands in Parma, Emilia-Romagna, and up to 965 €/ha for kiwi fruit. Payments based on the volume of water distributed to the user are increasingly common: INEA indicates an average of 0.54 €/m^3, varying from 0.04 to up to 6.3 €/m^3. In some cases, the payments take into account the multifunctional role of irrigation, in particular in groundwater recharge, a key environmental externality of irrigation.

The degree of coverage of the operating costs of water distribution has only been assessed for some Italian RICs. For a Sardinian RIC, Giraldo et al. (2014) evaluated the impact of applying a pricing system based on the cost of water distribution, as estimated by Dono et al. (2012). They found small efficiency gains, which could become negative by counting implementation costs, and major income redistributive effects among farmers. The current system is in fact designed to facilitate smaller farms, and with a lower income, which would then be disadvantaged by the redistribution of income that would follow by applying a volumetric price (Cortignani et al. 2018).

[11] Analogous schemes apply to the contribution to the land reclamation service.

Dono et al. (2012) also noted that operating costs respond more to the size of the irrigated land than to the volume of water supplied: so, a volumetric price would provide incomplete signals to optimize the water use. These authors also show that the RIC structures are currently underutilized, resulting in an average cost of distribution greater than that of efficient use. They attribute this under-use to the reduction in CAP support to many irrigated crops compared to when the facilities were designed. Hence, inefficiency does not depend on farmers' choices and would not be reduced by an increase in the price of water. On the contrary, full coverage of the average cost of underused facilities with irrigation tariffs would transfer to farmers the inefficiencies for which they are not responsible. Also, this increase could reduce the use of irrigation water and further reduce the cropping intensity.

Cortignani et al. (2018) examined the budget of the Sardinian Water Authority to compare farmers' irrigation payments to the costs of water supply in the same RIC. They showed that the diversion fees for the use of water as a public resource together with the cost of non-use imposed by minimum instream flow[12] cover 60% of the total water supply costs. Taxpayers and industrial users pay 90% of these costs by means of direct contributions and higher tariffs. Furthermore, farmers pay 60% of the RIC's distribution cost (i.e. 24% of total costs), while taxpayers cross-subsidize the remaining 40% with public contributions to the RIC's electricity costs. Overall, farmers pay 29.5% of the total water cost, taxpayers pay 22.5%, and the industrial sector pays 48%.

In brief, applying volumetric pricing in this RIC would not significantly increase the overall efficiency and would especially redistribute income among farmers. The latter actually pay only 30% of the costs of water supply, but the rest is paid mainly by industrial users, and not by taxpayers, by virtue of the greater environmental impact of their activity. This appears in line with the principles and objectives of the WFD.

6.4.3 The Example of the Capitanata Reclamation and Irrigation Consortium

The Capitanata Consortium for Reclamation and Irrigation (CBC) operates in Puglia on 441,000 ha, from the border with Molise to the border of the Foggia and Bari provinces (CBC 2017). Since the early 1960s, CBC planned and partly implemented three water schemes, Fortore, Sinistra Ofanto, and Carapelle, covering 200,000 ha. This was financially supported by CASMEZ. In the following decades, inter-sectoral allocation problems arose and made it necessary to adapt agricultural water use to meet the needs and demands of domestic, civil, and industrial uses. Water availability for agriculture was notably reduced with respect to the original

[12] The Sardinian Water Authority calculates this cost of nonuse based on the percentage of the volume returned to the water bodies to the total amount of water managed, i.e., attributing to these return flows a proportional share of its total water costs.

situation when farming was the only recipient of the water resource of the three schemes. Figure 6.2 presents the essential features of these water schemes that are briefly described below.

The Fortore water scheme, the northernmost area of the consortium, covers 140,000 ha with an estimated water requirement of 280 Mm^3 (based on 2000 m^3/ha irrigable). Yet, the distribution networks have not yet been completed and only serve 102,500 ha. Water is diverted from torrential streams and mainly from the Occhito dam, with a total capacity of 333 Mm^3, of which 250 can be used. 60 Mm^3 are used annually for drinking and 5 Mm^3 for industrial use: as a result, total need is higher than the availability in the reservoir, which creates social tension around the use of the resource. The projects being implemented – a dam on the Celone River (16 Mm^3) and a diversion weir on the Vulgano River (6.5 Mm^3) – will hardly be able to fill the gap. New works awaiting funding include two dams (with 48 Mm^3 of total capacity) and the reuse of wastewater from the main municipalities.

CBC also manages withdrawals from the Ofanto River for the irrigation of the Ofanto scheme. This was built with CASMEZ funding and includes two reservoirs,

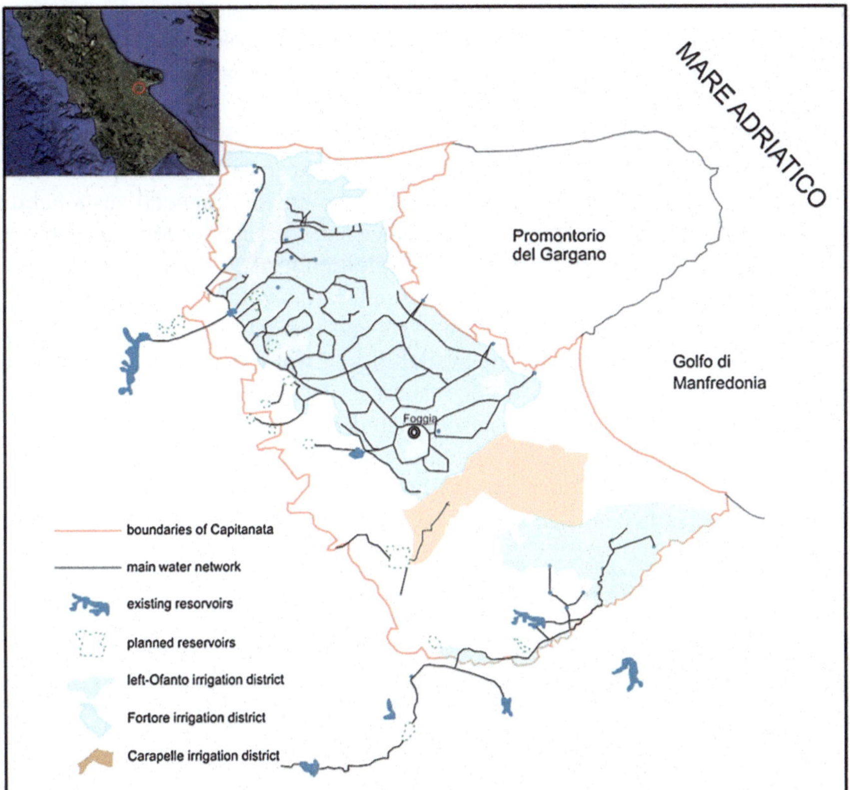

Fig. 6.2 Map of the Capitanata irrigation scheme. (Source: our processing on http://consorzio.fg.it/index.php?option=com_content&view=article&id=50&Itemid=136)

as well as two reservoirs of modulation and accumulation, and a bypass cross-branch. The available resources are estimated at 76 Mm^3 which, based on 2000 m^3/ha, cover an irrigable area of 38,000 ha. This theoretical potential is actually reduced to 56 Mm^3, with a 20 Mm^3 deficit that can also grow with the gradual transfer of a large part of this resource for drinking. The deficit can only be filled with new investment and the recovery and reuse of wastewater from the main municipalities nearby.

Finally, CBC manages the Carapelle scheme where the building of a dam is planned at Palazzo d'Ascoli with a usable capacity of 67 Mm^3: irrigation will receive 40 Mm^3 of this water, the remainder being allocated to industry. The project is awaiting funding. This resource is considered crucial to reduce groundwater over-exploitation that notably lowered the aquifer levels because of excessive well drilling over the past 20 years.

In recent years, agriculture in that area rationalized and significantly reduced the use of water on farms. In particular, surface irrigation (flooding, furrows) was abandoned, both for the high cost of water and the poor efficiency of the system. Sprinkler systems (low and high intensity) have also given way to micro-irrigation systems. At present, all trees and most horticultural crops are irrigated with drip. Even the fixed, low-intensity sprinkler systems, still used for artichokes, are being replaced by micro-irrigation. It is estimated that drip irrigation is now used on about 70% of irrigated crops (CBC 2017).

We can single out three important features of the CBC. First, its water network involves a lot of maintenance work that becomes more and more expensive with the ageing of the structures. This increasingly weighs on financial sustainability and weakens management, including vigilance and the supervision of the system. Second, the difficulty of containing the exploitation of groundwater is causing aquifer salinization. The large number of small farms with individual abstraction hinders the monitoring of these withdrawals and the enforcement of any restrictions or prohibitions. Increasing the availability of water in the consortium would relieve the pressure on groundwater resources. However, in some areas of the RIC, water sources are now scarce compared to demand, while in others, the water network has not yet reached the farms, and tapping groundwater is the only option. Third, the extension of the RIC network to areas where irrigation with groundwater prevails has been delayed due to a lack of sufficient public funding for many years.

6.5 Irrigable and Irrigated Land in Italy

6.5.1 Characterization of Irrigated Areas

This section describes the recent development of irrigation in Italy and the dissemination of its various techniques. Irrigable land[13] increased between 1970 and 1982 and leveled off up to the year 2000, to decline in the following decade (Table 6.4). The total utilized agricultural area (UAA) declined from 1970 as a result of the abandonment of the mountain and hilly areas and the rapid increase in the nonagricultural use of land (i.e. industrial buildings, infrastructure, and urbanization). However, irrigable land has not declined as much as UAA (Table 6.4). This suggests that the ability to irrigate may be one reason for keeping agricultural land utilized. Irrigation has been and still is an important activity. According to the last farm census, around 29% of the total UAA is irrigable and around 19% is effectively irrigated (Table 6.4).

However, the amount of irrigated land, when compared to irrigable land, has declined since 1970, and currently only two thirds of the irrigable land is irrigated due to factors such as low water availability and low profitability of irrigated crops. The 2005 reform of the Common Market Organization for sugar caused a large number of Italian refineries to shut down, removing the demand for sugar beet in many irrigated areas of the center and south of Italy (Severini and Valle 2007; Dono and Severini 2007). Similarly, the abolition of milk quotas led to a decrease in production (except in the more traditional areas of the north), and this reduced the demand for corn that used to make up a large part of the irrigated land. Another factor that reduced the irrigated area is the reduced profitability of fruit and vegetables. In particular, this was the result of a decline in the price of these products due to changes

Table 6.4 Evolution of irrigable and irrigated land in Italy since 1970

	1970[a]	1982[a]	1990	2000	2010
	(Millions of ha)				
Utilised Agricultural Area (UAA)	17.49	15.84	15.05	13.21	12.86
Irrigable area	3.35	3.86	3.88	3.89	3.75
Irrigated area	2.56	2.52	2.71	2.47	2.42
Irrigable land/UAA	19.1%	24.4%	25.8%	29.4%	29.2%
Irrigated land/UAA	14.6%	15.9%	18.0%	18.7%	18.8%
Irrigated/irrigable land	76.5%	65.3%	69.8%	63.6%	64.5%

Sources: For irrigable land, Fassò (2003). For the remaining data, ISTAT Agricultural Census
[a]Irrigable land data refer to 1969 and 1983

[13] ISTAT defines irrigable area as the maximum that could be irrigated during the year according to the potential of facilities and normal availability of water. Irrigated land is net of unproductive surfaces.

in the supply chain, where small operators were replaced by large companies often vertically integrated with big retailers (Petriccione and dell'Aquila 2011) and the growing competition from other southern Mediterranean countries that has caused a noticeable deterioration in the Italian trade balance in this sector (Malorgio and Hertzberg 2007; Nomisma-Unaproa 2016).

These factors have limited the use of the potentially irrigable land. Indeed, one third of Italian farms have irrigable land, but only a quarter of them effectively irrigate any of their land (Table 6.5).

Northern Italy has traditionally been more equipped with irrigation facilities, both in terms of the number of farms and the area (Tables 6.5 and 6.6). While in the north, around one third of the land is irrigated; the proportion for other areas is less than one eighth (Table 6.6).

More than half of Italy's irrigated land is supplied by RICs. However, the relative number of single farms is much higher in the center and in the south where only around 40% of the irrigated land relies on water from RICs. Water from single farm sources is mainly groundwater, especially in the south. By contrast, in the north, the two water sources allow about the same amount of land to be irrigated (Table 6.7). Unfortunately, no data exist on the specific sources of water RICs rely on.

Table 6.5 Farms with irrigable land and irrigated land. Italy and main regions

		Number of farms	
	Total	**with irrigable land**	**with irrigated land**
Italy	1,620,884	544,997	398,979
North	397,102	208,719	158,172
Center	252,012	54,792	33,002
South	971,770	281,486	207,805
	Irrigable/total	**Irrigated/total**	**Irrigated/irrigable**
Italy	33.6%	24.6%	73.2%
North	52.6%	39.8%	75.8%
Center	21.7%	13.1%	60.2%
South	29.0%	21.4%	73.8%

Source: ISTAT. Farm Census (2010)

Table 6.6 Irrigable and irrigated areas as a share of total UAA. Italy and main regions

	Irrigable/UAA	Irrigated/UAA	Irrigated/irrigable
Italy	29.2%	18.8%	64.5%
North	50.6%	34.8%	68.9%
Center	15.0%	6.6%	44.2%
South	18.2%	11.2%	61.5%

Source: ISTAT. Farm Census (2010)

Table 6.7 Prevalence of water sources in terms of irrigated land. Hectares and percentage

	Single farm sources				
	Groundwater[a]	Surface water[a]	Water users associations (RICs)	Other sources	Total
	Hectares of irrigated land				
Italy	616,330	364,623	1,348,406	89,562	2,418,921
North	247,973	248,596	1,050,916	44,261	1,591,746
Center	68,491	42,922	27,072	6617	145,102
South	299,866	73,105	270,418	38,684	682,072
	Percentages				
Italy	25.5	15.1	55.7	3.7	100
North	15.6	15.6	66.0	2.8	100
Center	47.2	29.6	18.7	4.6	100
South	44.0	10.7	39.6	5.7	100

[a]Inside and outside the farm. Source: ISTAT. Farm Census (2010)

6.5.2 *Sources of Irrigation Water*

Farmers access water (surface or groundwater) individually, or through water user associations, and sometimes other sources (Table 6.7).

6.5.3 *The Economic Importance of Irrigated Agriculture in Italy*

Irrigated agriculture is important from an economic point of view. A recent estimate suggests that a large share of the crop output in Italy is indeed generated by crops that are generally irrigated (i.e. that in most of the years and areas are irrigated) (Severini et al. 2016). This is coming from a classification of crops according to four categories described in Table 6.8, provided by a sample of national and regional experts who were asked to assess the situation in specific regions.[14]

The aggregated data show that the crops always irrigated represent more than half of the monetary value of the gross output generated by all crops in Italy. Considering also the output generated by the crops that are occasionally irrigated, the share of gross output reaches around two thirds of the total (Table 6.8).

The importance of the irrigated crops differs among regions.[15] In central Italy, irrigation is less important than in the north and south (Table 6.8), where the

[14]Given that the different Italian regions have different precipitation levels and irrigation requirements, the same crop can be classified in different classes in the different regions. For example, a crop that is always irrigated in a region of the South can be classified as irrigated only in some cases in a region of the North.

[15]The interested reader can get additional information regarding, for example, the specific crops within the four categories in Severini et al. (2016).

Table 6.8 Relative importance of crops classified according to the use of irrigation in terms of monetary value of gross output. All crops. Italy. 2011–2012. Percentages (%)

	Rainfed	Supplemental irrigation only	Always irrigated	In some cases irrigated in other cases rainfed: Total	of which: Irrigated	of which: Rainfed	Total
Italy	14.5	5.7	56.1	23.7	9.6	14.1	100.0
of which:							
North	9.2	3.1	53.3	34.4	14.9	19.4	100.0
Center	30.6	5.6	35.9	27.9	9.5	18.3	100.0
South	14.0	8.4	66.3	11.3	4.2	7.0	100.0

Source: own elaborations on ISTAT and questionnaire survey data

share of output generated by (non-supplemental) irrigated crops is around 70%. The south differs from the north because of the higher relative importance of the "always irrigated crops." The ability to grow high revenue crops (e.g., perennial and horticultural) strongly depends on the access to predictable, non-supplemental irrigation. This is true for southern Italy because of the nature of the crops and the climatic conditions, which means that restrictions in the access to irrigation could have a major impact on the output volume.

6.5.4 *Water-Saving Techniques*

The relative importance of different irrigation techniques has changed in the last century: the area with sprinkler irrigation almost doubled between 1961 and 2000 (Fassò 2003), and micro-irrigation also spread. These changes are due to several factors, including the increasing relative cost of labor, the declining availability of water, and the need for a more uniform and timely distribution of water. The first factor has caused a shift from labor-intensive irrigation technologies, such as flood/furrow and basin irrigation, to more capital-intensive irrigation technologies, such as sprinkler and micro-irrigation. The investments required by these latter technologies have been mainly supported by Rural Development Policies. This was done in the last programming period (i.e., 2007–2013), through Measure 121 of the "thematic axis" aimed at supporting the modernization of agricultural holdings. In particular, this measure provided state support for tangible investments which improve the overall performance of the agricultural holding, covering up to 60% of the amount of eligible investment in the case of young farmers located in mountain areas or other areas with handicaps or affected by environmental constraints. Indeed, benefiting from Measure 121 for investments related to irrigation is conditional upon achieving an improvement in farm-level water efficiency.

The need for a more uniform and timely distribution of water has been also driven by a need to ensure a high product quality standard especially in the fruit and vegetable sector where micro-irrigation is now used in a very large portion of the cultivated area. This change is due not just to the increasing availability of on-farm equipment but also to the shift toward on-demand delivery systems that allow the use of this technology (Measure 124). The support was also granted in order to increase water efficiency and, especially, to reduce water losses occurring in the delivery of water managed by the RICs.

According to data from the last census (Year 2010), sprinkler irrigation is the most used irrigation technique, accounting for around 40% of the irrigated area. However, flood/furrow and basin irrigation techniques together account for around the same share of irrigated land as sprinkler irrigation (Table 6.9). Despite the relative lack of water, basin irrigation is still used on slightly less than 10% of the irrigated land. Finally, drip irrigation accounts for less than 20% of the irrigated land.

Such national data hide huge differences between the north and south. Basin irrigation is mainly used in the north (rice and some forage crops use this technique extensively) but not in other parts of Italy. Similarly, the use of flood/furrow irrigation is higher in the north (40%) than in the other areas (only 10%) (Table 6.8). Contrastingly, the percentage of micro-irrigation is much higher in the south (40% of irrigated land) (Table 6.9).

Irrigation techniques use different amounts of water per ha. In 2010, the average was estimated at around 4500 m^3 per ha of irrigated land (Table 6.10). However, this figure hides a large heterogeneity. Farmers apply less water per unit of land in the center and south than in the north (Table 6.10). This is because of the nature of the crops, the type of irrigation method, and because the north is less affected by water shortages and can rely on a relative abundance of surface water coming from the Alps. The greater

Table 6.9 Relative importance of the different irrigation techniques in Italy (Year 2010)

	Flood/furrow irrigation	Basin irrigation	Sprinkler irrigation	Micro-irrigation	Other systems	Total
	Ha of irrigated land					
Italy	748,391	221,025	958,535	422,534	68,436	2,418,921
North	649,340	213,054	578,729	122,769	27,853	1,591,746
Center	14,707	686	89,733	31,990	7986	145,102
South	84,344	7284	290,073	267,775	32,596	682,072
	Percentages					
Italy	30.9	9.1	39.6	17.5	2.8	100
North	40.8	13.4	36.4	7.7	1.7	100
Center	10.1	0.5	61.8	22.0	5.5	100
South	12.4	1.1	42.5	39.3	4.8	100

Source: ISTAT. Farm Census (2010)

Table 6.10 Average amount of water used for irrigation by irrigation techniques (m³/ha)

	Flood/furrow irrigation	Basin irrigation	Sprinkler irrigation	Micro-irrigation	Other systems	Total
	Water per irrigated land (m³/ha)					
Italy	4034	17,492	3106	2528	2455	4588
North	3956	17,791	2543	1838	1805	5093
Center	4290	9529	3587	2831	2744	3473
South	4595	9497	4079	2809	2939	3648

Source: ISTAT. Farm Census (2010)

volume of water used per hectare and the higher proportion of the UAA that is irrigated explain why the north accounts for 73% of the estimated use of irrigation water in Italy.

The average unitary volume differs between regions and depends on the relative prevalence of the techniques previously discussed. However, there are also differences in the unitary volumes within the same irrigation technique. The volumes for sprinkler irrigation and micro-irrigation are on average higher in the south and the center than in the north. This suggests that, given the crops grown and higher irrigation requirements (due to the warmer climate and more limited precipitations), even the more water efficient techniques require relatively high volumes of water in the south (Table 6.10).

6.6 Future Challenges

Here we examine two major challenges for Italian irrigation: groundwater overexploitation and the possible impact of Climate Change (CC).

6.6.1 Groundwater Overexploitation

Groundwater overexploitation affects many coastal and non-coastal areas of Italy. In the former, summer peaks in water demand create depressions in the water table that invite sea water intrusion. The problem will be compounded by a rising sea-level possibly caused by CC. Agriculture not only contributes to salinization by using groundwater for irrigation but also suffers from it. In fact, soils are increasingly affected by the salinity of coastal waters used in irrigation, or rising by capillarity from salty aquifers (Napoli 2010; Ghiglieri et al. 2006). Recent data confirm the exacerbation of this quality problem (PTA 2009). In the CBC, withdrawals from aquifers vary annually according to the availability of (cheaper) CBC surface water supply (Giannoccaro et al. 2011). Usually each source provides around 50% of irrigation supply, but during long drought periods, groundwater can reach 100% of supply. Groundwater extraction is regulated through nonmarketable private licenses without clear access rules or quantity limits. The drilling of private wells is subject to public authorization, yet public authorities have proved to be incapable of preventing illegal or excessive water abstraction (ibid.). The difficulty of accurately monitoring a wide, scarcely populated area has also favored illegal drilling. The Act n.9/2008 of the Apulia region established groundwater rights requiring farmers to install water meters and to periodically monitor the quality status based on indicators for nitrates, organic carbon, and salinity. A fixed quota is attributed to each well, and farmers must send data records to the public authority, which may cancel the license in case of omission. New permits are banned, until a monitoring of the current situation is achieved. A system for the exchange of water rights could help reduce the inefficiency of overexploitation, although with distributional impacts on farmers. However, to implement a water market would require substantial investment as well as new organizational skills and application to enable an effective volume monitoring and control of illegal drilling.

Aquifer overexploitation for irrigation not only affects the south but also other areas, such as the Po Valley. Bazzani (2012) estimated that in the *Consorzio di Bonifica di Piacenza* (Emilia), 40% of irrigation water was sourced from aquifers. An increased use of groundwater, along with withdrawals from surface waterways, and reduction of rainfall in recent years, is accentuating the penetration of underground salt water wedge in the coastal areas of the Po Delta (Mainetti and Pecora 2013).

In some non-coastal central areas, the problem is related to the drawdown of aquifers where groundwater is crucial for water supply and there is competition among agricultural, urban, and industrial uses. Casini et al. (2012) have discussed the case of the ornamental nurseries in the Pistoia province, a key source of income for the area. Irrigation is provided by individual wells and has a strong impact on the aquifer. This is relevant because the Pistoia province (290,000 residents) also uses these aquifers for domestic water supply. Further increases in withdrawals by the nurseries and a succession of particularly dry years are expected to generate a dangerous imbalance for both groundwater and surface water, with the need to limit a sector that is important for the local economy.

6.6.2 Possible Impact of Climate Change

Climate Change (CC) could generate problems for Mediterranean agriculture due to a higher frequency of dry seasons and the need to find additional water resources. Several studies have investigated the production and economic effects of CC in the short-to-medium term, with a Change in Climate Variability (CCV) that increases water shortages in agricultural use, especially in areas where supply is currently insufficient. Impacts and possible mitigation measures vary according to context.

The typically Mediterranean case of the Cuga Hydrographic Basin, NW Sardinia, shows a combination of expected reduction in supply (lower autumn-winter rainfall reducing the multipurpose dam storage) (Dono et al. 2013), increased crop water needs in the summer, and priority given to domestic use that is expected to result in a reduction in the irrigated area. Improving the efficiency of collective irrigation systems is seen as more important than supporting individual farm investment.

In another Mediterranean zone, Oristano, CW Sardinia, part of the area is supplied by a dam whose storage capacity has been found sufficient to face the increase in water demand and sustain the current crops, but this situation could be threatened by further increases in priority domestic water needs, notably those associated with tourism on the island, which have priority in the use of water. In the remaining area that relies on groundwater, the increase in crop water requirements is expected to result in more pumping, with a worsening of the salinization problem (Dono et al. 2016). In Destra Sele RIC (Salerno), home to intensive greenhouse vegetable production, orchards, and buffalo farms, crop evapotranspiration demand will increase during the spring when water supply from the Sele River is already insufficient under present conditions. Groundwater pumping from private wells will also increase, with severe consequences on groundwater salinization (Dono et al. 2014b).

In five RICs that source water from the Po River and supply farms in Cremona and Piacenza (main centers of DOP Grana Padano cheese production), CCV is expected to result in future shortages of water in late spring-early summer when agriculture (forage and grain crop) increases its water demand, because of an insufficient storage capacity limited to minor streams (Dono et al. 2014a).

Increasing the efficiency of the RICs' services and, in some cases, extending supply to non-served areas may support adaptation to CC. The National RDP was planned to improve use efficiency in water networks serving 395,000 ha. There is a call for this program to finance irrigation investment, mainly by RICs, to improve reservoirs and complete and modernize pre-existing, and establish new, infrastructure. The call finances projects between 2 and 20 million euro (including VAT) with a 100% payment of the allowed capital expenses. The total budget is € 291 million, and the examination of the RICs' applications is now taking place (2018). It will be interesting to verify the need for financing that still remains unmet, with the ratio between requests and the allocated amount, since this action follows a long period without major investment in the sector. On the other side, the Regional RDPs support investment to increase, at the farm or very small collective level, the efficiency of irrigation, encouraging restructuring and modernization. This will contribute to the sustainability of agriculture by promoting CC adaptation and mitigation.[16] However, it should not be overlooked that in Italy it is often difficult for the approaches and choices of national and regional planning to result in coordinated action. The future challenge will therefore be to integrate the selection of investments in large irrigation infrastructure, to be borne by the national RDP managed by the minister, with that of irrigation investment in farms, financed by the RDPs of the various regions.

[16] In this regard, Dono et al. (2016) show that adaptation to CC could also exploit the effects of early maturation and CO_2 fertilization on various crops, generating possible water savings. Additionally, Cortignani and Severini (2011) showed that a kind of environmental payment could help spread deficit irrigation techniques.

Another viable strategy for meeting future increasing agricultural water needs, while preserving primary water sources, is the reuse of treated wastewater. This option was initially regulated by the Merli Law that banned the use for irrigation of untreated urban wastewater (Resolution February 4, 1977). A significant step forward was achieved with the Inter-Ministerial Decree June 12, 2003 n.185, which established that physical-chemical and microbiological quality requirements be strictly respected for domestic and agricultural reuse. Decree 152/06 defers to regions the task of defining the parameters to be mandatorily controlled and monitored and to fix limits in compliance with national regulations. The reuse of treated wastewater meets diverse needs in various regions depending on the availability of *primary* sources: in the south, reuse is intended for irrigation and is motivated by the need to mitigate water shortage, thus preserving the resource for domestic use; in the north, it is justified by environmental protection issues (pollution of the Po River, eutrophication of the Adriatic). According to Zucaro et al. (2012), even with the technological improvements needed to meet regulation standards, the investment costs in treatment plants and water networks would exceed the benefits generated by the value of the treated water produced.

6.7 Conclusions

Irrigation and land drainage are rooted in the Greek, Etruscan, and Roman civilizations. Water abundance in the north led local authorities to build reclamation and irrigation schemes and private landowners to come together to manage water facilities. In the south and the islands, with the exception of small-scale irrigation in the Arab and Norman periods, no major projects were built for centuries. Only in the twentieth century were major public irrigation projects financed and integrated with reclamation, and long-distance water transfer implemented. Italian water policy has acted simultaneously on the reclamation and irrigation fronts, integrating many measures aimed at these two interventions, and addressed issues related to infrastructure, the development of the marketing of agricultural products, and scientific research on agriculture and water use.

The legislation of the last 35 years has shaped irrigation by enacting policies of soil conservation, the environmental protection of water bodies and reforming governance systems. This was also affected by the decentralization and reorganization of the functions of the central state and regional and local authorities. A milestone was the definition of the river basin as the territorial unit for planning water use and protection, to be managed by AATOs. A system based on authorities at river basin level and ATOs certainly contributed to a better assessment and control of water use at the local level. However, it is interesting to note that the last normative acts have reinforced supra-regional forms of water governance. These can be crucial in the center-south where water use is based on large transfers between different watersheds and located in different regions. This type of governance is also crucial in the north, where water use is largely centered on the resource of the Po basin. In those contexts, tensions and conflicts between different uses and water areas suggest that water governance defined at river basin level is not always sufficient. The same is true of negotiations between regional authorities, which have not always adequately solved the problems of water transfers or the use of common resources. Increased competition between water uses in the context of CC will enhance the relevance of supra-regional governance to settle and manage disputes between regional administrations or technical bodies in due time.

Irrigation finds itself in a delicate phase: irrigable land increased between 1970 and 1982, leveled off up to 2000, and then declined in the following decade. Irrigated land decreased mainly due to the reduced availability of water, the modification of the CAP subsidy system, and the reduction in the competitiveness of the Italian fruit and vegetable sector. Still, agricultural production in nonirrigated areas declined even further, suggesting that irrigation helps support agriculture.

Adopting the volumetric pricing of water for the total recovery of costs could generate only limited efficiency gains, while major redistributive income impacts could mainly affect small farms with low income. In the south, charging farmers the full cost of water distribution would also make them pay for inefficiencies in the system and for higher fixed costs due to the underutilization of facilities linked to changes in market and political support conditions for which they are not responsible.

However, collective irrigation systems can play a key role in countering the impact of CC. Enhanced functioning and strengthening of these systems can restrain the use of groundwater, whose overexploitation is causing a diffuse salinization problem in Mediterranean regions. Extending RICs' facilities to adjacent areas could support them in adapting to the new conditions, which will be warmer and with less rain in key seasons. CC could also have a significant impact in northern regions where water needs are expected to increase. This would affect the area's dairy cattle industry that makes a large contribution to the country's agricultural income. Adaptation to CC will hence require integrating interventions on several fronts. Clearly, better results will be achieved by directing the available financial resources to zones where farm-level irrigation investment will cover large areas and economically important crops and where interventions will modernize and expand the collective water networks and reservoirs. Currently, the support for farm-level investment is planned by the regional RDPs, while the modernization and expansion of the networks are the responsibility of the national RDP, managed by the Ministry of Agriculture. The challenge over the coming years will be to make the best use of available financial resources by finding synergies between the intervention at the regional and national levels of Italian planning.

Acknowledgments The study was partially carried out under the MACSUR 1 and 2 projects (D.M. 2660/7303/2012 – www.MACSUR.eu) funded by the Italian Ministry of Agriculture, Food and Forestry (MiPAAF). MACSUR is funded as part of the JPI FACCE.

References

Barbero, G., & Bertoli, L. (Eds.). (1998). *L'influenza del deflusso minimo vitale sulla regolazione dei grandi laghi prealpini*. Milano: Guerini e Associati.

Bazzani, G. M. (2012). Analisi economiche e ambientali della scarsità idrica in presenza di fonti plurime con un sistema di supporto territorial. *Economia & Diritto Agroalimentare, XVII*, 429–447 2012.

Borelli, G. (2008). La governance di un territorio plurale. Sviluppo locale e nuove soggettività nella Media valle del Po. In G. Borelli (Ed.), *Tracce di Governance. Comunità e sviluppo locale nella Media Valle del Po* (pp. 13–67). Milano: Franco Angeli.

Braca, G. (2017). La Disponibilità della Risorsa Idrica Naturale e suoi Trend. In *3° Rapporto Generale sulle Acque: Obiettivo 2030*. Roma: Utilitatis.

Casini, L., Marone, E., Daniele, C., & Boncinelli, F. (2012). L'analisi dei costi della risorsa idrica: il caso del vivaismo pistoiese. *Economia & Diritto Agroalimentare, XVII*, 449–466.

Cassazione, Sezioni Unite, Dep. 11 febbraio 2008, Ord. n. 3171.

CBC. (2017). *Consorzio per la Bonifica della Capitanata*. http://consorzio.fg.it/index.php?option=com_content&view=article&id=50&Itemid=68

Conte, G. (2010). Verso una gestione sostenibile delle acque in Italia. *Gazzetta ambiente: rivista sull'ambiente e il territorio, 4*, 69–85.

Cortignani, R., & Severini, S. (2011). An extended PMP model to analyze farmers' adoption of deficit irrigation under environmental payments. *Spanish Journal of Agricultural Research, 9*(4), 1035–1046.

Cortignani, R., Dell'Unto, D., & Dono, G. (2018). Recovering the costs of irrigation water with different pricing methods: Insights from a Mediterranean case study. *Agricultural Water Management, 199*, 148–156. https://doi.org/10.1016/j.agwat.2017.12.016.

Council Regulation (EC) No 1698/2005 of 20 September 2005 on support for rural development by the European Agricultural Fund for Rural Development (EAFRD) Official Journal L 277/1, 21/10/2005, Brussels, Belgium.

CREA. (2018). *Consiglio per a la ricerca della agricoltura e l'analisi dell'economia agraria*. www.crea.gov.it/tag/sigrian/. Last accessed 4 Apr 2018.

Dono, G. (2008). Il recupero del costo pieno nella direttiva quadro delle acque: problemi per l'agricoltura italiana. In G. Bonati & C. Liberati (Eds.), *Uso irriguo dell'acqua e principali implicazioni di natura ambientale*. Roma: INEA and MATT.

Dono, G., & Severini, S. (2007). La riforma dell'OCM zucchero in due aree irrigue dell'Italia centro-meridionale. *PAGRI – Rivista di Politica Agricola Internazionale, 2*.

Dono, G., Liberati, C., Severini, S. (2003). La distribuzione dell'acqua d'irrigazione nell'Italia meridionale: un'analisi economica della gestione della risorsa nell'Italia meridionale. In *Istituto Nazionale di Economia Agraria. Indagine INEA sull'irrigazione nelle regioni meridionali: quadro conoscitivo e supporti alle decisioni*. Roma.

Dono, G., Giraldo, L., & Severini, S. (2012). The cost of irrigation water delivery: An attempt to reconcile the concepts of cost and efficiency. *Water Resources Management, 26*, 1865–1877.

Dono, G., Cortignani, R., Doro, L., Giraldo, L., Ledda, L., Pasqui, M., & Roggero, P. P. (2013). An integrated assessment of the impacts of changing climate variability on agricultural productivity and profitability in an irrigated Mediterranean catchment. *Water Resource Management, 27*, 3607–3622.

Dono, G., Cortignani, R., Dell'Unto, D., Mazzapicchio, G., Pasqui, M., Roggero, P. P., Mula, L., Lacetera, N., & Vitali, A. (2014a). *Effetti economici del cambiamento climatico nell'area Padana. L'agricoltura italiana è pronta ai cambiamenti climatici?* La parola al Progetto Agroscenari. Roma. 29–30 October.

Dono, G., Mazzapicchio, G., Cortignani, R., Dell'Unto, D., Pasqui, M., Alfieri, S. M., Basile, A., Bonfante, A., Roggero, P. P., & Mula, L. (2014b). *Effetti economici del cambiamento climatico nell'area del Destra Sele. L'agricoltura italiana è pronta ai cambiamenti climatici?* La parola al Progetto Agroscenari. Roma. 29–30 October.

Dono, G., Cortignani, R., Dell'Unto, D., Deligios, P., Doro, L., Lacetera, N., Mula, L., Pasqui, M., Quaresima, S., Vitali, A., & Roggero, P. P. (2016). Winners and losers from climate change in agriculture: Insights from a case study in the Mediterranean basin. *Agricultural Systems, 146*(1), 65–75.

Fassò, C. (2003). l'Irrigazione in Italia: ieri, oggi e domani, Conference proceedings I campi hanno sete, Associazione Nazionale Bonifiche ed Irrigazione and Food and Agriculture Organization, FAO-ANBI, Roma, 29 Ottobre 2003.

Galvani, A. (2009). La legislazione della bonifica e i Consorzi di Bonifica in Italia, AMS Acta, Università di Bologna, Biblioteca Nazionale di Firenze, ID 2750.

Ghiglieri, G., Barbieri, G., & Vernier, A. (2006). *Studio sulla gestione sostenibile delle risorse idriche: dall'analisi conoscitiva alle strategie di salvaguardia e tutela.* Roma: ENEA.

Giannoccaro, G., Prosperi, M., & Zanni, G. (2011). Economic effects of legislative framework changes in groundwater use rights for irrigation. *Water, 3*, 906–922. https://doi.org/10.3390/w3030906.

Giraldo, L., Cortignani, R., & Dono, G. (2014). Simulating volumetric pricing for irrigation water operational cost recovery under complete and perfect information. *Water, 6*, 1204–1220. https://doi.org/10.3390/w6051204.

HERA. (2012). Acqua. Chi la gestisce in Italia? http://www.gruppohera.it/gruppo/com_media/dossier_acqua/articoli/pagina25.html

INEA. (2008). *Agricoltura irrigua e scenari di cambiamento climatico – Stagione irrigua 2007 nel Centro Nord.* Roma: INEA.

INEA. (2009a). In R. Zucaro & A. Furlani (Eds.), *Rapporto sullo stato dell'irrigazione in Emilia-Romagna.* Roma: Istituto Nazionale di Economia Agraria.

INEA. (2009b). In R. Zucaro & A. Arzeni (Eds.), *Rapporto sullo stato dell'irrigazione nelle Marche.* Roma: Istituto Nazionale di Economia Agraria.

INEA. (2014). *Atlas of Italian irrigation systems* (R. Zucaro & A. Corapi, Eds.). Roma: Istituto Nazionale di Economia Agraria. Available at: www.inea.it

IRSA-CNR. (1999). *Un futuro per l'acqua in Italia.* Bozza per la discussione del corso del Convegno IRSA-30 anni. Roma.

Isenburg, T. (1986). *Acque e Stato. Energia, bonifiche, irrigazione in Italia fra 1930 e 1950.* Milano: Franco Angeli.

ISTAT. (2010). *6th agricultural census.* Rome: Istituto Nazionale di Statistica Available at: http://www.istat.it/en/agricultural-census.

ISTAT. (2017). *Giornata mondiale dell'acqua: le statistiche dell'istat* [World water day: Statistics ISTAT]. Rome. Available at: https://www.istat.it/it/archivio/198245

Mainetti, M., & Pecora, S. (2013). Difesa idrogeologica, serve un approccio proattivo. In *Ecoscienza, Rivista di ARPA (Agenzia regionale prevenzione e ambiente dell'Emilia-Romagna).* 5, IV.

Malorgio, G., & Hertzberg, A. (2007). *Competitiveness of the Southern Mediterranean Countries in the Italian Agri-Food Market.* Paper presented at the 103rd EAAE Seminar 'Adding Value to the Agro-Food Supply Chain in the Future Euromediterranean Space'. Barcelona, Spain, April 23rd–25th, 2007.

Margat, J., & Vallée, D. (2000). *Mediterranean Vision on water, population and the environment for the 21st century.* Blue Plan for the Global Water Partnership/Medtac in the programme of the World Water Vision of the World Water Council. http://www.gwp.org/Documents/The%20Library/GWP%20MED%20Publications/Studies%20on%20the%20Mediterranean%20region/Water%20Vision%20Mediterranean.pdf

Massarutto, A. (2008). *L'acqua.* Bologna: Il Mulino.

Napoli, R. (2010). *Valutazione del rischio di salinizzazione dei suoli e di intrusione marina nelle aree costiere delle regioni Meridionali in relazione agli usi irrigui.* INEA.

NBA-AM National Basin Authority of Appennino Meridionale, Liri-Garigliano and Volturno Rivers, Abruzzo, Basilicata, Calabria, Campania, Lazio, Molise Region, Puglia Regions. (2010). *Piano di Gestione delle Acque Distretto Idrografico dell'Appennino Meridionale.* www.ildistrettoidrograficodellappenninomeridionale.it. Last accessed on 9 July 2016.

Nomisma-Unaproa. (2016). *2° Rapporto Nomisma-Unaproa sulla Competitività del settore ortofrutticolo nazionale*. Roma: Agra Editrice Srl.

Novello, E. (2003). *La bonifica in Italia: Legislazione, credito e lotta alla malaria dall'Unità al fascismo*. Milano: Franco Angeli.

Petriccione, G., & dell'Aquila, C. (2011). Ortofrutta e catena del valore globale. *Agriregionieuropa, 7*(27), 10.

PTA. (2009). Piano di Tutela Acque. Regione Puglia. *Bollettino Ufficiale della Regione Pugli, 130*, 4–8.

Quaglia, T., Musco, F., & Camori, G. (2011). *Aree protette e sviluppo sostenibile un piano per la conservazione della biodiversità e lo sviluppo dell'economia locale nel delta del Po*. Regione Veneto: Università IUAV di Venezia.

Rainaldi, F. (2010). *Il governo delle acque in Italia e Inghilterra: un'analisi del Policy change*. PhD in Political Sciences, Ciclo XXII Settore scientifico-disciplinare di afferenza: SPS/04; Università di Bologna.

RBD (River Basin District) of the Po (Autorità di bacino del fiume Po). (2010). Piano di gestione del Distretto idrografico del fiume Po, http://www.adbpo.it/. Last accessed on 19 May 2015.

Regione Lombardia. (1996). Atti del Convegno "La regolazione dei Grandi Laghi Subalpini" Gardone Riviera, 2–3 Maggio 1996. Regione Lombardia Assessorato all'Agricoltura, Enti Regolatori dei Grandi Laghi Alpini – Giornate studi.

Santuari, A. (2013). *Profili giuridici e assetti istituzionali della gestione del servizio idrico integrato*. www.giustamm.it, 6-2013.

Severini, S., & Valle, S. (2007). La riforma dell'OCM zucchero in Italia: una valutazione di impatto sui produttori bieticoli di Veneto ed Emilia Romagna. *Rivista di Economia Agraria, LXII*(1), 91–122.

Severini, S., Cortignani, R., Dono G., & Vita E. (2016). Assessing the economic role of irrigation and the productivity of irrigation water in Mediterranean countries: The case of Italy. Paper presented at the 158th EAAE Seminar: "Euro-Mediterranean Cooperation in Sustainable Agriculture and Food Security", Chanià (Greece), 8–9 September 2016.

Zucaro, R. (Ed.). (2008). *Direttiva quadro per le acque 2000/60. Analisi dell'impatto sul settore irriguo e della pesca*. Roma: INEA.

Zucaro, R., & Pontrandolfi, A. (Eds.). (2007). *Agricoltura irrigua e scenari di cambiamento climatico – Stagione irrigua 2006 nel Centro Nord*. Roma: INEA.

Zucaro, R., Rago, C., & Vollaro, M. (2012). Valutazione tecnico-economica delle potenzialità di riutilizzo irriguo dei reflui depurati: il caso della Valpadana. Proceedings of the Towards a Sustainable Bio-economy: Economic Issues and Policy Challenges, Trento, 4–5.

Chapter 7
Turkey

Sevilay Topcu, Aysegul Kibaroglu, and Zeynep Kadirbeyoglu

Abstract This chapter reviews irrigation development and policy with specific references to the main water- and land-based regional socioeconomic development projects in Turkey. It analyzes the expansion of irrigation investment as well as institutional and technological changes in irrigation policy and development in parallel with policies of liberalization and decentralization in the late 1980s. The chapter also discusses institutional changes in the management of the irrigation systems as a result of (partial) transfer of management of large-scale irrigation systems to a variety of water user organizations. Finally, it describes current technological and institutional problems and the further challenges to the irrigation sector, such as infrastructure deterioration, risks of drought, environmental and ecological system degradation, and insufficient investment. It also notes the efforts to equip new irrigation schemes with modern technology, such as closed pipes for conveying water instead of open channels, and water-saving micro-irrigation methods rather than surface irrigation techniques.

Keywords Turkey · Irrigation technology · Irrigation policy · Irrigation management

S. Topcu (✉)
Department of Agricultural Structures and Irrigation, Cukurova University, Sarıçam/Adana, Turkey
e-mail: stopcu@cu.edu.tr

A. Kibaroglu
Department of Political Science and International Relations, MEF University, Istanbul, Turkey
e-mail: aysegul.kibaroglu@mef.edu.tr

Z. Kadirbeyoglu
Department of Political Science and International Relations, Bogazici University, Bebek, Istanbul, Turkey
e-mail: zeynep.kadirbeyoglu@boun.edu.tr

F. Molle et al. (eds.), *Irrigation in the Mediterranean*, Global Issues in Water Policy 22, https://doi.org/10.1007/978-3-030-03698-0_7

7.1 Introduction

Turkey, a middle-income country with a population of over 79.8 million (TUIK 2017), is one of the world's 20 largest economies as well as the world's 9th largest agricultural producer (Çakmak and Kasnakoğlu 2016). Although the pace of population increase has slowed in past decades, its net growth rate is still positive: 1.31% per year. The rural-to-urban population ratio decreased from 80% in 1927 to 8.3% in 2016 due to steadily increasing internal migration, mainly to the big cities. As a result of population growth, alterations in water consumption habits driven by increasing socioeconomic development and growing urbanization, water-demanding, and water-dependent sectors have increased and diversified in recent decades, such as energy production, industry, tourism, and environment, along with the traditional consumers like agriculture and domestic use.

While Turkey's agricultural output in the past century has managed to outpace population growth and keep up with growing demand from rising incomes, labor and capital were scarce, while land was relatively plentiful until the mid-twentieth century (Pamuk 2006). A concerted drive to expand agricultural land brought spectacular results after WWII. In keeping with the global Green Revolution, the 1950s, 1960s, and 1970s saw the rapid introduction of machinery, fertilizers, high-yielding varieties, and new inputs, while irrigation in particular increased rapidly (Zurcher 2004). Although Turkey's agriculture was self-sufficient in the past, there has been a sharp decline in agricultural production as a share of GDP in recent years. In 1923, agriculture made up about 43% of total value added. Although gross domestic production value of the agricultural sector increased from about 9 billion TRY (Turkish Lira) in 1998 to 161 billion TRY in 2016,[1] its share of GDP declined from 12.5% to 6.2% during the same time period (TUIK 2017). However, agriculture is still vital to the national economy as a large contributor to export revenues not only generated from agricultural products directly but also as a producer of raw materials for industrial export commodities.

In this chapter, we analyze irrigation policy and development in Turkey. We first present Turkey's geographic and climatic attributes with specific references to water and land resources. We then describe the role played by irrigation and water policy in the creation of the modern state in Turkey with special reference to the major water- and land-based regional socioeconomic development project, namely, the Southeastern Anatolia Project (Turkish acronym: GAP). Subsequently, we analyze the expansion of regional irrigation investments as well as institutional and technological changes in irrigation policy and development which started in the late 1980s in parallel with a policy of liberalization and decentralization. We find that irrigation investment is likely to continue in Turkey. There is a concerted effort to equip new irrigation schemes with modern technology, such as closed pipes for conveying the water instead of open channels and water-saving micro-irrigation methods

[1] Nine billion TRY in 1998 was worth 18.5 billion USD (28 billion in 2016 values), and 161 billion TRY in 2016 was worth 53 billion USD.

rather than surface irrigation techniques. We argue that stakeholders in the irrigation sector, namely, the state and farmers, should sustain institutional progress and reform programs for the sustainable and equitable management and use of water and land resources in the country.

7.2 Physical Setting and the Importance of Water in Turkey

In Turkey, water bodies such as rivers and lakes total around 10,000 km^2 out of a total area of 783,577 km^2. With an average altitude of about 1131 m above sea level, the country is predominantly mountainous, and lowlands are confined to coastal areas. There are three primary climate types (following the Köppen-Geiger climate classification) in Turkey. The Mediterranean and Aegean coastal areas of the country experience a temperate Mediterranean climate characterized by hot and dry summers and mild and wet winters. The Taurus Mountains' abrupt rise on the Mediterranean coast contributes to the irregular distribution of precipitation both in space and time. The Black Sea coastal region has a temperate maritime climate with warm and wet summers and cool and wet winters. This region receives on average 2250 mm of annual precipitation, evenly distributed throughout the year. The interior of Turkey, including central, eastern, and southeast Anatolia, is subject to a continental climate, which has hot summers and cold winters. In central and southeast Anatolia, rainfall is low throughout: less than 300 mm per year. Eastern Anatolia, which consists of several high mountain ridges and high plateaus, receives heavy snow (Türkeş 2010).

The long-term averages for mean temperature, annual precipitation, and evaporation in Turkey are 13.5 °C, 574 mm, and 1173 mm, respectively (see Fig. 7.1). In

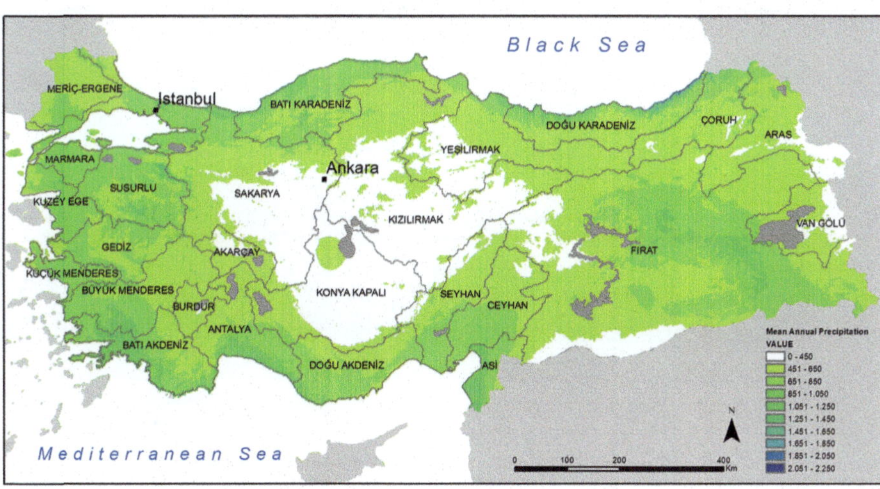

Fig. 7.1 Mean annual precipitation distribution over Turkey

all regions, except the eastern Black Sea region, approximately 70% of total precipitation falls between October and March. Besides the limited rainfall, evapotranspiration rates are high due to prevailing high temperatures during the summer months (Coşkun et al. 2017). Therefore, agricultural production, particularly in fertile regions, such as western, southern, and southeastern Turkey, can only be achieved through irrigation (Topcu 2011).

Comprehensive surveys of Turkey's land and water resources began around the 1930s and were carried out after the establishment of the General Directorate of State Hydraulic Works (Turkish acronym DSI) in the 1950s. Turkey's total annual precipitation amounts to 450 Bm^3, and, considering the water lost to evaporation from the surface, transpiration through plants, and seepage to aquifers, and gained from surface runoff coming from neighboring countries, the total surface runoff is about 172 Bm^3. Since not all the renewable water resources can be utilized, for topographical, geological, economic, and technical reasons, the exploitable surface runoff, which includes inflow from neighboring countries, only amounts to 94 Bm^3. The renewable groundwater potential is 18 Bm^3, hence the total exploitable surface and groundwater resource amounts to 112 Bm^3. In total, 54 Bm^3 per year (50%) of the usable water potential was diverted in 2016 (DSI 2016). The largest share (40 Bm^3) of freshwater resources (74%) was utilized by the agricultural sector, while the remaining 26% was shared equally between industry (7 Bm^3) and domestic use (7 Bm^3).

Turkey has 25 river basins with a wide range of catchment sizes, and large variations in the average annual precipitation, evaporation, and surface runoff parameters. Consequently, river discharges are irregular (see Fig. 7.2). Sixteen rivers rise in-land and reach the Marmara and surrounding seas within Turkey's territorial boundaries. Four of the 25 are closed basins, i.e. they have no outflow to the sea: the Konya, Akarçay, Burdur Lakes, and Lake Van basins. The other five are

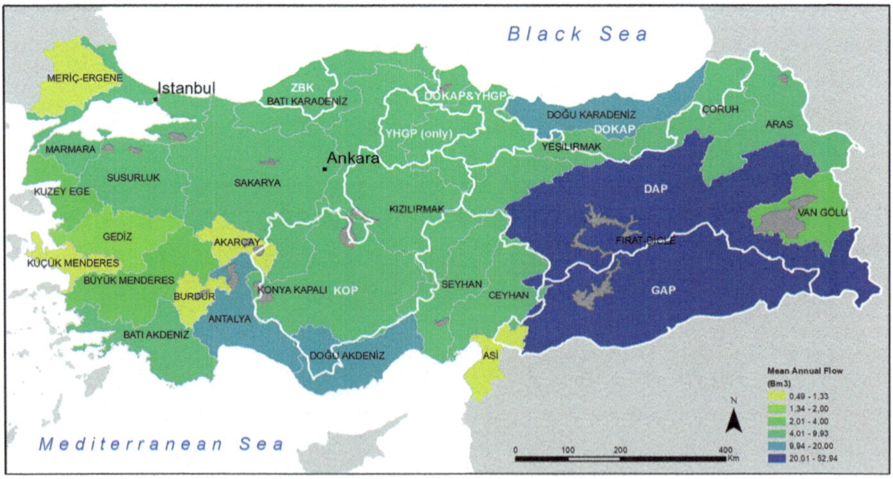

Fig. 7.2 Turkey's 25 river basins and their mean annual flows, Bm^3. (DSI 2016)

the main transboundary basins. Turkey is upstream of Syria/Iraq in the Fırat-Dicle/Euphrates-Tigris, of Georgia in the Coruh, and of Georgia and Iran in the Kura-Araks. Conversely, it lies downstream from Lebanon and Syria in the Asi/Orontes basin and downstream from Bulgaria in the Meriç/Maritza basin. These five transboundary rivers constitute about 35% of Turkey's water potential (DSI 2016). Some regions, such as Thrace, the Aegean Sea coastline, and central Anatolia, include about 20% of Turkey's cultivable area but only 10% of the country's surface water resources. Per-capita water availability is below 1000 m^3 in the Marmara, Kücük Menderes, and Asi basins. Together with the Sakarya basin, these three basins are threatened by severe water stress in terms of both quantity and quality.

Turkey's landscape is mainly composed of natural land cover (54%), dominated by transitional woodland and shrubs, and agricultural areas (42%), including nonirrigated and irrigated arable land, agriculture with natural vegetation, and complex cultivation patterns. The total cultivable land area is about 28 million hectares, 25.85 million of which is suitable for irrigation. Considering the water resource potential given by DSI, 12.5 million hectares can be irrigated, but, for technical and economic reasons, only 8.5 million hectares (7.9 and 0.6 million hectares from surface and groundwater resources, respectively) are planned to be equipped for irrigation by DSI by the year 2023 (DSI 2016).

Out of a total irrigated area of 6.09 million ha, an estimated 3 million ha of arable land is affected by waterlogging (i.e. poor drainage and topography) and a further 1.5 million ha, a quarter of the irrigated area, has yield limitations due to salinity and sodicity problems. In particular, irrigated lands of the Harran plain in the Euphrates river basin, the Amik plain in the Asi basin, Konya in central Anatolia, and the Lower Seyhan plain in the Seyhan river basin are threatened by salinity and sodicity (Küsek 2010). As much as 83.21% of agricultural land is currently under risk of severe water erosion. The hilly topography, soil conditions that facilitate water erosion (fine texture, low organic matter, poor plant coverage due to semiarid climate), inappropriate agricultural practices (e.g. excessive soil tillage and cultivation of steep lands, not using land in accordance with its capabilities), and forest fires are the main causes for intensified erosion in Turkey.

The most common crops in terms of percentage of irrigated area are maize, cotton, and cereals with a share of 25%, 14%, and 13%, respectively. These are followed by fruit trees (7%), fodder crops (6%), sugar beet (5%), vegetables (5%), sunflower (5%), legumes (3%), vineyard (2%), and other crops (12%). However, cropping patterns show regional variance depending on land-soil-climate conditions, as well as economic and social factors, such as input and output market prices and farmer habits. The benefit of irrigation in terms of yield increase as reported by the DSI (2017) was 273% for cotton, 147% for fruit trees, and 155% for citrus, whereas the largest increase (a sixfold increase in yield) was achieved for maize since irrigation was introduced.

Water is at once the most critical and limiting factor in agricultural production in many parts of the country including central and south Anatolia, the Aegean, and the Mediterranean regions. Intensive drought periods caused the loss of crops and animals and the migration of farmers to other areas over recent decades. The cost to the Turkish agricultural sector in a year of significant drought, 2007, was estimated at around €2.3–2.5 billion (see Oral 2008; MoFAL 2013).

The productivity in many agricultural areas equipped for irrigation is much lower than originally planned due to (i) a low *irrigation ratio* and (ii) inefficient *water use*. The irrigation ratio refers to the net area that is actually used for irrigated crop production within the irrigation scheme at least once a year. The low *irrigation ratio* means that a considerable amount of land within an irrigation scheme is not irrigated although it is served by irrigation networks. The average irrigation ratio is about 65% in the country (DSI 2016), varying between 45% and 90% in Water User Association-managed areas (Yavuz et al. 2006). Another important performance indicator for irrigation projects is *water-use efficiency* (Alizadeh and Keshavarz 2005). According to the latest records, the irrigation efficiency was around 45% in irrigated areas managed by Water User Associations (DSI 2016), meaning 55% of diverted water returned to the river/aquifer system.

7.3 State Consolidation and Irrigation Policy

Turkey's hydraulic history is characterized by urban waterworks rather than irrigation schemes, though potable water supply and irrigation often went together. Dam and channel building in Turkey has a millennial history, several dam types were premiered on its territory, and several of these structures are still in place. The Greeks, Romans, and Byzantines constructed a host of well-preserved waterworks: water conveyance systems, aqueducts, stone pipes and lead-pipe inverted siphons, tunnels, spring water collection chambers, reservoirs, cisterns, but also dams, such as the Faruk dam near Van, which collapsed as recently as 1988 (Ozis et al. 2005). The first of three dams constructed in Dara near Mardin during the reign of Justinian (527–565) is known as the oldest arch-type dam in the world (Cakmak et al. 2004). The Ottomans built many dam structures, several of which are still operational, and countless urban water structures. The first modern irrigation and drainage project was implemented in response to a severe drought devastating the Konya region: the Cumra Irrigation and Drainage Project in Central Anatolia (1907–1914) conveyed water to the Konya plain, mainly from Beyşehir Lake.

From the 1920s to the 1950s, Turkey was engaged in state consolidation efforts, which included the investigation and exploitation of water and land resources. New government institutions, namely the Ministry of Public Works (established in 1920) and the Electrical Power Resources Survey and Development Administration (established in 1935) were mandated to conduct hydrological surveys of the country's water resources and hydropower potential and to carry out related civil works (Tigrek and Kibaroglu 2011). Framework laws, such as the Village Law (1924), the Water Law (1926), and the Law on Municipalities (1930), were enacted, and various water projects were implemented. The earliest studies of the Euphrates-Tigris basin date back to the 1930s and eventually led (in the 1980s) to the Southeastern Anatolia Project (GAP). Poor water quality across the country was one of the urgent problems which the state had to address to improve public health. Responsibility for implementing the Water Law was therefore entrusted to the Ministry of Health and Social Aid. Similarly, the draining of swamps was seen as essential if certain water-borne diseases, such as malaria, were to be eradicated. Throughout this period, public investment in infrastructure was decided centrally on an ad hoc basis and thus in response to pressing needs.

Turkey had long followed a Keynesian development model, which assigns the state an active interventionist role in the economy, including the provision of public services (Kibaroglu et al. 2009). After World War II, state-led water resources development was fostered by the establishment of a central water bureaucracy, the DSI, which assumed a role similar to that of the US Bureau of Reclamation. One of the DSI's main objectives was to stimulate economic development through water-related infrastructure. It was believed that only the state was able to overcome underinvestment and realize economies of scale (Scheumann et al. 2011).

From the 1950s onward, Turkey adopted a river basin planning approach based on exploratory hydrological studies. Law No 6200 of 1953 ruled that the DSI should focus on major river basins and have regional directorates. Two laws enacted in this phase reinforced the status of the DSI as the main public water agency: the first was the Groundwater Law (1960), which mandated the DSI to grant licenses for the utilization of the country's groundwater resources (however, individuals and groundwater cooperatives would be entrusted with management issues); and the second was Act No 1053 (1968), which made the DSI responsible for water supply to cities with a population greater than 100,000.

Following the adoption of an import-substitution industrialization program, public investment in the water sector was arranged through the national 5-year development plans: major water infrastructure, such as irrigation systems including storage facilities and dams for hydroelectricity generation, was state-financed and state-managed. In recognition of the economic and political importance of the agricultural sector to the country, particular attention was also paid to rural development (Scheumann et al. 2011). Hence, Law no. 7457 of 1960 established the early General Directorate for Soil and Water (Toprak-Su), which was later reconstituted as the General Directorate for Rural Services (GDRS), under the Ministry of Agriculture and Rural Affairs, in 1985. While the DSI, under the Ministry of Energy and Natural Resources, was responsible for the construction and management of the large-scale public irrigation schemes, the services for on-farm irrigation, including the construction and management of irrigation facilities up to the 500 l/s, and supplying water to municipalities in rural areas below 3000 inhabitants, were the responsibility of the GDRS. Other on-farm services related to irrigated agriculture (e.g. surface and subsurface tile drainage, land leveling, land consolidation, soil and water conservation measures, research on soil-plant-water relationships) were also within the remit of the GDRS. As part of an economic program in 2005, the GDRS was abolished, and its duties and facilities (as well as the personnel) at all provincial levels were transferred to the Special Provincial Administrations (SPA) except for two metropolitan municipalities in Istanbul and Kocaeli. The former Bank of Provinces (renamed İlbank in 2011), under the Ministry of Environment and Urban

Planning, is another actor in the financing and providing of technical support to local authorities for the construction of water supply (also surface and groundwater for irrigation purposes) and waste water treatment units.

Under the guidance of the State Planning Organization, Turkey made considerable progress in augmenting water supply. This phase also witnessed the birth of GAP, one of the most significant projects in Turkey's history. With the stated aim of overcoming the relative backwardness of the southeastern Anatolia region, the development of water and land resources through public investment was regarded as an effective strategy (Kibaroglu et al. 2009). GAP is Turkey's largest integrated development project and is considered vital to the economy. It has the potential to meet the rising demand for hydropower caused by population growth, along with urbanization and the country's industrialization impetus. Upon the completion of the GAP project, 1.7 million ha of land will be brought under irrigation, equivalent to nearly one fifth of Turkey's irrigable land; energy production in the region will reach 27 billion kWh; per capita income is expected to rise by 209%; and employment opportunities will be created for around 3.8 million people. This will be accomplished through the construction of 22 dams, 19 hydropower stations, and extensive irrigation and drainage networks. By the end of 2016, 504 million ha of agricultural land were under irrigation in the GAP region. GAP's basic development objectives are defined as follows: to raise the income levels in the GAP region by improving the economic structure in order to narrow the regional income disparities; to increase productivity and employment opportunities in rural areas; to enhance the assimilative capacity of larger cities in the region; and to contribute to the national objective of sustained economic growth, export promotion, and social stability through the efficient utilization of the region's resources.

To these ends, the GAP was transformed from a pure infrastructure development project into a project in support of sustainable development with additional investment in urban and rural infrastructure, agriculture, transport, industry, education, health, housing, and tourism. The sporadic political and economic crises during the 1970s, 1980s, and 1990s prevented this investment from being completed in a timely fashion. From the very beginning, the GAP project and, in particular, its dam component has come in for some harsh criticism. The objections specifically concern resettlement issues, environmental and cultural aspects, and impacts on the rivers' riparian countries, Syria and Iraq (Scheumann et al. 2011).

The economy ran into a serious crisis with high inflation, a growing trade deficit, and high unemployment rates at the end of the 1970s (Kibaroglu et al. 2009). A balance-of-payments crisis emerged in late 1979, arising from insufficient exports and external debt service obligations (Scheumann et al. 2011). This meant that the state-led import-substitution model had to be restructured. This also had repercussions on irrigation planning, development, and management, as the next section outlines.

To summarize, until the 1980s, Turkey's water policy, including the irrigation development among the other water-related sectors, can be characterized as follows: (i) policy initiatives were basically shaped by national considerations, and governments decided on their agendas without the interference of foreign actors. This certainly does not mean that it was an entirely closed scene: professional relations existed with the US Bureau of Reclamation, and leading bureaucrats and politicians had been educated abroad. (ii) Managing a river's water resources was largely the mandate of the DSI: planning and investigation were focused on the 25 river basins, and decisions were taken by bureaucracies at the national level and dictated by national priorities. (iii) Public spending was the only instrument used to finance irrigation infrastructure projects.

7.4 Irrigation Development and Management[2]

In addition to the GAP project, which marks the post-1980 era, there are other important irrigation development projects (Fig. 7.2), such as the Zonguldak-Bartın-Karabük (ZBK) Regional Development Project (1995–1996), the Yeşilırmak Basin Development Project (YHGP) (1997), the Eastern Anatolia Project (DAP) (1999–2000), the Eastern Black Sea Regional Development Plan (DOKAP) (1999–2000), and the Konya Plain Project (KOP) (2011) (DSI 2016).

The UN Environment Program (UNEP) identifies Turkey as one of the first places where desertification will start in Europe, and, within Turkey, the Konya closed basin (central Anatolia) will face desertification by 2030 if nothing is done. While 17% of Turkey's irrigated area is located in the KOP region, that region has only 4% (4.36 Bm^3) of the available water resources, three fifths (2.44 Bm^3) of which come from groundwater resources. The KOP region covers about 3 million ha of cultivable land; however, with the existing water resources and even using water-saving irrigation techniques, only 1.1 million ha of this area can be irrigated. Hence, the KOP also envisages water transfer from the river Göksu through the Blue Tunnel into Konya, to be stored in reservoirs and used for irrigation. Agriculture is the major water-using sector (94%) in the Konya closed basin, and, as a consequence of groundwater overexploitation over recent decades, the groundwater level has fallen by 3 m a year on average. The KOP does not so much aim to expand irrigation in the area as to make it more efficient (DSI 2016).

The Eastern Anatolia Project (DAP) is to irrigate about 1.37 million ha. It covers 23% of Turkey's geographical area, involving 15 provinces and affecting 8% of Turkey's population. The DAP includes irrigation, drainage, and reclamation projects, as well as hydropower and domestic supply projects, and one third of the agricultural lands in the region are already under irrigation (DSI 2016).

The Eastern Black Sea Project (DOKAP) aims to restore degraded soils for agricultural production by means of irrigation and drainage projects and flood control, but also involves fishery, forestry, industry, tourism, and cultural heritage conservation. The planned area to be irrigated within the DOKAP region is around 287,000 ha, and irrigation projects were completed for almost two thirds of the planned total by the end of 2016 (DSI 2016).

[2] Some parts of this section are drawn from Kibaroglu et al. (2012:27–34).

Besides controlling the irregular flow regime which causes floods every year, the Yeşilırmak River Basin Development Project (YHGP), initiated in 1997, aims to prevent erosion and pollution as well as to promote irrigation and drainage in the project region (DSI 2016).

Following the above-mentioned regional priority projects, in 2012, Turkey began a project called "1000 reservoirs in 1000 days" to bring irrigated farming to rural areas that were outside planned large-scale irrigation schemes. Flood and erosion control were also among its goals. The project was completed in 2015, and the reservoirs, with a total storage capacity of 600 million m^3, are for irrigation and flood control in an area of about 1.7 million ha (DSI 2016).

7.4.1 Technological Change

During the 1950–1965 period, water was mostly conveyed using open channel distribution networks from the reservoirs to the fields. Since the 1970s, canalets (concrete raised parabolic flumes) were constructed, whereas from the 1980s onward, pipeline distribution networks were utilized in areas with irregular topography, high slopes, and limited water resources. Pipelines conveying and distributing water to the fields have become the rule, reaching 94% of irrigation schemes under construction by the end of 2016. However, in many old schemes, surface irrigation continues to be dominated by gravity-fed systems such as furrow, border, and even wild (uncontrolled) flooding; pumping systems only account for 22% due to their higher cost (DSI 2017); electricity can account for 80% of operation expenses.

The main canal, tertiary canals, and discharge canals are of trapezoidal cross section with concrete lining. The smaller backup, tertiary, and distribution canals, on the other hand, are generally prefabricated canals placed on supports above ground level at heights varying according to the topography. Water delivery systems (see Fig. 7.3) in irrigation schemes as a country average comprise:

- Classic trapezoidal open canals (37%)
- Canalets: precast concrete half ellipsoidal open canals (41%)
- Piped systems (22%)

Fig. 7.3 Typical open canal, canalet, and piped systems used in Turkey's irrigation schemes (DSI 2017)

As of 2016, an area of about 6.09 million ha was equipped with irrigation infrastructure, 63% of which was developed by the DSI and 37% by the now-abolished General Directorate of Rural Services (GDRS) and its successors. Although the DSI has continued to build irrigation systems, most irrigation schemes have now been transferred for operation and management to either:

- Irrigation cooperatives, e.g. groundwater irrigation cooperatives and water user associations
- Local authorities, e.g. village authority and municipality
- The public sector, e.g. universities, research institutions, and public agricultural enterprises: TIGEM

About 75–80% of Turkey's irrigated area is irrigated using surface water, and the remaining 20–25% using groundwater (see Box 7.1) (DSI 2017). To enhance flexibility and reliability in irrigation schemes, the conjunctive use of surface water (canal water) and groundwater is required, particularly in areas where surface water is insufficient.

Box 7.1 Resources used for abstracting irrigation water (DSI 2014)

- *Surface water resources, 80%*
  - Pumping: 18%
  - Gravity: 82%
- *Groundwater resources, 20%*
  - Supplementary to gravity: 16%
  - Abstraction by pumping*: 84%

**Groundwater irrigation cooperatives and individual (private) irrigation*

For this purpose, the DSI installs wells and manages groundwater abstraction from those wells into the irrigation schemes they operate. Nevertheless, the operation of these groundwater irrigation networks has been occasionally transferred to organizations such as water user associations and groundwater irrigation cooperatives. The use of groundwater resources more than 10 m deep requires a license issued by the DSİ, which solely covers the right to use it; it can neither be transferred nor sold.

About 14 Bm^3 of the total groundwater potential (18 Bm^3) has already been assigned to different water users including mainly the groundwater irrigation cooperatives, individuals (private), and public entities for different purposes, such as drinking and domestic use, industry, and irrigation. Agriculture (69%) is the main user of groundwater. About 4.4 Bm^3 is allocated to organizations such as groundwater irrigation cooperatives, the DSI (then on to the irrigation associations), and public entities for irrigation of about 684,000 ha. The Groundwater Irrigation Cooperatives founded from 1966 onward are a major user, and 72% of the total area irrigated with groundwater is under their control. Although 273,962 licenses for the use of around 5.1 Bm^3 of groundwater had been granted for individual (private) irrigations as of

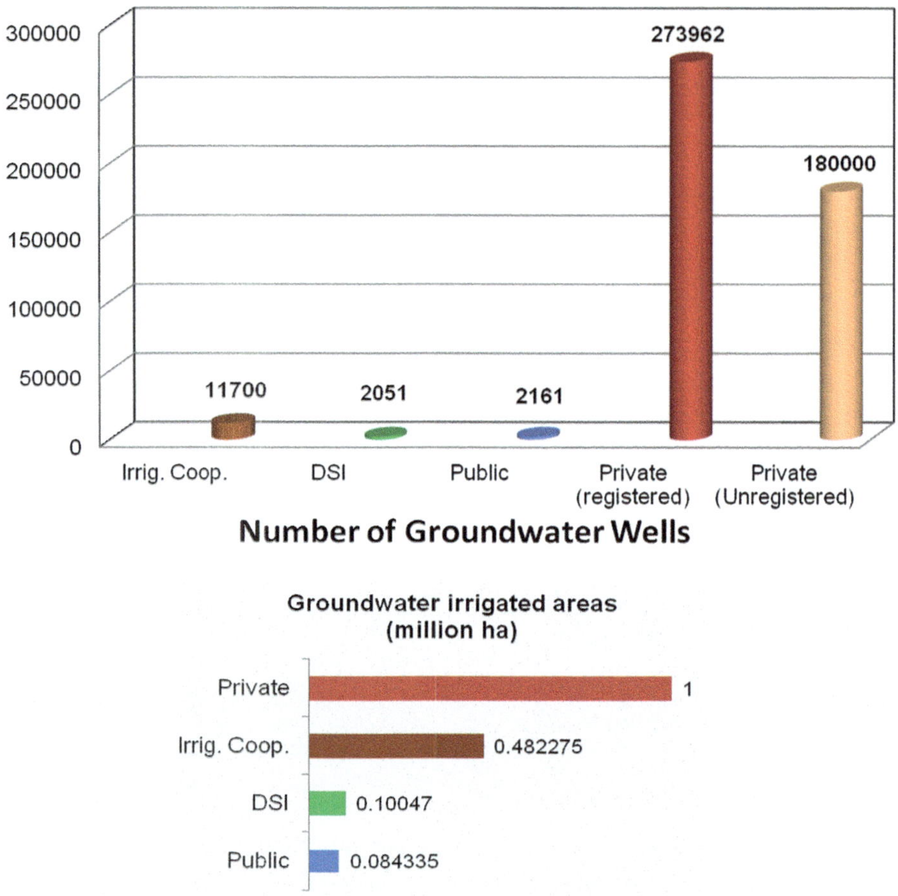

Fig. 7.4 Number of groundwater wells and irrigated areas (DSI 2014)

2014, there is no reliable data about the irrigated areas using these resources except for a rough estimate of around 1 million ha (Fig. 7.4).

DSI and the Ministry of Food, Agriculture, and Livestock (MoFAL, formerly the Ministry of Agriculture and Rural Affairs: MARA) have taken steps to encourage farmers to use sprinkler and drip irrigation methods, particularly in areas with high climatic vulnerability, such as the Aegean and Mediterranean regions as well as in central Anatolia (Konya closed basin), where resources are scarce due to overexploited surface and groundwater. Turkey has experienced prolonged droughts every 30 years, and the frequency appears to be increasing (Sen et al. 2012). The latest run of dry years was 2007–2009–2013. After the 2007 drought, the government prioritized short-term compensation measures to offset losses but also introduced economic instruments to tackle droughts. Farmers' debt repayments to state banks were postponed for 1 year, and interest-free loans and subsidies for implementing water-saving irrigation technology were provided, along with direct payments to farmers for mitigating drought effects.

These incentives successfully convinced farmers to switch from flooding to pressurized systems, such as drip and sprinkler, and even more so after the subsequent drought in 2009. Within the first 3 months following this initiative, more than 8000 farmers had changed their irrigation practices to pressurized systems (Topcu 2011). By the end of the year 2012, 205,919 farmers had benefitted from the incentives, and a total of 402,726 ha land was equipped with drip irrigation (Türker 2013). According to Turkey's fifth national communication under the UNFCCC report, a grant was provided for irrigation cooperatives and village authorities to transform on-farm irrigation systems to closed and pressurized systems through the Supporting Rural Development Investments Program implemented by MoFAL. The grants support 75% of costs for collective pressurized irrigation applications and 50% of within-parcel modern pressurized irrigation investments. The rest of the investment is provided as low-rate, long-term loans (Republic of Turkey Ministry of Environment and Urbanization 2013). Until 2017, the subsidy granted to individual farmers and collective entities could amount to a maximum of 200,000 TRY and 500,000 TRY, respectively. Although these ceilings were reduced to 50,000 TRY and 2000,000 TRY in 2017, the Ministry of Forestry and Water Affairs announced the preparation of a new law to enforce the compulsory use of water-saving irrigation methods, particularly in water-scarce areas in 2018. State loans (50% of the cost) and interest-free credit were provided to farmers (Akşam 2018).

As at 2017, approximately 67% of the total area in Turkey was irrigated by surface irrigation methods (furrow, border, etc.). The remainder was irrigated with pressurized irrigation methods such as sprinkler (19%) and drip (14%) (DSI 2017).

The shift from surface to micro-irrigation has generally taken place in fruit and vegetable growing areas in addition to those under protected cultivation. The use of sprinkler irrigation is also becoming common for field crops, such as sugar beet, potato and groundnut, particularly in drought-prone and water-scarce areas. The citrus, apple, grape, cherry, peach, fig, olive, and walnut orchards are the main fruit tree varieties to be irrigated with drip systems in the northwestern, western, and Mediterranean regions of the country, whereas apricot and pistachio orchards are irrigated mostly with drip systems in the eastern Mediterranean and eastern Anatolian regions. For vegetables, furrow irrigation methods have also been replaced with sprinkler and drip irrigation.

Water user associations calculate irrigation water charges per hectare according to balanced budget principles, generally dividing the cost of O&M and investment expenses by the total irrigated area. However, the fees may not be less than the operating and maintenance tariffs announced by the ministerial cabinet each year. In most of the country's irrigation schemes, irrigation water charges are based on the "area-crop" system in which the charge per hectare is determined by the type of crop. Farmers may pay less per hectare when using water-efficient irrigation technologies and increasing the production per area in many irrigation schemes. The charges showed a significant increase compared to those collected by the DSI before the irrigation management transfers to WUAs. Irrigation water charges vary significantly between schemes using gravity and pumping systems; however, there are also different cost categories within each system depending on the infrastructure and combined use of both systems. In general, farmers under pumping schemes pay around 2.5–5 times more than those under gravity schemes. Depending on the cultivated crop, the fees per hectare vary as well, e.g. farmers growing cereals, vegetables, and rice in a gravity-irrigated scheme with basic infrastructure pay 10.5 TRY, 20.5 TRY, and 40 TRY, respectively. The sample figures are selected from the tariff table published in the Official Gazette of 9/26/2017 (no 30192) for 2018.

Over the last 15 years, attempts at land reform could not prevent the Turkish agricultural sector from being beset by fragmentation, so that the size of farm holdings remains small. According to reports by the General Directorate of Land Reform and DSI, the average size of a farm holding is around 6 ha; however, these farms are generally fragmented into six or more parcels, each of approximately 1 ha or less. Despite incentives and increased farmer awareness, the use of water-saving technology is far behind the ultimate target due to increased costs from fragmentation. Consolidation is now part and parcel of regional development schemes, several of which are now in progress. During the 41 years between 1961 and 2002, only 450,000 ha was consolidated, whereas the consolidated area reached 6 million ha in 2016 (MoFAL 2016; DSI 2016).

7.4.2 *Irrigation Management Transfer*

In 1993, under an accelerated program of Irrigation Management Transfer (IMT) which was guided and partially financed by the World Bank, Water User Associations (WUA) were established to operate and maintain the secondary and tertiary levels of almost all large-scale public irrigation systems (Kibaroglu et al. 2009). Until 1993, the DSI and the General Directorate of Rural Affairs (KHGM) planned, built, operated, and maintained irrigation infrastructure, even though the small-scale decentralization of irrigation operation and maintenance (O&M) had taken place prior to this date. The DSI operated the irrigation schemes with a top-down approach that had low levels of farmer participation and very low cost-recovery rates, at around 10% (Akuzum et al. 1997: 552). In fact, Law 6200 of 1954 granted the DSI the right to transfer the management of state-owned infrastructure. However, the rate of transfer of irrigation O&M was insignificant prior to 1993. The transfer of irrigation O&M was caused by "(a) rapidly escalating labor costs, (b) a hiring freeze on government agencies, and (c) consequent concern over the agency's ability to operate and maintain systems serving the expanding irrigated area for which it was responsible" (Svendsen and Nott 1999a). This model of irrigation management transfer was inspired by the Mexican example (Palerm-Viqueira 2004).

Following the transfer of O&M, the DSI maintains only the ownership of the resource infrastructure. The responsibility for the secondary and tertiary canals is transferred

Table 7.1 Organizations governing transferred irrigation schemes

	Number	%	Area (ha)	%
Villages	201	21	33,671	1.5
Municipality	124	13.0	89,551	3.9
Water user association	383	39	2,036,836	88.6
Irrigation cooperative	246	25	130,105	5.7
Other	17	2	9727	0.4
Total	755	100.0	2,299,890	100.0

Source: DSI (2016)

to the WUAs or irrigation cooperatives.[3] WUAs and cooperatives took over O&M in cases where the irrigation scheme covered more than one village (Svendsen 2001). In other cases, it was the village administration or municipality which took over the O&M responsibility. As can be seen from Table 7.1, in recent decades, the management of over 2 million ha of irrigation schemes has been transferred.

The legal standing of the WUA should in principle be guaranteed by an enabling law which authorizes its establishment and the transfer agreement between the state agency and the WUA (Salman 2002). However, in Turkey, the accelerated transfer program progressed much faster than planned, and there was no opportunity to prepare such a law. The associations were established by reference to three laws: the Village Act (No. 442), the Local Government Act (No. 1580), and the Provincial Governance Act (No. 5442). The need for a new law to determine the principles of WUA functioning was articulated by different agencies. In 2005, WUAs were brought under the jurisdiction of legislation pertaining to Local Administrative Unions (Law No. 5355, *Mahalli Idare Birlikleri Kanunu*, May, 26, 2005). That legislation did not bring about major change.

WUAs finally gained public legal authority status following the legislation of the 2011 Water User Association Law[4] (Özerol 2013). Many changes are brought to the structure and functioning of WUAs. According to the 2011 Law, WUAs are set up by the local authorities in an irrigation zone and apply to the DSI in order to sign the transfer agreement and protocol which gives them the right to collect fees and assigns them the responsibility of distributing water and maintaining the canals. According to the law, WUAs are responsible for their operation, management, maintenance, and repair.

Each association has a chairperson, a council, an executive committee, and an audit committee. Prior to the 2011 Law on Water User Associations, the headmen – the principal elected authority in a village – and the mayor (if there is a municipality within the borders of the WUA zone) occupied the permanent seats in the council.

[3] A cooperative is different from WUAs as it is "owned and operated by its members who share its profits or benefits" (Svendsen and Nott 1999a:17). All members of a cooperative participate in the general assembly meetings, and there is direct democracy rather than the representative structures of the WUAs.

[4] Law No. 6172, 08.03.2011.

Although councilors were to be elected by the farmers who owned or rented land within the irrigation zone of an association, this practice was abandoned in some localities due to tension and conflict. In some places, the representatives were not elected from the water users at large but appointed by the headmen. Therefore, it was possible for large landowners to become chairs or councilors and to favor the interests of large farmers (Unver and Gupta 2002). This kind of elite capture led to allegations of corruption and embezzlement in some of the WUAs.

The Law on WUAs changed the "one farmer, one vote" principle by increasing the weight of those farmers who own or rent tracts larger than the average in their WUA. Currently, the number of votes in the election of councilors depends on the amount of land a farmer owns or rents (for a period of more than 5 years) – with a maximum of five votes per farmer.

According to the 2011 law, the chairperson of the association is elected by the members of the WUA assembly (parliament) for a 4-year term and is the head of the executive committee, which decides on matters related to the management of the associations. Technical staff are hired to operate the system (Palerm-Viqueira 2004). According to the law, the associations are not allowed to spend more than 30% of their annual budget on personnel expenditure.

The revenues of the association consist mostly of fees collected from users. The fees (per donum[5]) depend on the crop that will be cultivated and are set by each association (Unver and Gupta 2002). Self-auditing mechanisms for WUAs existed but were not widely used. Prior to the 2011 law, a group of councilors could be selected to audit the accounts, question the chair, and scrutinize the annual report submitted to the council by the chair (Naik and Kalro 2000). The law established an audit committee selected from the councilors. However, the extent to which the committee can perform its duties depends on the local power dynamics. There are also external checks and balances in the system, in that it is the responsibility of the governor's office to monitor the activities of the WUAs and approve their fees and budgets. The governor's office is responsible for establishing an audit commission to scrutinize the finances and administration of the associations.

WUAs collect water-demand forms (based on the cropping pattern and evapotranspiration) before the start of each irrigation season (usually in April), and forward the total amount required to the DSI, which allocates the water from the reservoir. During the off-season (November to April), the WUAs clean the secondary and tertiary canals. The associations' technicians carry out repair work, if necessary, in order to ensure a healthy distribution of water. Another key duty of the association is to clean the drainage canals.

During the irrigation season, the association is responsible for making sure that all users get the necessary water to irrigate their crop. The distribution of water is organized in a variety of ways. The most common is the rotation system for different tertiary canals combined with a distribution order among farmers set by technicians. Ideally, farmers submit water demand forms to the field technicians 3 days prior to irrigating their fields. The number of siphons they can use to divert water from the

[5] Donum is the measurement unit used among farmers and corresponds to approximately 919.3 m^2.

canal to their field is determined by the amount of land under irrigation, and the field technician monitors the process. The irrigation order is sometimes determined randomly by a lottery. In either case, field agents monitor whether farmers over-irrigate their land or take water when it is not their turn. There are serious questions as to whether this monitoring is effective in some contexts.

Harris (2005) claims that state agencies in Turkey had high expectations of the WUAs, which were expected to increase efficiency, promote the sustainability of irrigation resources, and establish horizontal networks. However, there are divergences in farmer satisfaction from the services they receive from the WUAs (Kadirbeyoglu and Ozertan 2015). Furthermore, WUAs were unable to implement participatory irrigation Operation and Management in some local contexts characterized by power asymmetries (Kadirbeyoglu 2008). In such settings, the associations are sometimes captured by powerful and large landowners, who can use the association resources for their own benefit. In other contexts, the local, participatory management of irrigation enabled a more efficient co-management of irrigation, especially at times of drought. The state agency and associations were able to devise new payment mechanisms to reduce the amount of irrigation without endangering the crops (Kadirbeyoglu and Ozertan 2015).

While the fee collection rate of the state agency prior to the transfer was on average 38% between 1989 and 1994, it had already reached an average of 72% in 1995 under the governance of WUAs (Svendsen and Nott 1999b). However, it should be noted that there was an undercover farmer support system which was at the root of the state agency's inability to collect irrigation fees. The Public Debt Act stipulated that debt would incur a 10% penalty only if the farmer failed to pay on time. Subsequently, even if the farmer paid 5 years later, (s)he would still owe the original amount plus 10%. In an economy where annual inflation rates were on average 70% during the 1980s and 1990s, this left no incentive to pay on time. Since WUAs depend on the fees they collect to survive financially and pay their personnel, they had to collect their fees and prevent those who did not pay from accessing water the following season. There is variance in the proportion of fees that the associations can collect from their users: whereas some can collect only 60% of their budgeted water fees, others approach 100% (Unver and Gupta 2002:13; Kadirbeyoglu 2008).

The sustainable use of irrigation water and the elimination of over-irrigation, which causes problems such as salinity and waterlogging, were not addressed by the establishment of the WUAs. These problems relate more to the type of infrastructure constructed: especially with open canals and insufficient drainage systems, we cannot expect WUAs to bring about sustainable water use.

7.5 Challenges for the Irrigation Sector in Turkey

Since irrigation is likely to continue to be a bulk consumer of water, serious challenges need to be tackled: scarce water resources, erratic and unevenly distributed precipitation, the overexploitation of aquifers, the effect of climate change, decreasing water quality, competition among water user sectors, and increased demand for irrigation. In addition to poor water use efficiency, irrigated agriculture in Turkey has several serious structural problems, including the small size of agricultural holdings, fragmented and scattered fields, the deterioration of infrastructure, insufficient and/or inappropriate drainage, and shallow water table management, as well as a lack of investment. These factors compromise the potential benefits of irrigation. Some are briefly discussed below.

7.5.1 *Infrastructure Deterioration*

Most irrigation schemes in Turkey were built about three or four decades ago. The continuous deterioration of irrigation infrastructure (particularly supply canals) has led to substantial conveyance losses during the transportation and diversion of water. These mainly consist of operational losses, evaporation, and seepage, the latter being by far the main loss. Average conveyance losses of up to 8.6% in main canals, 5% in trapezoidal secondary canals, and 7% at the tertiary level have been reported for several irrigation schemes (e.g. Akkuzu et al. 2007). The most common reasons for high seepage losses include low concrete quality, poor joint construction, inadequate compaction, wear and tear, and consequent cracking. Furthermore, the inappropriate construction of water intake structures, such as constant head orifices and outlets on the main canal, as well as the misuse or damaging of the canals by breaking and/or dumping all types of waste/rubbish into the canals may often cause slippage of the fixed concrete legs and/or blockage of the flow, consequently flooding the roads and fields (Yenigun and Aydogdu 2010). Another source of loss from the canals is weeds and plants growing on the slopes of the canals, on which the DSI and WUAs spend considerable amounts of money every year. Besides infrastructure deterioration, the lack of automation at control structures, insufficient flow monitoring and measuring structures also cause the overuse of irrigation water, consequently lowering water-use efficiency and resulting in salinity and shallow water table problems.

7.5.2 *Climate Change Impacts on Water Resources and Irrigation Sector*

Recent analyses of Turkey's climate data indicate that the average summer temperature of the 2000s was about 1.5 °C higher than the 1960s or 1970s. While only a moderate decrease in the total amount of precipitation has been reported, the temporal distribution pattern of winter rainfall has significantly changed; the decreasing trend in precipitation, particularly in winter months, is correlated with drought events (Sen et al. 2012). Global climate change is expected to worsen climate conditions by causing lower and more erratic rainfall combined with increased temperatures, resulting in higher rates of evaporation, more severe and frequent drought events, and increasing water stress and water requirements for crop production (Sen et al. 2012). Evaluation of annual minimum, maximum, and mean stream flows of all the Turkish rivers showed significant decreases in most basins in the western part of the country (e.g. Topaloglu 2006). A study conducted in the Seyhan river basin showed that as a consequence of lower rainfall, both surface and groundwater resources would decline drastically in the Mediterranean region (Tezcan et al. 2007).

The snow-fed Euphrates and Tigris rivers will undergo surface temperature increases across the entire basin. The increase is comparatively greater in the highlands in winter, which would mean a reduction in the snow cover and change in the seasonality of surface runoff. The annual surface runoff is projected to decrease by 26–57% on average in Turkey by the end of the present century. All other countries in the basin are expected to feel such stress (Bozkurt and Sen 2013). Similarly, other transboundary river basins, such as the Maritza (Meric) and Orontes (Asi), as well as rivers in western Turkey, are at great risk of climate change-induced extreme weather events, both droughts and floods.

Irrigated agriculture is inherently extremely vulnerable to climate variability and change, due to the natural connections and dependencies that exist between climatic conditions, water availability, and plant development. A variety of climate drivers can impact agricultural productivity, both directly and indirectly. Frequently occurring flooding of low-lying areas of the river deltas and coastal cities has resulted in damage to physical assets and disruption to operations in large agricultural areas. Therefore, insurance premiums may increase, or insurance may become unavailable, and the value of exposed assets may decrease.

Several studies suggest that climate change may lead to moderate increases in irrigation water use at the global scale but to larger changes at the local and regional scale. Crop water requirements will increase in many agricultural districts of Turkey, including but not limited to the Aegean, Mediterranean, and central and southeastern Anatolia regions (e.g. Sen et al. 2012).

7.5.3 Environmental and Ecological System Degradation

The natural hydrology of watersheds is disturbed during the storage and conveyance of water allocated for irrigation as well as during the discharge of drainage and return flows from irrigated areas. Overexploitation of groundwater, changing flow regimes of rivers, and rising water tables (which may trigger salt accumulation), soil erosion due to poor agricultural practices, particularly the surface irrigation of sloped fields and/or deforestation of the upstream areas, are some of the environmental issues to be addressed. Furthermore, large bodies of water stored in dams and diversion channels can cause significant changes in local climate, and the extent of wetlands, which have the important ecological functions of biodiversity, nutrient retention, and flood control, can be reduced. The Ereğli and Eşmekaya reedbeds are only two examples of wetlands which are "natural sites" and "wildlife protection areas" (Karadeniz et al. 2009). Due to the excessive use of agrochemicals, such as fertilizers and pesticides, in irrigated agriculture, surface and groundwater resources in some regions are contaminated with heavy metals and nitrates. In west Turkey (Bornova-Izmir) and middle Anatolia (Niğde), groundwater resources are polluted mainly by the industrial sector, municipality, and agriculture, while the rivers including but not limited to the Gediz, Ergene, Sakarya, Kızılırmak, Yeşilırmak, Seyhan, Asi, and Tigris have also been affected by the discharge of wastewater from those three sectors (e.g. Harmancioglu et al. 2001; MoEU 2011; Varol 2011; Muluk et al. 2013). Horticultural production in the irrigated areas of the Marmara, Aegean, and Mediterranean regions accounts for over 70% of Turkey's total pesticide use; however, the intensity of pesticide use in the country is low compared with that in other OECD countries (OECD 2011). Note that drainage systems are used not only to remove excess irrigation return flows from agricultural lands but also wastewater from the industrial sector and municipalities. Moreover, the poor maintenance of drainage systems reduces the effective removal of excess water.

The European Water Framework Directive (WFD), in which Turkey has taken great interest since its adoption in 2000, emphasises water quality and ecological aspects; however, neither sufficient networks for monitoring nor trained personnel for measuring and analyzing the data are available yet. Building and sustaining such an extensive network would also require additional funding (Çiçek 2010).

Overall, due to excess water application and excess irrigation, in addition to a lack of drainage facilities, about 3 and 1.5 million ha of irrigated land is at risk from water logging and salinization, respectively, or is already affected (Küsek 2010). The hilly topography, soil conditions that facilitate water erosion (i.e. fine texture, low organic matter, poor plant coverage due to semiarid climate), inappropriate agricultural practices (e.g. excessive soil tillage and cultivation and irrigation of steep lands), and forest fires/deforestation are the main causes of intense erosion in Turkey. Soil erosion caused by land degradation is the main reason for the loss of fertile agricultural soil in Turkey; it is estimated that around 83% of agricultural land is currently at risk from severe water erosion (Ozsahin and Uygur 2014).

Regional development projects, such as GAP and KOP, will also cause various forms of degradation of the relevant watersheds if timely adequate measures are not taken. For example, waterlogging and salinity, largely due to unsustainable irrigation methods, insufficient drainage, and poor land management in the GAP areas have caused an increase in soil salinity. Hence, the salt-affected areas in some parts of the plains, mainly in Akcakale in the lower part of the Harran plain, totaled 5500 ha in 1987, 7498 ha in 1997, and 11,403 in 2000 (Cullu et al. 2002; Kapur et al. 2009). Furthermore, most of the irrigation projects in the Tigris basin are situated in areas that are subject to a significant risk of erosion.

Given the levels of environmental degradation in Turkey, it is important to ask whether the government and civil society mobilize around issues of environmental protection and biodiversity conservation. Although Turkey has had a body of environmental legislation and a ministry since the 1990s, the legislation has not been very efficacious, and environmental degradation and biodiversity loss have been rampant in recent decades (Sekercioglu et al. 2011). Aydin (2005) states that lip service is being paid to problems such as air/water pollution, deforestation, and desertification in official documents, and NGOs are tolerated as long as they do not challenge the development policies. Overall, an emphasis on the need for economic growth has overshadowed concerns for the environmental impact of development projects (Karapinar 2010). The post-1990 period saw growing numbers of environmental NGOs and social movements in Turkey. Although there is a participatory discourse at the governmental level, in reality participation is mostly on paper, is ineffective, and becomes impossible when NGOs take up issues such as water regimes, nuclear energy, mining, and international waters (Paker et al. 2013).

In the field of agriculture and irrigation, a stakeholder survey conducted with private sector associations, international NGOs, media, university and research institutions, domestic NGOs, political parties, and governmental institutions revealed that they think agriculture in Turkey is not sustainable and that there is environmental degradation (Karapinar 2010). The same study, however, shows that although these stakeholders agree that the growing intensity of agriculture tends to aggravate environmental degradation, there is a clear distinction between the domestic NGOs and government representatives and private sector associations, whereby only the former think that environmental protection should be prioritized (Karapinar 2010). The differing priorities cannot be incorporated into policymaking unless there is a more bottom-up approach, favoring the participation of different groups which are affected by such policies. The WFD requires stakeholder participation, which is as yet very limited. Only the Environmental Impact Assessment procedure has institutionalized forms that allow direct public participation in decision-making on water resource development (Scheumann et al. 2011). However, even then participation tends to be treated as just a formality to be fulfilled and in fact does not allow for input from the community into the project design or implementation (Kadirbeyoglu and Kurtic 2013). EIA applies only to individual projects, not to the setting up of river basin management plans or the totality of its infrastructure components.

7.5.4 *Investment and Financing*

The cost of irrigation development in Turkey varies between US$ 5000/ha for small schemes and US$ 15,000/ha for large schemes (including pumps). The costs of operation and maintenance (O&M) vary from US$ 330/ha for schemes smaller than 1000 ha to US$ 180/ha for schemes larger than 1000 ha (including dams) (DSI 2012). There is a general consensus among stakeholders – the state bureaucracies, WUAs, chambers of agricultural engineers, and other relevant NGOs – that irrigation investment should continue to reach the potential of economically irrigable land, i.e. 8.5 million ha. However, there is also an emphasis on irrigating with modern technology under proper land consolidation and reform programs.

Since the 1920s, water for irrigation has always been subsidized. Pricing policies have aimed at recovering service costs (O&M&R) and, to a lesser extent, capital costs. If levied per area and crop, irrigation costs are only a proxy of the amount of water used, and incentives to save water are difficult to introduce. Given the technical difficulties of volumetric measuring, it is not easy to implement charges as a stimulant for water savings. While transferring the O&M of irrigation systems to WUAs has improved O&M cost-recovery, state financial transfers are still needed for the rehabilitation and maintenance of irrigation and drainage systems (Topcu 2011).

Although almost 44% of the DSI's total investment (3.48 billion out of 7.98 billion Turkish Lira) was allocated to the regional development projects GAP, KOP, and DAP in 2012, a lack of investment has significantly impeded agricultural development in general. In 2009, following the GAP Action Plan of June 2008, the allocated budget for irrigation investment to the GAP Project increased tenfold compared to 2008 (DSI 2009). The annual share of the GAP within the country's investment was 7% in 1990–2007, 12% in 2008, and 14% in subsequent years. The new action plan for the period 2018–2022 envisages continuing the investment at a similar pace (GAP 2016).

GAP is an expensive project: the total cost has been estimated at US$ 32 billion, half of which has been spent so far. Due to the sensitivity of the transboundary flows involved, the Turkish government was unable to secure international finance (an exception being German and Swiss credit which was obtained to purchase equipment). The severe economic and budgetary crisis in Turkey, along with, for example, the slow pace of land redistribution, caused considerable delays in the project implementation. Despite these drawbacks, Turkey is persistently pursuing its plans to harness the Euphrates and Tigris rivers (Tigrek and Kibaroglu 2011).

Urgent rehabilitation and modernization of irrigation facilities are needed not only to increase the irrigation ratio (share of net irrigated area within the area equipped for irrigation) but also to raise irrigation efficiency. Hence, 80% of the total cost of this has been covered by the DSI and the remaining 20% by the WUAs. As per Law No. 6200, 80% of the total cost incurred by the DSI (within the tertiary canal levels) must be reimbursed by the WUAs. The DSI would only put the projects in its investment programs if the WUAs provided an advance payment of about 5% of the total cost of rehabilitation and rejuvenation (DSI 2012).

7.5.5 Prospects and Pitfalls in Irrigation Management

About two thirds of Turkey's crop production relies on irrigation; consequently, irrigation will continue to be a bulk consumer of water. There is significant potential for improving water use efficiency, and options might cover a combination of physical (engineering), agronomic, environmental, institutional, and managerial, as well as sociopolitical measures (Howell 2001).

Physical measures include the rehabilitation and modernization of the irrigation network, and the completion of on-farm works to increase the irrigation ratio and decrease conveyance losses, converting from an open canal distribution network to a piped system using flow meters. Diverting and applying irrigation water can prevent water wastage and increase efficiency (Topcu 2011). Introducing new irrigation methods in order to increase irrigation efficiency requires greater assistance from the extension services as well as close collaboration between farmers and the agricultural extension services, which has never been a great success story in Turkey.

In recent decades, notwithstanding several projects aimed at combating desertification and drought, adapting to climate change, and improving water use efficiency in irrigated agriculture, irrigation development in Turkey nonetheless still mainly focuses on the construction of new dams, pools, canals, and other infrastructure. Less and/or mostly delayed attention has been paid, and limited investment has been made into developing a comprehensive and integrated management approach that is environmentally sound and sustainable. Integrated River Basin Management (IRBM) takes a much broader approach than traditional water management and includes significant attention to land-use planning, erosion control, land consolidation, agricultural policy, environmental management, and the continuous monitoring of water quality and quantity. IRBM, however, is a new concept in Turkey, and existing political/institutional structures do not facilitate the necessary change in mentality; its implementation will therefore be a major challenge. Obstacles that threaten the successful implementation of IRBM in Turkey are (i) incomplete water resources development, (ii) administrative problems, and (iii) lack of participatory approaches (Divrak and Demirayak 2011). The DSI made preparations so that IRBM could be implemented in 11 river basins, while 4 river basin plans are under revision.

Industrial and municipal waste, as well as drainage water from agricultural areas, is degrading the water quality of the rivers and groundwater resources. This in turn threatens the irrigated land in terms of salinity and also heavy metals. In order to prevent the negative effects of excessive organic and inorganic fertilizer use, soil analysis has been made obligatory for farmers owning agricultural landholdings of more than 5 ha in order to benefit from fertilizer/manure subsidies. Wastewater treatment plants have been increasingly installed in cities. Most of this water is flowing into rivers and lakes, while a significant volume of untreated wastewater is discharging into lakes and ponds, and onto land. The use of wastewater for irrigation is so far limited to arid/semiarid areas, such as the central and southeastern regions. According to the very few available records and reports, about 200,000 ha of land was irrigated using wastewater in various provinces by the year 2004 (Gokcay 2004), and some small wastewater irrigation projects comprising a few thousand hectares were initiated in the Konya and Ergene basins, as well as in Afyonkarahisar, some of which were implemented for urban and suburban agriculture (e.g. Duman 2017). Irrigation return flows can also be reused in the lower part of the irrigation schemes, like in the Seyhan and Harran plains, where water is not sufficient and/or irrigation infrastructure has not yet been completed.

7.6 Conclusion

Agriculture is the biggest water user in Turkey. Most of the country is very dry during the summer months. Therefore, agricultural production, particularly in fertile regions, such as western, southern, and southeastern Turkey, can only be achieved by supplying irrigation water. State authorities assert that an additional two million jobs will be generated once the targeted areas have been developed for irrigation by the year 2023.

Given agriculture's significant share of water consumption, this chapter has provided a closer examination of the irrigation sector. We found that Turkey is eager to continue with investments in large-scale irrigation systems, particularly within the context of regional development projects: the GAP, KOP, DAP, DOKAP, and YHGP. We also observed that technological change has been under way in the irrigation sector since the late 1990s. To illustrate, the share of pipelines conveying and distributing the water to the fields in irrigation schemes increased substantially. However in many old schemes, surface irrigation in Turkey continues to be dominated by gravity-fed systems, such as furrow, border, and even wild flooding.

We observed that there has been drastic institutional change in the management of irrigation systems. The (partial) transfer of the management of large-scale irrigation systems to a variety of organizations from the early 1990s had mixed results. Key policy entrepreneurs in the devolution of irrigation management in Turkey have been national and international bureaucrats, namely those of the DSI and the World Bank as donor agency. Successive governments have seemed quite content with the "reform process" in irrigation management, and the World Bank has pronounced Turkey a "success" (Kibaroglu et al. 2012). A coalition against the accelerated irrigation management transfer process, however, has leveled considerable criticism. The Chamber of Agricultural Engineers has led the objections, focusing on a number of issues. The participatory aspect of the transfers in particular has been questioned, owing to the exclusion of irrigators from WUA general assemblies and boards. In addition, the top-down approach that was adopted rather than generating grassroots involvement from farmer interest and involvement has caused fierce debate over the characterization of the associations as democratic. Critics stress that the maintenance, rehabilitation, and modernization of the irrigation canals, some of which are 40 years old, cannot be accomplished due to deficiencies in the WUAs' technical, administrative, and legal capacities.

In addition to the technological and institutional problems, further challenges for the irrigation sector include infrastructure deterioration, risks of drought, environmental and ecological system degradation, including that associated with large dams and small run-of-the-river hydropower projects, insufficient investment, and the financing of continuing projects, as well as the maintenance of existing infrastructure.

References

Akkuzu, E., Ünal, H. B., & Karataş, B. S. (2007). Determination of water conveyance losses in the Menemen open canal irrigation network. *Turkish Journal of Agriculture and Forestry, 31*, 11–22.

Akşam, G. (2018) *Tarlada yağmurlama veya damlama zorunluluğu*. Akşam Gazetesi. Ref. No. 75004794, www.aksam.com.tr/ekonomi/tarlada-yagmurlama-veya-damlama-zorunlulugu/haber-710835. Accessed on 24.02.2018.

Akuzum, T., Kodal, S., & Çakmak, B. (1997). Irrigation management in GAP. *International Journal of Water Resources Development, 13*(4), 547–561.

Alizadeh, A., & Keshavarz, A. (2005). *Status of agricultural water use in Iran. Water conservation, reuse, and recycling: Proceedings of an Iranian-American workshop* (pp. 94–105). Washington, DC: The National Academic Press Open Book.

Aydin, Z. (2005). The state, civil society and environmentalism. In F. Adamanve & M. Arsel (Eds.), *Environmentalism in Turkey: Between democracy and development?* Aldershot: Ashgate.

Bozkurt, D., & Sen, O. L. (2013). Climate change impacts in the Euphrates-Tigris basin based on different model and scenario simulations. *Journal of Hydrology, 480*, 149–161. https://doi.org/10.1016/j.jhydrol.2012.12.02.

Cakmak E.H., Cetik M., Tüzün M., Baysan N., Karaca Bilgen G., Terzi M., Kanber R., Yazar S. (2004). Participatory irrigation management and water cultural heritage: Turkey country report. In : Hamdy A. (ed.), Tüzün M. (ed.), Lamaddalena N. (ed.), Todorovic M. (ed.), Bogliotti C. (ed.). *Participatory water saving management and water cultural heritage*. Bari : Ciheam, p. 263–285 (Options Méditerranéennes : Série B. Etudes et Recherches; n. 48)

Çakmak, E., & Kasnakoğlu, H. (2016). *Türkiye Ihracatçılar Meclisi Tarım Raporu* (p. 160). Istanbul: Küçük Mucizeler Yayıncılık ve İletişim Hizmetleri.

Çiçek, N. (2010). *Water framework directive and EU and Turkey approach within the example of Büyük Menderes River basin management plan*. Selçuk University.

Coşkun, M., Sumer, U. M., Ulupinar, Y., Şensoy, S., Demircan, M., Boluk, E., Arabaci, H., Eskioglu, O., Acar, Y., Ekici, M., Yagan, S., & Kervankiran, S. (2017). *2016 Yılı İklim Değerlendirmesi*. Ankara: T.C. Orman ve Su Işleri Bakanliği Meteoroloji Genel Müdürlüğü.

Çullu, M. A., Almaca, A., Şahin, Y., Aydemir, S. (2002). *Application of GIS for monitoring soil salinization in the Harran Plain, Turkey*. SSST International Conference on Sustainable Land Use and Management (pp. 326–331). Çanakkale-Turkey.

DSI. (2009). *Devlet Su İşleri Genel Müdürlüğü 2008 Yılı Faaliyet Raporu*. Ankara: T.C. Çevre ve Orman Bakanlığı, Devlet Su İşleri Genel Müdürlüğü.

DSI. (2012). *Water and DSI*. Ankara: Ministry of Forestry and Water Affairs, General Directorate of State Hydraulic Works 1954-2012.

DSI. (2014). *Türkiye'de Yeraltı Sulama Faaliyetleri Yeraltısuyu Sulama Kooperatifleri*. Ankara: T.C. Orman ve Su İşleri Bakanlığı, Devlet Su İşleri Genel Müdürlüğü, Jeoteknik Hizmetler ve Yeraltısuları Dairesi Başkanlığı.

DSI. (2016). *Devlet Su İşleri Genel Müdürlüğü 2016 Faaliyet Raporu*. Ankara: T.C. Orman ve Su İşleri Bakanlığı, Devlet Su İşleri Genel Müdürlüğü.

DSI. (2017). *2016 Yılı DSİ'ce İşletilen ve Devredilen Sulama Tesisleri Değerlendirme Raporu*. Ankara: T.C. Orman ve Su İşleri Bakanlığı, Devlet Su İşleri Genel Müdürlüğü İşletme ve Bakım Dairesi Başkanlığı, DSİ Destek Hizmetleri Dai. Başkanlığı Basım ve Foto – Film Şube Müdürlüğü.

Divrak, B. B., & Demirayak, F. (2011). NGOs Promote Integrated River Basin Management in Turkey: A Case-Study of the Konya Closed Basin. In A. Kramer, A. Kibaroglu, & W. Scheumann (Eds.), *Turkey's Water Policy* (pp. 161–176). Berlin, Heidelberg: Springer.

Duman, H. (2017). Arıtılmış Kentsel Atıksuların Sulamada Yeniden Kullanımı; Kayseri Atıksu Arıtma Tesisi Örneği. Uzmanlık Tezi. Ankara: T.C. Orman ve Su İşleri Bakanlığı. 144pp.

GAP. (2016). *Güneydoğu Anadolu Projesi Son Durum*. T.C. Kalkınma Bakanlığı Güneydoğu Anadolu Projesi, Bölge Kalkınma İdaresi Başkanlığı. Mayıs 2016.

Gokcay, C. (2004). *Evaluation of the Turkish reuse standards and the compliance status*. International Workshop on Implementation and Operation of Municipal Wastewater Reuse Plants. 11–12 March 2004, Thessaloniki, Greece.

Groundwater Law, Act No. 167 relative to groundwaters. (1960). Date of text: 16 December 1960

Harmancioglu, N., Alpaslan, N., & Boelee, E. (2001). *Irrigation, health and environment: A review of literature from Turkey, IWMI working paper no. 6*. Colombo: International Water Management Institute (IWMI) 21pp.

Harris, L. (2005). Negotiating inequalities: Democracy, gender, and the politics of difference in water user groups of Southeastern Turkey. In M. Arsel & F. Adaman (Eds.), *Turkish environmentalism: Between democracy and development* (pp. 185–200). Aldershot: Ashgate.

Howell, T. A. (2001). Enhancing water use efficiency in irrigated agriculture. *Agronomy Journal, 93*, 281–289.

Kadirbeyoglu, Z. (2008). *Decentralization and democratization: The case of water user associations in Turkey*. Unpublished PhD Thesis. Montreal: McGill University.

Kadirbeyoglu, Z., & Kurtic, E. (2013). Problems and prospects for genuine participation in water governance in Turkey. In J. Goldin, L. Harris, & C. Sneddon (Eds.), *Contemporary water governance in the Global South: scarcity, marketization and participation*. Oxon: Routledge.

Kadirbeyoglu, Z., & Özertan, G. (2015). Power in the governance of common-pool resources: A comparative analysis of irrigation management decentralization in Turkey. *Environmental Policy and Governance, 25*(3), 157–171.

Kapur B, Kapur S, Akça E (2009) *Global climate change and effects on soil, crop productivity and water resources*. International Workshop for the research project on the impact of climate change on agricultural production System in arid areas, Adana, Türkiye, (pp. 100).

Karadeniz, N., Tiril, A., & Baylan, E. (2009). Wetland management in Turkey: Problems, achievements and perspectives. *African Journal of Agricultural Research, 4*(11), 1106–1119.

Karapinar, B. (2010). "Sustainability" in agriculture: Stakeholders' perceptions in Turkey. In B. Karapinar, F. Adaman, & G. Ozertan (Eds.), *Rethinking structural reform in Turkish agriculture*. New York: Nova Science Publishers.

Kibaroglu, A., & Baskan, A. (2011). Turkey's water policy framework. In Kibaroglu et al. (Eds.), *Turkey's water policy, national frameworks and international cooperation* (pp. 3–25). Berlin: Springer Verlag.

Kibaroglu, A., Baskan, A., & Alp, S. (2009). Neo-liberal transitions in hydropower and irrigation water management in Turkey: Main actors and opposition groups. In D. Huitema & S. Meijerink (Eds.), *Water policy entrepreneurs. A research companion to water transitions around the globe* (pp. 287–304). Cheltenham: Edward Elgar.

Kibaroğlu, A., Sumer, V., & Scheumann, W. (2012). Fundamental shifts in Turkey's water policy. *Méditerranée, 119*, 27–34.

Kusek, G. (2010). Tarım Sektörü Mevcut Durum Değerlendirmesi Raporu, 2. Taslak-Web, http://iklim.cob.gov.tr/iklim/Files/Tarim_Sektoru_Mevcut_Durum_Degerlendirmesi_Raporu.pdf. Accessed on 03.09.2014.

Law on Municipalities no. 1580, 03.04.1930, Official Gazette No. 1471. http://www.resmigazete.gov.tr/arsiv/1471.pdf.

MoEU. (2011). *2011 Türkiye Çevre Durum Raporu*. Ministry of Environment and Urbanisation (www.csb.gov.tr).

MoFAL, (2013). *T.C. Gıda Tarım ve Hayvancılık Bakanlığı 2013 Yılı Faaliyet Raporu*. Ankara: T.C. Gıda Tarım ve Hayvancılık Bakanlığı.

MoFAL, (2016). *T.C. Gıda Tarım ve Hayvancılık Bakanlığı 2016 Yılı Faaliyet Raporu*. Ankara: T.C. Gıda Tarım ve Hayvancılık Bakanlığı.

Muluk, Ç. B., Kurt, B., Turak, A., Türker, A., Çalişkan, M. A., Balkiz, Ö., Gümrükçü, S., Sarigül, G., Zeydanli, U. (2013). Türkiye'de Suyun Durumu ve Su Yönetiminde Yeni Yaklaşımlar: Çevresel Perspektif. İş Dünyası ve Sürdürülebilir Kalkınma Derneği – Doğa Koruma Merkezi.

Naik, G., & Kalro, A. H. (2000). A methodology for assessing impact of irrigation management transfer from farmers' perspective. *Water Policy, 2*, 445–460.

OECD. (2011). *Evaluation of agricultural policy reforms in Turkey* (p. 7). OECD Publishing, 2011. https://doi.org/10.1787/9789264113220-en, (02 February 2014).

Oral, N. (2008). *Agriculture in Turkey in 2007*. www.zmo.org.tr/resimler/ekler/20a7ce2a627ba83_ek.pdf?dergi=139 (in Turkish, accessed 14 Feb 2011).

Özerol, G. (2013). Institutions of farmer participation and environmental sustainability: A multi-level analysis from irrigation management in Harran Plain, Turkey. *International Journal of the Commons, 7*(1), 73–91. https://doi.org/10.18352/ijc.368.

Ozis, U. A., Turkman, Y., Ozdemir (2005). Historical waterworks in Turkey, ninth international water technology conference, IWTC9 2005, Sharm el-Sheikh, Egypt.

Özşahin, E., & Uygur, V. (2014). The effects of land use and land cover changes (LULCC) in Kuseyr plateau of Turkey on erosion. *Turkish Journal of Agriculture and Forestry* (2014), *38*, 478–487.

Paker, H., Adaman, F., Kadirbeyoglu, Z., & Özkaynak, B. (2013). Environmental organisations in Turkey: Engaging the state and capital. *Environmental Politics, 22*(5), 760–778.

Palerm-Viquiera, J. (2004). *Irrigation institutions typology and water governance through horizontal agreements*. www.geocities.com/jacinta_palerm.

Pamuk, S. (2006). Agricultural output and productivity growth in Turkey since 1880. International Economic History Congress, 21–25 August 2006, Helsinki.

Salman, M. A. S. (2002) Regulatory Framework for Participatory Irrigation Management: An Overview; in Slavko Bogdanovic, (ed.) Legal Aspects of Sustainable Water Resources Management, (YuAWL and IWLRI) 343.

Scheumann, W., Kibaroglu, A., & Kramer, A. (2011). Trajectories of Turkey's water policy. In Kibaroglu et al. (Eds.), *Turkey's water policy: National frameworks and international cooperation* (pp. xxiii–xxxxi). Berlin: Springer Verlag.

Şekercioğlu, C. H., Anderson, S., Akçay, E., Bilgin, R., Can, Ö. E., Semiz, G., Tavşanoğlu, Ç., Yokeş, M. B., Soyumert, A., Ipekdal, K., Sağlam, İ. K., Yücel, M., & Nüzhet Dalfes, H. (2011). Turkey's globally important biodiversity in crisis. *Biological Conservation, 144*(12), 2752–2769.

Sen, B., Topcu, S., Türkeş, M., Sen, B., & Warner, J. F. (2012). Projecting climate change, drought conditions and crop productivity in Turkey. *Climate Research, 52*, 175–191.

Svendsen, M. (2001). *IMT case study: Turkey*. Paper read at International e-mail Conference on Irrigation Management Transfer, June-October 2001.

Svendsen, M., & Nott, G. (1999a). Irrigation management transfer in Turkey: Process and outcomes. In D. Groenfeldt & M. Svendsen (Eds.), *Case studies in participatory irrigation management, EDI Participatory Irrigation Management Services* (Vol. 273). Washington, DC: IBRD.

Svendsen, M., & Nott, G. (1999b). *Irrigation Management Transfer in Turkey: Process and Outcomes, Economic Development Institute (EDI) Participatory Irrigation Management Case Studies Series*. Washington DC: World Bank.

Tezcan, L., Ekmekci, M., Atilla, O., Gürkan, D., et al. (2007). *Assessment of climate change impacts on water resources of Seyhan River Basin. The final report of ICCAP. Impact of climate changes on agricultural production system in the Arid Areas, Kyoto*, (p 59–72).

Tigrek, S., & Kibaroglu, A. (2011). Strategic role of water resources for Turkey. In Kibaroglu et al. (Eds.), *Turkey's water policy, national frameworks and international cooperation* (pp. 27–42). Berlin: Springer Verlag.

Topaloglu, F. (2006). Trend detection of stream flow variables in Turkey. *Fresenius Environmental Bulletin, 15*(6), 644–653.

Topcu, S. (2011). Water for agriculture: A major but inefficient consumer. In Kibaroglu et al. (Eds.), *Turkey's water policy, national frameworks and international cooperation* (pp. 93–115). Berlin: Springer Verlag.

Tuik. (2017). TUIK Haber Bulteni, Sayı: 24569, 11 Aralık 2017, Saat: 10:00, www.tuik.gov.tr. Accessed on 03.02.2018.

Turker, M. (2013). International conference on policies for water and food security in dry areas 24th to 26th of June 2013, Cairo, Egypt.

Türkeş, M. (2010). The mediterranean climate regime. In: *Lecture notes of the TSMS-NOAA-USAID-WMO training workshop on climate variability and predictions for the Mediterranean Basin*. 27 July – 4 August 2010, TSMS-WMO Training Center, Alanya.

Unver, O. I. H., Gupta, R. K. (2002). *Water management. Cross cutting issues*. Middle East Technical University, 210pp.

Varol, M. (2011). Assessment of heavy metal contamination in sediments of the Tigris River (Turkey) using pollution indices and multivariate statistical techniques. *Journal of Hazardous Materials, 195*, 355–364. https://doi.org/10.1016/j.jhazmat.2011.08.051 Epub 2011 Aug 22.

Village Law, Act no. 442, 18.03.1924, Official Gazette no. 68 http://www.mevzuat.gov.tr/MevzuatMetin/1.3.442.pdf

Water Law no. 831, 28.04.1926, Official Gazette no. 368 http://www.mevzuat.gov.tr/MevzuatMetin/1.3.831.pdf

Yavuz, M. Y., Kavdır, I., & Delıce, N. L. (2006). Performance evaluation of water user associations: A case study of the lower Seyhan basin. *Journal Agricultural Faculty HR.U, 10*(3/4), 35–45.

Yenigun K, Aydogdu MH (2010) Scientific Research and Essays 5(21), 3237–3253., 4 November, 2010, www.academicjournals.org/SRE, ISSN 1992-2248 ©2010 Academic Journals.

Zurcher, E. J. (2004). *Turkey: A modern history* (3rd ed.p. 424). London: I.B.Tauris (September 4, 2004).

Chapter 8
Israel

Doron Lavee, Eran Feitelson, and Hadas Joseph-Ezra

Abstract Israel has succeeded in advancing irrigated agriculture on a wide scale on arid and semiarid lands, with an intensive use of technology and capital, and a firm state-led irrigation policy. This chapter describes the evolution of the Israeli irrigation sector since the British Mandate period, distinguishing four distinct eras linked to contrasting political and technological frameworks. It also explores the changes in the institutional framework of the irrigation sector (from the local community level to the state structure). The chapter underscores the critical role of technology in Israeli irrigation, most particularly in the development of micro-irrigation and monitoring systems, and the shift towards desalination and wastewater use for irrigation. Finally, it identifies the major challenges to be overcome, considering water quantity and quality problems, and the impact of climate change, possibly compounded by tensions with neighboring countries.

Keywords Water scarcity · Water policy · Intensive farming · Wastewater recycling · Desalination

8.1 Introduction

8.1.1 Geography, Climate, and Water Resources

Israel is a small country, located at the junction of Europe, Africa, and Asia on the southeastern coastline of the Mediterranean Sea and the northern tip of the Gulf of Eilat/Aqaba. With an area of approximately 20,770 km^2 that includes 445 km^2 of

D. Lavee (✉)
Faculty of Social Sciences and Humanities, Tell-Hai College, Qiryat Shemona, Israel
e-mail: doronl@telhai.ac.il

E. Feitelson
Hebrew University of Jerusalem, Jerusalem, Israel

H. Joseph-Ezra
Pareto Group, Natanya, Israel

F. Molle et al. (eds.), *Irrigation in the Mediterranean*, Global Issues in Water Policy 22, https://doi.org/10.1007/978-3-030-03698-0_8

inland water (U.S. Library of Congress 1988),[1] Israel is divided into four main geographical regions: the coastal plain, the central hills, the Jordan Rift Valley, and the Negev Desert. The Jordan River,[2] which terminates in the Dead Sea, flows along the northern part of the Rift Valley. The Sea of Galilee, which is the only lake in the region, serves as Israel's main reservoir. The arid south of Israel is the Negev Desert, covering more than half of the country's total land area.

Israel has two major seasons, summer and winter, with short transition periods. Rainfall occurs during the winter season alone. The difference between the major climatic regions in Israel is very prominent. The northern and coastal regions of Israel are characterized by a Mediterranean climate, with cool, rainy winters (400–1200 mm rain per year) and hot, dry summers, whereas the Negev and Judean deserts are characterized by an arid climate, with very little rainfall (less than 200 mm per year). Between these two areas there is a narrow strip of a hot, dry climate (200–400 mm per year) (Fine et al. 2007).

According to estimates, the total replenishment amounts to 1.9–2 billion m^3 (Kislev 2011; Feitelson et al. 2011). Until the advent of large-scale seawater desalination in 2005, this constituted the total available water supply to households, agriculture, and industry for both Israel and the Palestinians.

Water scarcity is a major concern in Israel. High population growth and rapid economic development, as well as obligations made in water agreements with Jordan and the Palestinians, have placed growing demand on Israel's scant water resources (OECD 2011). Irrigation is essential for intensive farming throughout the year, even in the areas with relatively high rainfall, due to the long and dry summer. Furthermore, the existing inventory of fresh water in Israel cannot satisfy all of the domestic, industrial, and agricultural needs.

The country's water planning, policies, and management are highly challenged by climatic fluctuations and the resulting uncertainty regarding water availability, due to multi-year droughts (Amiran 1994). The main role of the water sector is thus to store water from winter to summer and from rainy years to drought years, utilizing the aquifers (Gvirtzman 2002). Over 40% of the agricultural lands are located in the semi-arid northern Negev, while the highest rainfall occurs in the north. The large-scale water conveyance scheme (the National Water Carrier – NWC) completed in 1964 and extended in 1969 as part of Israel's nation-building effort, allows water to be conveyed from the Sea of Galilee in the north to the densely populated center, and further to the south – where modern agriculture cannot exist without irrigation (Fig. 8.1). Since the late 1980s, Israel has experienced several multi-year droughts, during which water consumption exceeded the natural rate of replenishment. Due to the unsustainable utilization of groundwater, aquifers are threatened with substantial pollution and salinization problems, particularly in the coastal aquifer which has been over-pumped since the 1950s (Furman and Abbo 2013).

[1] In discussing Israel in this chapter, we refer only to the area within the pre-1967 Armistice Line and the Golan Heights. We do not discuss the West Bank and Gaza, though we refer to the water relations between the West Bank and Israel wherever necessary to understand the Israeli water situation.

[2] The Jordan River has five riparians: Israel, Lebanon, Jordan, Syria and the Palestinian Authority.

Fig. 8.1 Israel's main freshwater supply resources and National Water Carrier (NWC)

Nonetheless, Israel has succeeded in advancing agriculture on a wide scale also on its dry lands. This was made possible by damming the outlet of the Sea of Galilee,[3] the construction of the NWC, water conservation, crop substitution, and Israel's national water policy. In particular, water supply has been expanded through wide-scale wastewater recycling, making it a world leader in such recycling, and more recently also in large-scale seawater desalination.

Today, Israel's main natural freshwater supply resources are the Sea of Galilee, parts of the Mountain Aquifers, the Coastal Aquifer, and a number of smaller aquifers (50%) (Fig. 8.1). Additional sources of water include treated wastewater, which is largely used for irrigation in the agricultural sector (20%), desalinated seawater (24%), and brackish water (mainly from several small brackish-water aquifers).

8.1.2 Agriculture in Israel: Past and Present

Agriculture occupied a prominent place in the Zionist enterprise and in the early years of the state. During the late Ottoman and early British Mandate period, farming was seen as central to the (re)connection of Jews with the land of Israel, as part of the nation-building efforts. Yet the nature of the Jewish rural settlements changed. While in the initial period, during the late nineteenth century, settlements were mostly on privately owned land, during the British Mandate era (between the two world wars), settlements were mostly cooperative on public land. These collectives were either Kibbutzim (fully collectivized) or Moshavim (where production and consumption are private, but distribution and financing are collective). To provide a "decent" level of living, the settlements were largely based on irrigated agriculture, thereby making water resource development an integral part of the settlement efforts.

Following the Peel Commission[4] of 1936, which suggested a two-state solution in Palestine where the borders between the states would be delineated on the basis of the (largely rural) settlement areas, geo-political factors came to dominate the Zionist rural settlement efforts (Reichman 1979). These continued to be central to rural development in the early state period (Reichman 1990). However, at this time, food security and the settlement of the massive immigration wave that flowed into the country[5] became major factors in agricultural development due to severe food shortages (which led to food-rationing) and the desire to absorb

[3] The outlet of the Sea of Galilee, also called Lake Kinneret or Lake Tiberias, was dammed as a side-product of the hydroelectricity project at the confluence of the Jordan and Yarmouk rivers, built in 1931.

[4] The Peel Commission was a Royal Commission of Inquiry established by the UK Government to investigate the causes of unrest in Palestine under the British mandate, following the 6-month-long Arab general strike and to recommend future measure.

[5] Over 700,000 immigrants arrived between 1949 and 1952, severely stressing the country's meagre resources in the early Fifties (Hacohen 1994).

the new immigrants into the economy as quickly as possible (Hacohen 1994). Hence, large-scale water projects were seen as central to nation-building, securing the borders, providing food security, and immigration absorption, thereby making them highly visible in the public agenda, and a priority in development budgets (Feitelson 2013). To advance these goals, irrigation was heavily subsidized. Not only did farmers not pay the capital cost of water supply, they enjoyed subsidized water rates that were much lower than those paid by the urban and industrial sectors.

Since the early 1960s, the relative share of agriculture to the country's economy has declined, due to rapid industrialization followed by a shift to a high-tech and service economy in the late twentieth century. Moreover, farming shifted away from food self-sufficiency goals toward an export-oriented agriculture, not least due to the realization that Israel's water resources could not meet the needs of food self-sufficiency, meaning food supply must be based on imports.[6] Today, agriculture contributes only about 1% of the country's GDP (Central Bureau of Statistics 2016), though it also serves as a beta-site for the agricultural technology industry. Still, until recently, agriculture continued to enjoy lower freshwater rates due to the power of the agricultural lobby in the Knesset (Israel's parliament).

As a result of the economic shifts, the loss of protective tariffs, the elimination of protective practices, and the gradual increase in water rates, there has been a decline in the number of self-employed farmers in Israel. Currently, agriculture is moving towards economies of scale, whereby fewer farmers operate larger farms. The decline in the relative importance of agriculture for the national economy and in the share of agriculture in the workforce has political ramifications, as the power of the agricultural lobby in the Knesset has gradually declined. In recent elections, only a scant number of Knesset members hailed from the rural sector (which was traditionally heavily over-represented).

As a result of the decline in the power of the agricultural lobby, the shift in geopolitical concerns,[7] and the rise of neo-liberal ideologies, the state's support for agriculture has diminished in the last three decades. The previously highly regulated agricultural production, whereby quotas were set for each crop to each farmer, has been largely de-regulated. Water rates have risen gradually and agricultural water subsidies for fresh water have decreased. Whereas farmland protection was a central tenet of land-use regulations, open space considerations, infrastructure, and urban development are now the main factors in planning decisions. Thus, agriculture is largely viewed today as a secondary economic sector, with diminishing political and symbolic capital.

Water is a fundamental factor in agricultural production and constitutes 11% of the average farmer's costs[8] (in the year 2016). Therefore, the change in water pricing is a key element of the change in farmers' decisions, as can be seen in the next paragraphs.

---

[6] From a water perspective this shift implies that Israel increasingly imports 'virtual water', the water embedded in food, thereby alleviating much of its water scarcity (Gilmont 2014).

[7] Following the rise of the Likud to power in 1977, the geo-political concerns shifted to the Occupied West Bank, where most Jewish settlements are not based on agriculture.

[8] www.cbs.gov.il/shnaton68/diag/19_05.pdf

8.2 Historical Development of Irrigation in Israel

Israeli water history can be seen as consisting of four eras (Feitelson 2013) preceded by a pre-state era. The historical development of irrigation can be related to these four eras.

Prior to the establishment of the state, irrigation was largely based on local sources, either surface water (mainly springs) or wells, managed at the village level. Until the 1930s, the wells utilized only shallow aquifers, as the main technology available, the Antilia (animal-driven water wheel), restricted pumping to a depth of 10 m (Avitsur 1991). Surface and shallow well water rights were local.

The latter part of the British Mandate period and the first state period – termed by Feitelson (2013) the hydraulic imperative era – were characterized by the expansion of water conveyance and irrigation systems, from the local to the national level, as part of the nation-building and agricultural development efforts of the time (Feitelson and Fischhendler 2009), as well as by the utilization of deeper aquifers. In the late 1930s, irrigation schemes were planned and built at the regional level, mainly in the northern part of the country, where water was available within the region (Feitelson et al. 2014). Such schemes were built in the Yizrael valley and later in the northern Negev, as part of the pre-state Zionist settlement efforts (Reichman 1979; Seltzer 2010). The Kishon project, supplying the western Yizrael valley, was the first large-scale project built by Mekorot, the national water company of Israel, thereby providing it with the expertise that was later to serve it in its struggle to become the national water company it is today (Seltzer 2010). This was enabled by deeper drilling to groundwater, thereby producing "new" water and circumventing the existing water rights system.

Similar schemes were carried out after the establishment of the state, perhaps most notably in the Beit Shean region (Nir 1989). These were driven in part by geopolitical concerns – particularly the fear that Israel might be forced back to the 1947 UN resolution lines (Reichman 1990). The Beit Shean project, as well as others of its kind, was facilitated by the annulment of local water rights after 1948, a move that was finally formalized in the 1959 Water Law, which nationalized all the water (surface, groundwater, and effluents). This law reflected a shift in attitude toward water resources, from a discourse of plenty to be discovered and utilized, to one of scarcity, to be utilized and managed judiciously (Alatout 2008). More importantly, this law reflected the shift in power from local water associations, controlled by local agricultural interests, to the national level (the newly formed Water Commissioner, Mekorot, and the then-state owned water planning company, Tahal[9]). Opposed by private farmers, this shift was enabled by the collective farming interests' (the moshavim and kibbutzim movements) control over the Ministry of Agriculture and all centralized water bodies. Still, existing regional farmer-controlled water associations continued to supply water to farmers at very low rates. Such associations exist mainly in regions that were settled

[9] Tahal was privatized in 1996, as part of the neo-liberalization of the Israeli economy.

by Zionist bodies before the state was formed, that is, in the northern and central parts of the country.

The main purpose of the spatial extension of waterworks was the conveyance of water to the arid northern Negev, which was seen to have the greatest potential for agricultural development once irrigation could be introduced, and hence important from a food-security perspective. Indeed, in this part of the country, the rapid expansion of irrigation could be observed at any point connected to the NWC (Feitelson et al. 2014). To this end, water was first conveyed from the Yarkon River,[10] and later from the Jordan River basin, once the NWC was completed in 1964. These projects had widespread deleterious consequences for both rivers (as well as for the coastal streams that were also captured as part of these projects), since their freshwater flows were replaced by sewage and brackish water. The capture of surface water and the subsequent pollution of rivers and streams (mainly in the coastal plain) reflect the power of the agricultural interests at the time, which largely dominated the main ruling party (Mapai).

The main beneficiaries of the massive water projects in this era were Kibbutzim, followed by the Moshavim. Private farmers and Arab farmers received hardly any additional water (Yunger et al. 1993). This reflects the political power in the irrigation sector at the time. All the water planning and development bodies were controlled by a small elite that was part of the collective agriculture establishment, largely dominated by the kibbutzim. These initial inequities were later institutionalized in the water allocations set under the 1959 Water Law, leaving kibbutzim with more water per dunam than moshavim, and much more than the Arab sector (Kislev 2011).

Once the NWC was completed, all the main streams and rivers were closed (i.e. hardly any water reached the sea). Moreover, the potential for further groundwater utilization was found to be less than originally thought, meaning the ability to further develop new freshwater resources was largely curtailed. The second era of the Israeli water system focused, therefore, on the improved utilization of existing resources (Feitelson 2013). With respect to irrigation, the emphasis shifted from the expansion of irrigation systems to water conservation in irrigation. The development of drip irrigation and its rapid dissemination proved crucial in this respect. Moreover, Israeli agriculture shifted in this period from the self-provision of food to a market orientation, whereby the goal was to derive the maximal income from it. Hence, the emphasis shifted to increasing the product value per unit of water. Figure 8.2 shows the success of this policy, which combined technology (mainly drip irrigation) and crop substitution. It shows that, since the early 1970s, the value of agricultural production was largely decoupled from irrigation (Gilmont 2014), rising faster than the agricultural product, thereby producing substantially higher income per unit of water. These shifts were facilitated by heavy investments in agricultural know-how,

[10] The Yarkon River, originating north-east of Tel-Aviv, was captured at the source in 1955 and diverted to the western Negev through the Yarkon-Negev pipeline, which was later incorporated into the NWC.

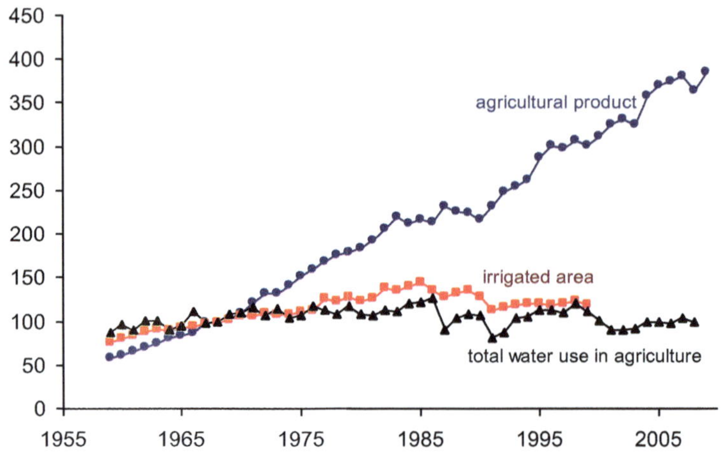

Fig. 8.2 Agricultural product, irrigated area (data on irrigated area were only collected until 1999) and total water use in agriculture (total water use in agriculture includes fresh water, brackish water and recycled wastewater) in Israel 1959–2008 (index: 1968=100). (Source: Central Bureau of Statistics, Annual Yearbooks)

mainly in the state-owned Vulcani Institute and by the ministry's agricultural extension service.

From an institutional-political perspective, this period was marked by the domination of a small, technocratic elite concentrated in the two often-competing national companies – Mekorot and Tahal. These were regulated by the Water Commissioner, who either came from within this technocratic elite or from the agricultural sector, and hence gave precedence to irrigation considerations (Feitelson, Fischhendler, and Kay 2007). Regardless, the Water Commissioner was appointed by the Agriculture Minister who invariably came from the collective agriculture sector (usually the Kibbutzim). Thus, irrigation interests held sway over water policies in Israel throughout this period. This was manifest in the late 1980s, when the Water Commissioner allowed deep drawdowns of the aquifers in order to prevent or reduce the cuts in irrigation quotas.

The brinkmanship policy, whereby excessive abstractions were allowed under the assumption, advanced by water commissioners from the agricultural sector, that a high-rainfall year would replenish both the Sea of Galilee and main aquifers once in a while, combined with several multi-year droughts, led to a series of crises from 1990 to the end of the first decade of the twenty-first century (Feitelson, Fischhendler, and Kay 2007). This is the third era identified by Feitelson (2013). The agricultural sector, and irrigation water in particular, were viewed as a buffer sector. That is, in drought periods, water for irrigation was to be cut and compensation paid to farmers. Such cuts were subject to fierce political struggles between the agricultural lobby and the Water Commissioner, as the agricultural sector utilized the power vested in the Minister of Agriculture to postpone such cuts in lieu of higher compensation (Fischhendler 2008). But, as the urban population grew rapidly in this period,[11] fresh water increasingly had to be shifted from agriculture to the domestic sector. To this end, cheaper recycled wastewater was offered to farmers in lieu of freshwater quotas.[12] This was enabled by the increasing availability of recycled wastewater resulting from the advanced (i.e. secondary+) treatment of wastewater due to pressure from and the increasing power of the Ministry of Environment.

During this third era, a policy impasse ensued, as the agricultural lobby in the Knesset prevented the raising of water tariffs, while the Treasury blocked suggestions to advance desalination, as it viewed higher water tariffs as a crucial instrument to reduce the demand for water, particularly in agriculture (Feitelson 2005). Moreover, as fresh water was increasingly shifted to the domestic sector, and water agreements were signed with the Palestinians and Jordan, resulting in inflexible international obligations for water supply,[13] agriculture's potential as a buffer sector was eroded, forcing the water commissioner to reduce the "red lines" on resource exploitation (Feitelson et al. 2005). To mitigate this impasse and its adverse implications for fresh water for irrigation, a significant effort was made to shift agriculture to recycled water. As discussed below, this had widespread implications for irrigation. Increasingly, farmers rely on recycled water for irrigation, rather than fresh water. To this end, drip irrigation systems had to be adapted to recycled wastewater. This had the additional benefit of largely decoupling irrigation from the vagaries of precipitation (Friedler 2001), and hence could also be viewed as an adaptation measure to climate change. Still, the shift away from freshwater led to several crises in agriculture, as discussed in the next section.

In 2005, the first large-scale seawater desalination plant came on line in Ashkelon, marking the beginning of a new era, the desalination era (Feitelson 2013). Through desalination, the total amount of water available in Israel has risen for the first time since the mid-1960s. Yet, the use of desalinated seawater for irrigation has proved problematic due to the high concentration of Boron, which is toxic to many crops, and the lack of ions that are essential for plant growth (Yermiyahu et al. 2007). Additionally, desalination is associated with several external costs, such as significant energy consumption leading to air pollution and GHG emissions, damage to marine life and resources through the discharge of residual salt to the sea, and the utilization of land along coastal areas. According to a study conducted in Israel, air

[11] In 1990 a massive immigration wave from the ex-soviet Union began. As a result of the immigration wave and the high birth rates Israel's population grew from less than 5 million in 1989 to 8.6 million at the end of 2016, a rate of growth unparalleled in any developed country.

[12] Subsidizing wastewater fits well with the government policy in the field of positive externalities. Using treated wastewater is a solution that prevents the disposal of contaminated wastewater into streams and into the Mediterranean. Furthermore, the use of treated wastewater increases the available water sources and therefore reduces the need for desalination of sea water and the negative externalities associated with them.

[13] The current obligations under the two treaties amounts to about 100 Mm^3, out of the 1200 Mm^3 considered safe long-term withdrawals (taking into account climate change projections). However, in the unratified Israeli 2010 draft masterplan the quantity reserved is 143 Mm^3, as it takes into consideration informal obligations and a realization of the amounting needs of both Jordan and the Palestinians.

pollution and GHG emissions resulting from desalination are associated with an external cost of 0.131 NIS/m^3 and the utilization of land is associated with an external cost of 0.135 NIS/m^3 (Lavee et al. 2011). However, desalination also changes the composition of wastewater, as it reduces its average salinity, allowing the wider use of recycled wastewater. Thus, from an irrigation perspective, this era is marked by an increasingly reliable supply of less saline wastewater for irrigation in the central and southern parts of the country. As a result of large quantities of desalinated water becoming available along the Mediterranean coast, the need to convey water from the Jordan River southward through the NWC diminished, leaving more fresh water within the basin. Hence, while irrigation in the central and southern parts of the county increasingly shifted to recycled wastewater, agriculture within the Jordan River basin continues to be irrigated with fresh water. This pattern is further justified when non-market benefits are taken into account (Becker et al. 2012), as it allows more water to be retained for nature.

Due to the increasing price of fresh water for irrigation supplied by Mekorot, the discrepancy in irrigation costs increased between farmers served by water associations and those served through Mekorot. This has a spatial pattern, as the lowest cost accrues to farmers in the north, who are supplied by local water associations from surface sources, while farmers in the south, to whom water must be conveyed inter-regionally, are served through Mekorot and hence pay increasingly higher rates. To address this, a new pricing formula has been advanced that will standardize the rates paid by farmers state-wide. This reform, proposed by the Treasury, is strenuously opposed by the water associations and farmers in the northern part of the country. The Ministry of Agriculture was caught in the cross-fire between the two groups of farmers, leading it to zigzag between antagonistic positions. Although the legislation standardizing water rates was passed, it has not yet been implemented at the time of writing (early 2018), and thus the full implications cannot be assessed at present.

8.3 Agricultural Water Demand

Water has long been the main constraint in Israel's agriculture. Due to the country's Mediterranean and semi-arid climate, all summer crops require irrigation.[14] Thus, irrigation is unequally distributed between the months of the year, with peak demand occurring in the dry summer months. At the beginning of the twentieth century, agricultural water supply was limited to the surroundings of local water sources. As water conveyance widened, irrigation expanded, and the amount of water supplied to agriculture increased. As a result, the extent of irrigated areas grew, until the mid-1980s (Fig. 8.2). By this time, all the available freshwater sources had been utilized. During those years, water for agriculture was mainly provided from natural

[14] Agriculture in the southern Negev Desert (mainly the Arava Valley) requires irrigation all year round.

freshwater sources. In 1986, Israel began to cut back on freshwater allocations for agriculture, due to increasing water shortages. The cuts were made mainly in the areas supplied by the NWC (Israel Water Commission 2002). Since the completion, in 1972, of the Shafdan – the wastewater treatment plant of the Tel Aviv metropolitan area (the Dan Region) – the availability of reclaimed wastewater for agriculture rose. By 2010, recycled wastewater made up about 40% of the water used for irrigation, compared to less than 4% 40 years earlier (Fig. 8.3). As can be seen in Fig. 8.2, for the last four decades, the total amount of water used in the agricultural sector (fresh water + recycled wastewater) has remained basically stable, while crop output has risen steadily. Currently, Mekorot provides about 60% of the water consumed in the agricultural sector, while about 40% is supplied by regional and local water associations (Kislev 2013). Since Israel's agriculture is based on kibbutzim and moshavim, they constitute the major water consumers in the agricultural sector.

The increase in the agricultural output (by weight) per unit of water is due to the introduction and widespread adoption of water-saving irrigation technology (fertigation), the substitution of fresh water by recycled wastewater, and a shift to water-efficient crops and to those allowing the use of saline water and recycled wastewater. Saline water, for example, is utilized for the irrigation of salinity-tolerant crops, such as cotton. For a number of crops, such as melons and tomatoes, brackish water improves product quality, although yields are lower (The Israel Export International Cooperation Institute 2013). However, the substitution of basic food crops by export-oriented crops implies that Israel cannot supply all of its food requirements. Since the mid-1960s, it has been realized that the available water in Israel is not sufficient for the production of the country's entire food intake. This

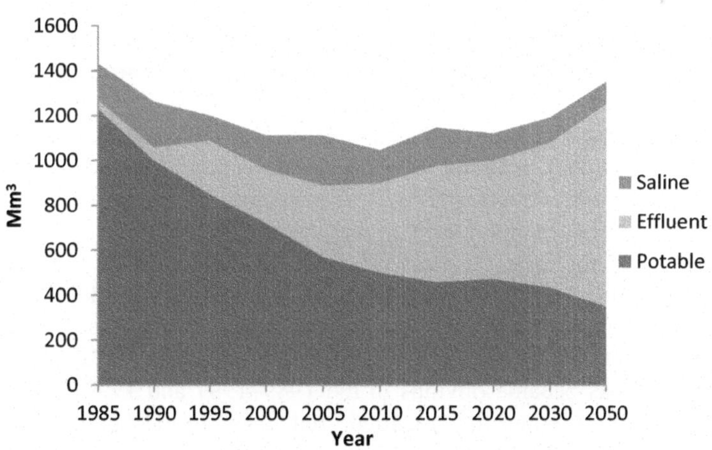

Fig. 8.3 Agricultural water use in Israel by water source, data for 1985–2015 and predictions for 2015–2050. (Source: Data for 1985–2010: Ministry of Agriculture and Rural Development 2010. Data for 2015: Israel Water Authority 2016a. Predictions for 2020–2050: Israel Water Authority 2010)

means that Israel is dependent on the import of water-intensive food, mainly grain and beef (Kislev 2011).

In addition to the transition to water-saving crops and the use of reclaimed wastewater and brackish water, farmers are also required to optimize their operations. Hence, they continually search for cost-effective management and marketing practices in order to reduce their expenditure, and, consequently, agricultural output and the value per unit of production factors, land, water, and labor, have been growing. Today, water-use efficiency in Israeli agriculture is among the highest in the world as a result of research, guidance, and the development of advanced irrigation methods.

According to Israel's national master plan for the water sector (IWA 2012), which constitutes the overall policy framework for water production and use in Israel by 2050, the sector will slightly reduce the amount of water for agriculture in the short term (2020); however, with the growth of desalination for domestic use, the amount of recycled wastewater will increase and thus the constraints on the use of recycled wastewater for agriculture should be alleviated.

In the following sections, we examine in greater detail the issue of agricultural irrigation in Israel under conditions of a water crisis.

8.4 Policy and Practice: Present Situation and Recent Developments

8.4.1 Water Governance and Institutions

Under the 1959 Water Law, all water resources are public property. The law established the office of the Water Commissioner, in order to manage the resources, and provided it with the power to authorize all abstractions and uses of all water resources. This legislation has been complemented by the earlier Supervision of Water Drilling Law (1955), the Water Measurement Law (1955), and the subsequent Water and Sewage Corporation Law (2001), as well as secondary legislation (rules and regulations) intended to regulate the water sector with emphasis on water use and management. Based on these laws, metering is required and applied. The Water Law has been amended over the years.

However, despite the seemingly wide-ranging powers of the Water Commissioner, there was widespread criticism of the management of the water resources, and their systemic over-exploitation, particularly the coastal aquifer (Gvirtzman 2002). As a result, the ability to accommodate multi-year droughts was severely eroded, as became obvious in the 1989–1991 and 1999–2001 droughts (Fischhendler 2008), when the so-called "red lines" in the Sea of Galilee were exceeded. This over-exploitation was the consequence of a 20-year policy impasse between the agricultural lobby, which prevented an increase in water rates for agriculture, and the Treasury, which blocked desalination (Feitelson 2005). This state of affairs led to several government and parliamentary inquiry commissions being set up.[15] Following their recommendations, a major reform was enacted in May 2006, in accordance with which the Water Commissioner was replaced by a Water Authority and the Water Authority Council to provide oversight. Within the authority's purview were placed all government bodies involved in water management, with the aim of creating a professional body with a comprehensive view of the needs of the water and sewage sectors, and the ability to properly manage and supervise all aspects of water and sewage in Israel.

The establishment of the Water Authority, as well as the decision to embark on widespread desalination, resulted from a shift in the position of the Treasury, which sought to reduce the power of Mekorot as part of its drive to introduce competition into the water sector (Feitelson and Rosenthal 2012), and in the power balance between the Treasury and the agricultural lobby (Singer 2011). Today, approximately 35% of the fresh water in Israel is provided by four private desalination plants (out of five in total), (calculated on the basis of data from the Israel Water Authority [2016b]).

Today, several institutions supervise and manage the Israeli water sector (EMWIS 2008; Gelpe 2010; Rejwan 2011). These include:

The Water Authority is responsible for the management, operation, and development of the water sector. It is responsible for the preservation and restoration of natural water resources, the development of new water sources, and the supervision of water consumers and producers. This authority is today within the Ministry of Energy and Water (in contrast to the Water Commissioner, who was originally under the Minister of Agriculture), reflecting the shift in water policy away from irrigation.

The National Water Company "Mekorot", Israel's main water supplier, is a corporate entity, originally established by the Zionist organizations and today a government company. Its main responsibility is to operate and manage the NWC and to supply water. The emergence of desalination somewhat diminished its power; however, it supplies most of the water for irrigation and much of the recycled wastewater. Moreover, it stands to gain from the recent legislation standardizing water rates across water associations.

Water associations are regional cooperatives whose members are kibbutzim and moshavim. Their purpose is to provide water for agricultural needs to members of the associations or to the local authorities in which they operate. They provide much of the water supply in the upper Jordan basin. The water associations are also platforms for political activity, handling negotiations with public officials. The associations' representatives bring their members' requests and needs to the table, giving the organized farmers a collective voice and power. The recent legislation to standardize water rates undermines these associations and severely reduces the voice of farmers in the areas they serve, which explains their strenuous objections.

---

[15] Most notable among them were the Arlosoroff Commission of 1997, and the Magen Commission in 2001.

8.4.2 Water Resource Management Policies

The primary focuses of Israel's water policy are: enhancing water-use efficiency by greater use of economic and market-based instruments, such as raising water prices; enforcing regulatory tools, such as water allocation and quotas; reducing the overall support for water in agriculture; and encouraging the use of non-potable water (effluent or brackish water) for irrigation. In addition, the policy supports research and development, training and technology.

8.4.2.1 Water Rights and Allocations

As mentioned, Israeli law provides that water sources are controlled by the state and are publicly owned. This includes water that may be drawn from wells located at farmers' properties, even if it is intended exclusively for the owner's use. Thus, such water can be drawn only with a water production license. Once allocated, the water is supplied to the farmers mainly by Mekorot and the Agricultural Water Associations.

Since the enactment of the Water Law, the Water Authority issues annual licenses to each of the water producers and suppliers, specifying the amount of water permitted for production, supply, or consumption during that year, as well as production conditions and restrictions.

The main goal of the Water Authority is to ensure water quality and the reliability of supply in the long run. For this purpose, it issues annual extraction licenses that may vary with local conditions. In addition, a levy was set on extractions by water producers, which varies according to region, water quality, use, and source (groundwater or surface water). The purpose of the levy is to reflect water resource scarcity, and thus to internalize externalities, so that, along with production and distribution costs, water prices will reflect more accurately the true value of water. Thus, the extraction levy is intended to encourage water conservation in the agricultural sector as well as to encourage farmers to switch to recycled wastewater, and to provide the Water Authority with an economic instrument to manage water both nationally and regionally (Zaide 2009). Currently, changes are being made to the extraction levy and, under Amendment 27 of the Water Act, a uniform water rate for agriculture was set for each type of water used. This much-contested amendment seeks to standardize water prices between farmers in different parts of the country, since, before the amendment, farmers in the north supplied by water associations enjoyed lower freshwater tariffs than those in the south.

The decisive factors for the allocation of water to various consumers combine political, economic, social, and environmental objectives. Economic efficiency is achieved by maximizing the total economic welfare from water use, whereas environmental efficiency is achieved by allocating water in a manner that will meet environmental objectives. As the vast majority of farmers are connected to water meters, two policy instruments used to achieve these objectives are quotas and pricing. Nonetheless, the issue of water allocation is often strongly linked to internal and external politics, as well as political pressure, rather than to a clear overall policy (Hadas and Gal 2012). Over the years, politicians tended to favor short-term considerations over long-term responsibilities (Feitelson 2005). The reform of 2006 and the creation of the Water Authority were designed, inter alia, to eliminate the political pressure over water allocation (Gilmont 2014). The social objective of the Water Authority is making sure that water is allocated to all Israeli farmers, while an important goal is the allocation of water for agriculture in the peripheral areas (OECD 2010).

Water in Israel is allocated to the various consumers according to type. In the first decades after the establishment of the state, agriculture was the main freshwater consumer. Freshwater allocations to the agricultural sector peaked in the mid-1980s. Over the years, fresh water has been directed toward the urban sector at the expense of the agricultural sector, partly replaced with treated wastewater. Since 1999, following a period of severe drought, major cuts were made in the allocation of fresh water to agriculture. In some cases, the reductions were compensated by payments to farmers; in others, farmers were compensated with treated wastewater as a substitute. The additional water sources (mainly treated wastewater and desalinated seawater) enabled the Water Authority to increase the total allocation of water (fresh water + recycled wastewater) for agriculture in recent years (Israel Water Authority 2016c).

Today, water allocation is made according to the following descending order of priority: domestic consumption, industrial consumption, agricultural consumption, and other needs (including environmental needs). For domestic and municipal uses, only potable water is allocated, while for industrial, natural, and landscape uses, both potable and brackish water are allocated. The agricultural sector receives treated wastewater, brackish water, and potable water (mainly in the eastern part of the country).

The urban sector does not have limits on use. Households are free to use any amount of water but face block-rate tariffs determined by the Water Authority. In the agricultural sector, the initial water allocation is administrative – that is, each consumer has a basic quota, and pays a subsidized tariff. But, due to the technological changes and the dismantling of protective measures, small-scale agriculture faces increasing difficulties. Consequently, farms are consolidated, de-facto, through largely unauthorized lease agreements. Water allocations are also informally leased by farmers who cease to farm their own land to those that farm both their land and that of others.

In Government Decision 828 (dated 01/06/2000), the quota for agriculture was set at roughly 1150 Mm^3 per annum as of 2010, of which approximately 40% (450–530 Mm^3) is fresh water and the remainder is marginal waters – saline water and treated wastewater, unfit for domestic consumption (Israeli Ministry of Agriculture and Rural Development 2013). As can be seen in Fig. 8.3, the use of fresh water is expected to drop to 350 Mm^3, accounting for 25% of the expected total in 2050. The fresh water is mainly designed for livestock farming, crops/areas that are not permitted for irrigation with treated wastewater,[16] or areas with insufficient treated wastewater (mainly in the Jordan basin). Since 2001, the use of fresh water in the urban sector has exceeded that in agriculture (Israel Water Authority 2016d).

8.4.2.2 Water Tariffs and Levies

The Water Authority is responsible for setting tariff levels to recover supply costs and reflect water scarcity. The charges for all water in Israel are determined according to the metered water volume, and municipalities are subject to fines if unbilled water quantities exceed 12% of that supplied by the local authority. Water prices for farmers vary according to the type of water, i.e. fresh water, treated wastewater, and surface or groundwater. The price differences reflect water quality and supply costs, as well as the government's objective to encourage the use of treated wastewater for agriculture.

Following the change in the dominant discourse within Israel, from a statist-socialist to a neo-liberal, market-oriented agenda, the water subsidy policy came to be viewed as a major economic burden. According to this narrative, Israel's water policy guidelines for agriculture were characterized by underpricing, discriminatory and distorted pricing, and misallocation amongst the various water users (Plaut 2000). In 2006, a water agreement was signed with the farmers, according to which the price of water for irrigation was to increase, so as to reflect the average cost of water supply. According to the agreement, new water tariffs would increase gradually, so that, as of 2017, the agricultural sector would pay the full cost of water production, including its share of desalination. As can be seen in Table 8.1, the price of fresh water increased by 111% over a 12-year period, thus achieving the goal. Similarly, increased water rates were expected for brackish water adjacent to fresh water. In addition, although not part of the Water Agreement, the Water Authority increased the tariffs for treated wastewater (Israeli Ministry of Agriculture and Rural Development 2013). As can be seen in Table 8.1, the price of wastewater increased by 56% over a 7-year period. We would expect such an increase to lead to lower levels of water use, but, in practice, farmers did not reduce their use of water (Table 8.1); instead they switched from fresh water to wastewater, the latter being much cheaper (Fig. 8.3).

The price hikes only affect farmers supplied by Mekorot, since water associations are subject to extraction levies, and can shift this cost onto farmers. Although, as farmers control these associations, the cost increase is much lower than for farmers supplied through Mekorot (Kislev 2011). This increasing discrepancy is at the heart of growing discord within the agricultural sector regarding the new legislation proposing to standardize irrigation prices across all farmers (essentially, increasing the cost to farmers supplied by water associations to somewhat reduce the average cost

[16] These are areas which are deemed vulnerable according geo-hydrological or health criteria.

Table 8.1 Water tariffs for agriculture use, 2005 until 2017

Year	Fresh water for agriculture (NIS)*	Change in prices (%)	Treated wastewater tariffs	Change in prices (%)	Use of water for agriculture (Mm³)	Change in land use (%)
2017	2.93	0.0%	1.36	5.1%		
2016	2.93	5.0%	1.33	5.1%		
2015	2.79	9.8%	1.3	10.0%	1096	−0.5%
2014	2.54	11.4%	1.22	11.1%	1102	−6.8%
2013	2.28	9.6%	1.11	9.9%	1183	11.2%
2012	2.08	10.1%	0.98	10.1%	1064	4.1%
2011	1.89	18.9%	0.92	5.7%	1022	−5.2%
2010	1.59	−3.0%	0.87	–	1078	8.2%
2009	1.64	2.5%			996	−9.4%
2008	1.6	7.4%			1099	−5.6%
2007	1.49	0.7%			1164	6.9%
2006	1.48	6.5%			1089	
2005	1.39					
2005–2017		111%		56%		1%

Source: Israel Water Authority (2018)
3.5 NIS = 1$ at December 2017
*Prices do not include VAT

and hence the cost to farmers supplied through Mekorot). As at 2018, this topic was still under discussion, one option being that Mekorot would supply fresh water to all farmers.

Along with the water agreement, farmers were given support to adjust to the new costs. During 2007–2009, farmers in the communities listed in appendix B of the water agreement were given direct refunds, while the rest were required to invest in water use efficiency. Since 2010, all communities have been required to improve their water systems, reducing waste and achieving more efficient water use (Israeli Ministry of Agriculture and Rural Development 2015).

Until recently, an integrated system was practiced for the agricultural sector: each farmer had a basic quota, and the payment for water was determined by block rate pricing – that is, the price rose with the increase in the relative amount of water consumed of the allocation, according to three price levels. An example for Mekorot's former tariffs for fresh water to agriculture, according to block rate prices, is shown in Table 8.2. However, as of January 2014, a uniform tariff was set for each type of water used in agriculture. This is one of several actions promoted recently by the Ministry of Agriculture together with the Water Authority, and compensatory measures include the allocation of fresh water for agriculture for 3 years in advance and a substantial increase in quotas for farmers. For example, the allocation of water from Mekorot to agriculture increased between 2013 and 2016 from 334 to 459 Mm³ (Israel Water Authority 2013; Israel Water Authority 2016e).

Table 8.2 Mekorot's tariffs for fresh water to agriculture, according to block rate prices (2013)

Price (NIS/ m^3)[a]	Present from total quota (%)	Block
2.61	50%	Block I
2.96	30%	Block II
3.66	20%	Block III
2.93 (average)	100%	Total

Source: Israel Water Authority (2018)
[a]Prices do not include VAT

These measures give greater certainty regarding the quantities of fresh water available for use by farmers, allowing them to plan crops in a more optimal and efficient manner. They were made possible by the desalination facilities along with a massive amount of treated wastewater for irrigation.

In practice, most farmers use the first and second blocks. On average, in 2016–2017, farmers used only 70–75% of the quota (compared to 82% in 1999), because of the rise in the average price of fresh water (Mecler 2018).

8.4.2.3 Cross-Subsidization

Cross-subsidization refers to charging one group of consumers higher prices in order to offer lower prices to another. In Israel, domestic and other water consumers in the urban sector cross-subsidize the water prices for agriculture, and farmers near the water sources cross-subsidize farmers in areas where supply costs are high – mainly the south and mountainous areas (Becker and Lavee 2002). The system is heavily criticized by the Treasury and the urban sector, as well as by many economists. This led to the shifts in pricing noted above. Hence, in contrast to the situation in the 1970s, the agricultural sector today covers most of the operating and maintenance costs, as well as the fixed capital costs of water supplies. As mentioned, freshwater prices for agriculture will continue to rise and inter-sectorial cross-subsidies will be reduced (Kislev 2013).

8.4.3 Development of Technologies for Coping with Water Scarcity

The decoupling of agricultural product from total water use since the early 1970s, seen in Fig. 8.2, can be largely attributed to the advent of drip irrigation. While the idea has a long history, the breakthrough was the invention of a new emitter by Simcha Blass in the mid-1960s. As irrigation systems were pressurized to allow sprinkler irrigation, the Blass emitter reduced the velocity to allow timed, low-volume irrigation in the root zone. Blass partnered with Kibbutz Hatzerim to establish Netafim, the leading drip irrigation company, which, in 1970, acquired the rights to the turbulent water passage patent, developed by Rafi Mahoudar to resolve the in-line laminar dripper's clogging problems. Once these were resolved, drip irrigation became available to a wide array of crops, and consequently became one of Israel's best known exports.[17]

Further improvements of the drip-irrigation technology include sub-surface drip irrigation and micro-spray heads. Sub-surface irrigation reduces evaporation and damage to the drip irrigation systems, and allows its use for recycled wastewater. This required further development, in particular, methods for cleaning slime within the tubes. An additional use of drip irrigation is for introducing fertilizers, as liquid fertilizers can be mixed with the irrigation water, thereby allowing more precise fertilization. To allow more precise irrigation, monitoring systems were added. These provide growers with real-time, computerized field display. Based on data collected from an array of sensors, such systems alert the grower to any deficiencies in moisture or nutrients. Further development allows the automatic release of different nutrients into the irrigation water, based on the data collected by the sensors.

While drip irrigation may have reduced the recharge of the coastal aquifer, it has little effect on other sources, as the recharge areas of all other aquifers, as well as the upper Jordan River, are in mountainous areas where there is little drip irrigation. The coastal aquifer has indeed been over-utilized, as seen in the decline in water level and the intrusion of saline water (Gvirtzman 2002). But, as this is also where much of the urban development and heavy pumping for urban consumption took place, it is difficult to identify the marginal contribution of the shift to drip irrigation to this decline.

An additional direction of technological development is the identification and development of salt-tolerant crops that can be grown on brackish, saline, or recycled water (Pasternak et al. 1986). The research and development of such crops is ongoing at present (Ventura et al. 2015, for example).

The advent and widespread adoption of new technologies in Israel can be attributed not only to the high level of research and development, but also to the human capital of farmers. As farming in Israel, for the most part, is not a traditional sector, and much of it was organized in a cooperative manner (whether in fully collectivized Kibbutzim or semi-collective Moshavim), farmers display a high level of education. Moreover, with exposure to world markets, farmers were forced to compete under increasingly challenging circumstances (Kislev 2013),which forced them to innovate. Hence, farmers were willing, and were often instrumental in developing and testing new technologies.

---

[17] For a timeline of Netafim's history and further drip irrigation innovations of the company see www.netafimlegacy.com/timeline.

8.4.4 Transition from Potable Water Irrigation to Treated Wastewater, and the Impact on Israeli Regulation and Water Quality

Until the 1990s, the water used for agriculture was mainly fresh water, with the exception of the western Negev, which was supplied with treated wastewater from the Shafdan, and the western Yizrael valley, which was supplied from the Haifa wastewater treatment plant. Since the 1990s, wastewater recycling for agriculture has greatly expanded. Due to the increasing power of environmental bodies and following the establishment of the Ministry of Environment in 1988, as well as the wider incorporation of environmental considerations into land use planning, any new development must be connected to a wastewater treatment plant (since 1992). These, in turn, face increasingly stringent standards. As a result, the supply of high-quality treated wastewater has increased dramatically.

While wastewater can be reused for urban and industrial uses, stream rehabilitation, and artificial groundwater recharge, most is actually recycled for irrigation. The use of treated wastewater has two main advantages: first, it may be considered a new water resource and substitute for conventional water (potable water) for irrigation and other purposes (Friedler 2001). Second, by switching to recycled wastewater the pressure on conventional water resources is lessened (Lavee 2011). However, the extended use of recycled wastewater has adverse effects on soils and aquifers, particularly the coastal aquifer (Jueschke et al. 2008).

8.4.4.1 Treated Wastewater Quantity

The significant increase in wastewater treatment and use in recent years is shown in Fig. 8.4. As a result of utilizing treated wastewater for irrigation, a considerable reduction in potable water use in agriculture has occurred (Fig. 8.3).

Today, 95% of Israel's domestic wastewater is treated by approximately 135 treatment facilities. Around 86% of the treated wastewater is reused, mostly in agriculture, which receives roughly 520 Mm^3 per year, representing around 45% of its total supply (Israel Water Authority 2015a, 2016a). The remaining treated wastewater is either lost through evaporation or discharged to sea, since some treatment plants lack the infrastructure to convey it to irrigable land. Some treated wastewater is also used for stream rehabilitation projects.

National policy calls for treated wastewater to be eventually fully utilized by agriculture, and it is estimated that, by 2020, virtually all municipal wastewater will be reused, and treated wastewater will provide 50% of Israel's agricultural needs (Brenner 2012; Inbar 2007). This would be accomplished by enlarging the existing wastewater treatment facilities as well as constructing new ones.

The largest player in Israel's treated wastewater industry is Mekorot, as its facility (the Shafdan) provides the largest amount of treated wastewater, approxi-

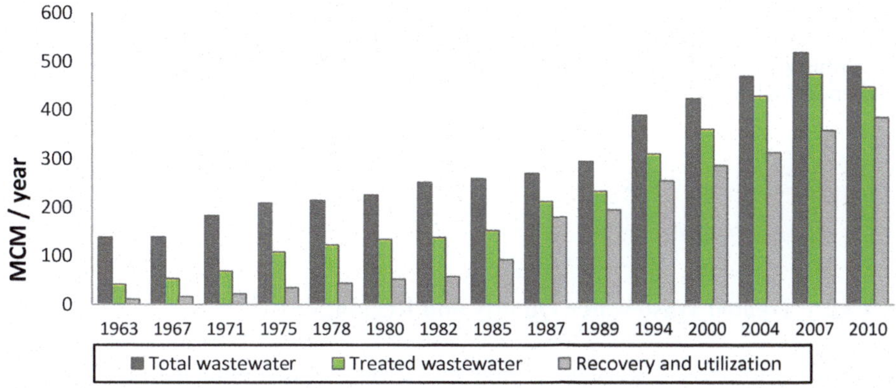

Fig. 8.4 Wastewater treatment, recovery, and utilization in Israel, 1963–2010. (Source: Israel Nature and Parks Authority 2012)

mately 130 Mm^3 per year. The wastewater treated at Shafdan is then conveyed to infiltration fields, where it is recharged into the confined aquifer, thereby further purifying the effluents. This also allows large amounts of water to be stored and made available for irrigation during "dry" seasons. To prevent the leakage of this water into drinking water wells in the area, the treated wastewater is pumped by a peripheral system of 150 recovery wells. Subsequently, the reclaimed wastewater is conveyed for agricultural irrigation in the south by Mekorot through its so-called Third Line. This is separated from the two other lines that convey potable water to the Negev and is intended to provide only treated wastewater for agricultural use.

In addition to the 130 Mm^3 of the Shafdan, through other recovery plants Mekorot provides roughly 63 Mm^3 of treated wastewater (of varying quality) each year. In addition, there are private wastewater treatment facilities, providing approximately 200 Mm^3 per year. By 2050, production is expected to reach approximately 900 Mm^3 per year (Table 8.3).

In 2000, an assistance plan for sewage treatment plants was initiated. Under this plan, financial assistance is granted for converting from the use of potable water to treated wastewater, as well as for improving the quality of effluents to the level required by regulations. The plan provided guidelines for the conveyance of excess wastewater from densely populated areas, which produce high volumes of wastewater but have little agricultural land, to sparsely populated peripheral areas that are based on agriculture and are in need of treated wastewater (Israel Water Authority 2011).

In the next few years, a significant increase in the amount of land irrigated with wastewater is expected, mainly due to reductions in the quotas of fresh water and the many wastewater recycling projects of recent years (Israel Ministry of Health 2011). The increase in total freshwater supply thanks to desalination will also lead to the additional availability of wastewater through recycling.

Table 8.3 National objectives and timeline for improving wastewater use in the agricultural sector

Year	% of irrigation water from effluent	Volume of effluent per year (Mm^3)
2010	38	400
2015	43	464
2020	51	587
2050	67	900

Source: Israel Water Authority (2011)

8.4.4.2 Treated Wastewater Quality and Regulations

Irrigation with treated wastewater may cause various environmental and sanitary hazards, such as water and soil pollution, as well as a decline in crop yields (Lavee 2011). These risks are largely a function of wastewater quality. Therefore, it is necessary to set standards and requirements for wastewater treatment and use.

Initially aimed at preventing health risks, a regulatory framework began to evolve alongside the development of wastewater reuse. Water quality standards for irrigation were set in 1977 by the Shelef Commission (Sagi and Shisha 1999). Subsequently, in 1981, wastewater irrigation was limited to crops specified by the Public Health Law, and permits were required for all wastewater reuse projects. In 1992, a secondary standard was set by the Ministry of Health in response to demands for the improvement of wastewater quality for irrigation and the prevention of the pollution of streams and further environmental risks (State Commission of Inquiry on water management in Israel 2010). While this standard reduced the environmental and health impacts arising from the use of wastewater, treatment plants in Israel continued to discharge effluents containing various pollutants and high levels of salt, raising various issues (Israel Ministry of environmental protection 2005; Lavee 2010).

Consequently, a special committee (Inbar Committee) was established in the year 2000 in order to set new standards for wastewater quality. The committee's objectives were to address various aspects of health, soil, plants, and hydrology, so that wastewater use would not constitute a hazard, on the one hand, and would be economically viable on the other (Pareto Engineering Ltd. 2003). On the basis of a cost-benefit analysis, the committee advanced a tertiary standard, to allow the use of treated wastewater for unlimited irrigation and improve the quality of wastewater released into the sea (Inbar 2007). The recommendations were approved in 2005 and took effect on 7/25/2010. Currently, most (59%) of the treated wastewater used for irrigation is of high quality and can be used to irrigate specific crops (Israel Ministry of Health 2011).

8.5 Future Challenges

8.5.1 *Water Supply*

Over the past three decades, water supply to Israel's agriculture has been characterized by climate vagaries, leading to uncertainty in the agricultural sector. Following the advent of seawater desalination, the increased use of treated wastewater, and the higher quality of this wastewater, it appears that the supply of water for agriculture in forthcoming years will be sustained and even increased. This is true, however, only for the central and southern parts of the country. Irrigation in the northeast, largely within the Jordan River basin, still depends on precipitation patterns. Due to increasing competition from tourism, the worrying situation in Jordan, and a series of below-average rainfall years, the future of irrigation in this part of the country remains uncertain.

Irrigation, particularly with treated wastewater, may cause changes in the irrigated soil properties and its environment due to high salt concentrations. In addition, recently, there has been increased awareness of the presence of organic pollutants, such as pharmaceuticals, cosmetics, and hormones, in treated wastewater and their effect on the environment and agricultural crops. Organic pollutants and salinity may affect the food chain and water resources, although these effects are not yet sufficiently known (Shaviv et al. 2011). Therefore, it will be necessary to address these impacts.

8.5.2 *Climate Change*

Despite the greater certainty of the water supply in the coming years, due to the combination of desalination and wastewater recycling, there are concerns that climate change may affect the water sector.

In the last few decades, a warming trend has been observed in the Middle East. While the trends in precipitation are ambiguous, scenario analyses have suggested that levels may decrease and the incidence of droughts will increase (Alpert et al. 2008). According to the A1B scenario of the IPCC, a rise of 1.5 °C in Israel's average temperature is expected by 2020 compared to 1960–1990. According to IPCC scenarios A2 and B2, between the years 2071 and 2100, average temperatures are expected to rise by up to 5 °C by the end of the century compared to 1960–1990. Furthermore, by 2020, a 10% decrease in precipitation is projected in Israel, and expected to reach a 20% decrease by 2050 (Israel Ministry of Environmental Protection 2010). The rising sea level is likely to lead to further intrusion of sea water into the coastal aquifer. Melloul and Collins (2006) estimate that the storage capacity of this aquifer will diminish by 16.3 Mm^3 per 1 km of coast for each 50 cm rise in sea level, thereby reducing the available fresh water.

According to the OECD (2010), by 2020, climate change will benefit agriculture in Israel, thanks to its ability to meet the demand of international markets earlier in the season. However, in the long term, Israel will need to supply the expected increase in total water consumption due to population growth and higher evapotranspiration rates, while the availability of freshwater resources is expected to be reduced due to the higher frequency and severity of extreme events (particularly droughts) and the rising sea level (Alpert et al. 2008). This will adversely affect rain-fed agriculture, thereby requiring additional irrigation.

To address this challenge, Israel will need to continually improve water savings, inter alia by using more efficient irrigation methods, expanding the use of water-saving crops, and increasing the use of brackish water and treated wastewater, whose availability will continue to rise as additional desalinated seawater is used. However, these measures will not be available in some parts of the country, mainly the northeastern part, where there are few sources of wastewater. Moreover, these measures will not be available to Israel's eastern neighbors, Jordan and the Palestinians. Hence, if additional water will be conveyed to these water-stressed neighbors, Israel will have to do with less fresh water, thereby furthering its reliance on desalination and wastewater recycling.

8.6 Summary and Conclusions

Irrigation enjoyed wide public and political support in the early years of the state, as it was deemed essential for state-building, food security, and immigration absorption. The agricultural lobby was closely connected to the centers of power. Consequently, major water infrastructure was built, at substantial cost, creating a national-level water system. One result of this, as well as of the trans-regional conveyance of recycled wastewater, is a spatial discrepancy between the sources of water and the areas irrigated. As the collective farming sector controlled the Ministry of Agriculture and the bodies that planned and implemented these water development projects, it was also the main beneficiary of these investments. Moreover, the control of the collective farming sector by the Ministry of Agriculture enabled the state to nationalize all water resources and form a centralized water management structure, as the management was entrusted to the ministry. The Water Commissioners that came from the agricultural sector prioritized irrigation, allowing groundwater levels to decline in order to increase the water supply to agriculture and reduce allocation cuts in drought years.

After almost 20 years of relatively abundant precipitation levels, a series of drought years began in the late 1980s. Due to the unsustainable utilization of water, mainly for irrigation, that preceded these droughts, Israel experienced severe water crises in the 25 years that followed. These were exacerbated by a combination of rapid urban population growth, rising standards of living, and new obligations under water agreements between Israel and its neighbors. This led to increasing pressure on the scant water resources, in terms of both quantity and quality.

Due to the technological and economic shifts in agriculture, the number of farmers declined dramatically, as did the status of the Ministry of Agriculture within the governmental system. As a result of these trends, and the Likkud party coming to power, the influence of the agricultural lobby declined. Along with the rise of neoliberal ideology and the increasing power of the Treasury within the governmental system, cuts were made to freshwater allocations and subsidies for irrigation, thereby raising the cost of fresh water for farmers. Moreover, with the marketization of agriculture and the decline of the central institutions that underlay the agricultural lobby, the previously unitary voice of the agricultural sector fragmented, further weakening the political power of irrigation interests.

The ability of the Israeli agricultural sector to withstand these adverse trends can be largely attributed to the high human capital in the Israeli agricultural sector, as well as to improvements in technology and the management of water resources. In particular, the substitution of fresh water by treated wastewater allowed irrigation to retain much of its water allocation, despite the sharp decrease in freshwater allocations, with the prospect of even increasing it in coming years. The availability of high-quality treated wastewater can be attributed to the growing power of environmental interests, which led to the enactment of more stringent standards for wastewater treatment, the demand by planning authorities that all new residential developments be connected to treatment plants, and the advent of seawater desalination that led to increasing quantities of higher quality (i.e. fresher) water flowing through urban systems. The increasing use of both treated wastewater and desalinated water has relieved much of the pressure on freshwater resources, thereby reducing the need for further cuts in the total water allocated for irrigation.

A second factor that allowed Israeli agriculture to continuously increase productivity without increasing the total water use (and decreasing freshwater allocations) is the development and application of advanced water-saving technologies, most notably drip irrigation. Thanks to these technological developments, Israel became a lead exporter of water-saving irrigation technology. Within Israel, drip irrigation has been central to improving irrigation efficiency, i.e. increasing the amount of agricultural production per unit volume of water consumed. Water conservation, along with crop-substitution and the substitution of fresh water by recycled water, allowed Israel to largely decouple agricultural output from (fresh) water. Indeed, since the 1950s, Israel's agricultural output has grown tenfold (volume), while the amount of freshwater for agricultural use has decreased significantly.

But Israel's water policy reforms were not only driven by technology. Rather, they resulted from the decline in support for the subsidized supply of water for agriculture, due to the shifting power balance between the agricultural lobby and the Treasury. This resulted in the introduction of real water prices for farmers, giving them an incentive to increase the efficiency of water use in agriculture, and to substitute fresh water with treated wastewater. The technological improvements allowed them to do so while increasing the agricultural output. Nonetheless, the falling profit from agriculture, not least due to the increase in water prices, has led most of the young generation of farmers to abandon agriculture. This is manifest in the rising median age of active farmers. Hence, the future of Israeli agriculture is uncertain. This may lead to a decline in the demand for recycled wastewater. Israel may face, therefore, an excess in treated wastewater, particularly in its northwestern parts. The widespread recycling of wastewater might have adverse effects on soils, the extent of which requires further study. Therefore, the success of Israel's irrigation systems may be more tenuous than they seem from past trajectories.

References

Alatout, S. (2008). States of scarcity: Water space and identity politics in Israel 1948–59. *Environment and Politics D, 26*, 959–982.

Alpert, P., Krichak, S. O., Shafir, H., Haim, D., & Osetinsky, I. (2008). Climatic trends to extremes employing regional modeling and statistical interpretation over the E. Mediterranean. *Global and Planetary Change, 63*, 163–170.

Amiran, D. (1994). *Rainfall and water policy in Israel*. The Jerusalem Institute for Israel Studies, No. 55 (Hebrew).

Avitsur, S. (1991). Water of affliction. In J. Aviram & A. Shachar (Eds.), *David Amiran volume, Eretz-Israel: Archeological, historical and geographical studies* (Vol. 22). Jerusalem: Israel Exploration Society (Hebrew).

Becker, N., & Lavee, D. (2002). The effect and reform of water pricing: The Israeli experience. *International Journal of Water Resources Development, 18*(2), 353–366.

Becker, N., Lavee, D., & Tavor, T. (2012). Desalinate or divert? Coastal non-market values as a decision tool for an integrated water management policy: The case of the Jordan River basin. *Ocean and Coastal Management, 64*, 27–36.

Brenner, A. (2012). Limitations and challenges of wastewater reuse in Israel. In *Clean soil and safe water* (NATO science for peace and security series C: Environmental security, pp. 3–9). Dordrecht: Springer.

Central Bureau of Statistics. (2016). www.cbs.gov.il

EMWIS (2008). Water resources management – institutions. Israeli EMWIS Focal Point. www.emwis il.org/en/Water_legislation/legislation_06.htm

Feitelson, E. (2005). Political economy of groundwater exploitation: The Israeli case. *International Journal of Water Resources Development, 21*, 41–423.

Feitelson, E. (2013). The four eras of Israeli water policies. In N. Becker (Ed.), *Water policy in Israel: Context, issues and options* (pp. 15–32). Berlin: Springer.

Feitelson, E., & Fischhendler, I. (2009). Spaces of water governance: The case of Israel and its neighbors. *Annals of the Association of American Geographers, 99*, 728–745.

Feitelson, E., & Rosenthal, G. (2012). Desalination space and power: The ramifications of Israel's changing water geography. *Geoforum, 43*, 272–284.

Feitelson, E., Gazit, T., & Fischhendler, I. (2005). *The role of "red lines" in safeguarding the Sea of Galilee (Lake Kinneret)*. The Jerusalem Institute for Israel Studies (Hebrew).

Feitelson, E., Fischhendler, I., & Kay, P. (2007). The role of a central administrator in managing water resources: The case of the Israeli water commissioner. *Water Resources Research, 43*(11), 1–11.

Feitelson, E., Tamimi, A., Bein, A., Laster, R., Marei, A., Rosenthal, G., & Salahout, S. (2011). *Defining water needs for fully exploited resources: A necessary step for Israeli-Palestinian reconciliation*. Jerusalem: The Jerusalem Institute for Israel Studies, Green Cross International and Peres Institute for Peace.

Feitelson, E., Seltzer, A., & Almog, R. (2014). Water history facets of landscape change in Israel/ Palestine 1920–1970: A question of scale and periodization. *Water History, 6*, 265–288.

Fine, T., Segev, S., Lavie, R. (2007). *Precipitation and climatic zones in Israel*. CET: Center for Educational Technology (Hebrew). http://lib.cet.ac.il/pages/item.asp?item=9591

Fischhendler, I. (2008). Institutional conditions for IWRM: The Israeli case. *Ground Water, 46*(1), 91–102.

Friedler, E. (2001). Water reuse – An integral part of water resources management: Israel as a case study. *Water Policy, 3*, 29–39.

Furman, A., & Abbo, H. (2013). Groundwater management in Israel. In N. Becker (Ed.), *Water policy in Israel: Context, issues and options* (pp. 125–136). Dordrecht: Springer.

Gelpe, M. (2010). Legal framework for allocation of water and for protection of water quality in Israel. In A. Tal & A. A. Rabbo (Eds.), *Water wisdom: Preparing the groundwork for cooperative and sustainable water management in the Middle East*. New Brunswick: Rutgers University Press.

Gilmont, M. (2014). Decoupling dependence on natural water: Reflexivity in the regulation and allocation of water in Israel. *Water Policy, 16*, 79–101.

Gvirtzman, H. (2002). *Water resources in Israel*. Jerusalem: Yad Izhak Ben Zvi.

Hacohen, D. (1994). *Immigrants in Turmoil: The great wave of immigration to Israel and its absorption 1948–1953*. Jerusalem: Yad Izhak Ben-Zvi.

Hadas, E., & Gal, Y. (2012). Inter-sector water allocation in Israel, 2011–2050: Urban consumption versus farm usage. *Water Environment Journal, 28*, 63–71.

Inbar, Y. (2007). New standards for treated wastewater reuse in Israel. In M. K. Zaidi (Ed.), *Wastewater reuse—Risk assessment, decision-making and environmental security* (pp. 291–296). Dordrecht: Springer.

Israel Ministry of Environmental Protection. (2005). *Effluent quality standards – Inbar Committee* (Hebrew). http://old.sviva.gov.il/bin/en.jsp?enPage=BlankPage&enDisplay=view&enDispWhat=Object&enDispWho=Articals%5El3576&enZone=vaadot_tkina

Israel Ministry of Environmental Protection. (2010). *Israel's second national communication on climate change*. Submitted under the United Nations Framework Convention on Climate Change.

Israel Ministry of Health. (2011). *Final activity report for 2010*. Israel Nature and Parks Authority, environmental unit, the water quality domain (Hebrew).

Israel Nature and Parks Authority. (2012). *Wastewater collection and treatment and effluent utilization for agricultural irrigation – A national survey – 2010*. Prepared for the Water Authority, May 2012 (Hebrew). www.water.gov.il/Hebrew/ProfessionalInfoAndData/Water-Quality/DocLib1/seker-kolhin-2010.pdf

Israel Water Authority. (2010). *A long-term master plan for the water sector – Policy document of the State of Israel*. Ministry of National Infrastructures, Water Authority, Planning Department.

Israel Water Authority. (2011). *The state of Israel: National water efficiency report*. Planning Department of the Israeli Water Authority, April 2011.

Israel Water Authority. (2012). *Master plan for the National Water Sector – Policy document of the State of Israel*. Ministry of National Infrastructures, Water Authority, Planning Department. www.water.gov.il/Hebrew/.../09-Israel-Water-Sector-Master-Plan-2050.pdf

Israel Water Authority. (2013). *Water allocations for 2016*. Water division regulation, licensing department. www.water.gov.il/Hebrew/ProfessionalInfoAndData/Allocation-Consumption-and-production/20131/water-quantity-2013.pdf

Israel Water Authority. (2015a). *Book of water and sewage rates – Rates for water and sewage consumers*. Economics Division. www.water.gov.il/Hebrew/Rates/DocLib1/water-prices-book-1.1.15.pdf

Israel Water Authority. (2016a). *Water consumption by purpose, 1998–2015*. http://water.gov.il/Hebrew/ProfessionalInfoAndData/Allocation-Consumption-and-production/Pages/Consumer-survey.aspx

Israel Water Authority. (2016b). *Desalination plants in Israel*. www.water.gov.il/Hebrew/Planning-and-Development/Desalination/Pages/desalination-%20stractures.aspx

Israel Water Authority. (2016c). *Annual report for the year 2015*. The Government Water and Sewage Authority.

Israel Water Authority. (2016d). *Water consumption for 2014 summary report*. http://water.gov.il/Hebrew/ProfessionalInfoAndData/Allocation-Consumption-and-production/20143/hakdama.pdf

Israel Water Authority. (2016e). *Water allocations for 2016*. Water division regulation, licensing department. www.water.gov.il/Hebrew/ProfessionalInfoAndData/Allocation-Consumption-and-production/20162/water-allocations-end-of-2016.pdf

Israel Water Authority. (2018). *The government water and sewage authority* (Hebrew). www.water.gov.il/Hebrew/Rates/Pages/prices-archive.aspx

Israel Water Commission. (2002). *Water in Israel – Consumption and production 2001*. State of Israel, Ministry of National Infrastructures, Water Commission, Demand Management Division, December 2002. http://gwri-ic.technion.ac.il/pdf/wcom/demand.pdf

Israeli Ministry of Agriculture and Rural Development. (2010). *Irrigated agriculture – The Israeli experience*. www.water.gov.il/Hebrew/ProfessionalInfoAndData/2012/13_Israel_Water_Sector-Water_and_Agriculture.pdf

Israeli Ministry of Agriculture and Rural Development. (2013). *Agriculture in Israel – Term summary* (Hebrew). www.moag.gov.il/NR/rdonlyres/0917C695-A29C-41A5-9CB3-5662E6C7B2B0/0/shulhan_mankal_23.pdf

Israeli Ministry of Agriculture and Rural Development. (2015). *Special agreements and administrative benefits of the development policy*. Agricultural Investment Administration (Hebrew). https://bit.ly/2GbevxL

Jueschke, E., Marschner, B., Tarchitzky, J., & Chen, Y. (2008). Effects of treated wastewater on the dissolved and soil organic carbon in Israeli soils. *Water Science and Technology, 57*, 727–733.

Kislev, Y. (2011). *The water economy of Israel* (Policy Paper No. 2011.15). Taub Center for Social Policy Studies in Israel. http://taubcenter.org.il/wp-content/files_mf/thewatereconomyofisrael.pdf

Kislev, Y. (2013). Water in agriculture. In N. Becker (Ed.), *Water policy in Israel: Context, issues and options* (pp. 51–64). Dordrecht: Springer.

Lavee, D. (2010). The effect of water supply uncertainty on farmers' choice of crop portfolio. *Agricultural Water Management, 97*(11), 1847–1854.

Lavee, D. (2011). Cost-benefit analysis of alternative wastewater treatment standards: A case study in Israel. *Water Environment Journal, 25*(4), 504–512.

Lavee, D., Ritov, M., & Becker, N. (2011). Is desalination the most sustainable alternative for water-shortage mitigation in Israel? *International Journal of Sustainable Economy, 3*(4), 410–424. https://doi.org/10.1504/IJSE.2011.042796.

Mecler, B. (2018). *Water allocation for agriculture*. www.mekler.co.il/hozrim/item/3602-%D7%94%D7%A7%D7 %A6%D7%90%D7%AA-%D7%9E%D7%99%D7%9D-%D7%9C%D7%97%D7%A7%D7%9C%D7%90%D7%95%D7%AA-2016-2018.html

Melloul, A., & Collins, M. L. (2006). Hydrological changes in coastal aquifers due to sea level rise. *Ocean and Coastal Management, 23*, 2015–2029.

Nir, D. (1989). *Beth-Shean Valley: The region and its challenges on the fringe of the desert* (3rd ed.). Tel Aviv: Hakibbutz Hameucahd (Hebrew).

OECD. (2010). *OECD review of agricultural policies: Israel 2010*. OECD Publishing. ISBN:978-92-64-07934-2.

OECD. (2011). *OECD environmental performance reviews: Israel 2011*. OECD Publishing. https://doi.org/10.1787/9789264117563-en.

Pareto Engineering Ltd. (2003). *Final report of the committee for establishing wastewater treatment standards* (Inbar Committee) (Hebrew). http://www.sviva.gov.il/InfoServices/ReservoirInfo/DocLib2/Publications/P0301-P0400/P0321.pdf

Pasternak, D., Aronson, J. A., Ben Dov, J., Forti, M., Mendliger, S., Nerd, A., & Sitton, D. (1986). Development of new arid zone crops for the Negev desert of Israel. *Journal of Arid Environments, 11*, 37–59.

Plaut, S. (2000). *Water policy in Israel*. Institute for Advanced Strategic and Political Studies, Policy Studies, 47.

Reichman, S. (1979). *From foothold to settled territory, 1918–1948*. Jerusalem: Yad Itzhak Ben-Zvi.

Reichman, S. (1990). Partition and transfer: Crystallization of the settlement map of Israel following the war of independence 1948–1950. In R. Kark (Ed.), *The land that became Israel*. New Haven/Jerusalem: Yale University Press/Magnes Press.

Rejwan, A. (2011). *The state of Israel: National water efficiency report*. Prepared by the Planning Department of the Israeli Water Authority. www.water.gov.il/Hebrew/ProfessionalInfoAndData/2012/04-The-State-of-Israel-National-Water-Efficiency-Report.pdf

Sagi, G., & Shisha, A. (1999). Advanced wastewater treatment, in modular field systems, for unrestricted irrigation. *Water and Irrigation, 390*, 11–13 (Hebrew).

Seltzer, A. (2010). *Mekorot: The story of the Israel national water company – The first 75 years*. Jerusalem: Yad Ben-Zvi.

Shaviv, A., Zaida, M., Tarchitzky, J., & Goldratt, T. (2011). *Water and agriculture: Summary and recommendations of the discussion forum water, no. 2*. Haifa: The Samuel Neaman Institute (Hebrew).

Singer, G. (2011). *Over troubled water: The establishment of the water authority as a case of implementation of government decisions in Israel*. Unpublished MA Thesis, School of Public Policy, The Hebrew University of Jerusalem (Hebrew).

State Commission of Inquiry on water management in Israel. (2010). *Committee Report* (Hebrew). http://elyon1.court.gov.il/heb/mayim/doc/sofi.pdf

The Israel Export International Cooperation Institute. (2013). *Israel's agriculture*. www.moag.gov.il/agri/files/Israel's_Agriculture_Booklet.pdf

U.S. Library of Congress. (1988). *Israel: A country study*. Washington, DC: GPO for the Library of Congress.

Ventura, Y., Eshel, A., Pasternak, D., & Sagi, M. (2015). The development of halophyte-based agriculture: Past and present. *Annals of Botany, 115*, 529–540.

Yermiyahu, U., Tal, A., Ben-Gal, A., Bar-Tal, A., Tarchintzky, J., & Lahav, O. (2007). Rethinking desalinated water quality and agriculture. *Science, 318*, 920–921.

Yunger, M., Carmon, N., & Shamir, U. (1993). *Who benefits from Israel's water resources?* Haifa: Technion, Center for Urban and Regional Studies (Hebrew).

Zaide, M. (2009). *Drought and arid land water management*. United Nations Commission on Sustainable Development (CSD)-16/17 National Report Israel.

Chapter 9
Egypt

François Molle

Abstract In no other country, perhaps, is the identification between water, irrigation, and agriculture as strong as in Egypt. Several phases of agrarian and technical change throughout its long history have taken Egypt to a situation where farmers grow between one to three crops a year on 3.78 million ha after lifting water from canals and drains, and sometimes from aquifers. This chapter first reviews the extraordinary cost of maintaining the delta 'irrigation machine', the history of irrigated agriculture expansion, and the evolution of public policies. It then examines key issues and challenges, including irrigation modernization, the search for water savings, the multiple experiences with participatory management at various scales, inter-sectoral integration, the water-food-energy nexus, the recent masterplan, and finally transboundary threats.

Keywords Irrigation policy · Efficiency · Water reuse · Water balance · Nile Delta

9.1 Introduction

In no other country, perhaps, is the identification between water, irrigation, and agriculture as strong as in Egypt. With the exception of the oases in the Western desert and the Red Sea coast, the 5% of Egypt's territory where the population is settled (divided between the Nile Valley and Nile Delta) corresponds to either irrigated agriculture or urban settlements, both sustained with Nile water. Like several other countries in the region, Egypt's irrigated agriculture is responsible for 76% of total water diversions and more than 85% of water *consumption* (NWRP 2017). Irrigated agriculture has played a crucial role in Egypt's history, social fabric, and national economy, as well as in the livelihoods and food security of a population of roughly 100 million. Hence, the intertwining of water and agricultural policies is key in understanding the country's evolution.

F. Molle (✉)
UMR-Geau, Institute of Research for Development (IRD), University of Montpellier, Montpellier, France
e-mail: francois.molle@ird.fr

F. Molle et al. (eds.), *Irrigation in the Mediterranean*, Global Issues in Water Policy 22, https://doi.org/10.1007/978-3-030-03698-0_9

Egypt (as at 2015) boasts an irrigated area of 3.78 million ha (or around 9 million feddan[1] [Mfed]), while rain-fed agriculture is negligible (CAPMAS 2018). This includes 2.25 million ha of Old Lands, the traditional farm land of the Valley and Delta, and 1.53 million ha of New Lands, which include land reclaimed from the desert adjacent to the Old Lands or in the oases. The social landscape of the new lands east and west of the Nile Delta, as well as the oases, is a complex and intricate mix of investors and settlers, including small farmers coming from the Delta, landless tenants, previous state farm workers, high school and university graduates, small and large investors, large companies, and state and other army-operated large farms. All types of irrigation technology can be found, from gravity and sprinklers to drip and pivot, using water pressurized by individual pumps or collective pumping stations.

Following an historical overview of irrigation/agricultural development, this chapter focuses on post-WWII irrigation policy, and examines how irrigation in the Valley and Delta is sustained and how it is adapting to changing conditions. The next sections are devoted to analyses of "horizontal" agricultural expansion, the interactions between agricultural policy and water/irrigation, and finally the key issues and challenges currently faced by the irrigation sector, including technological and institutional innovations. The chapter closes with a reflection on the link between irrigated agriculture and the wider policy and geopolitical contexts.

9.2 Brief Historical Overview of the Development of Irrigation

For millennia, the water management and irrigation of the Nile Valley and Delta consisted of guiding the rising tide of the flood—and its load of fertilizing silt—into interconnected compartments or basins (*houd*). The success and hazards of cultivation reflected the (mis)match between the timing, duration, and height of the flood, partly reshaped through a system of dikes and the planting of particular crops. Between 1810 and 1840, during the reign of Mohammed Ali, the dikes were heightened along the two branches of the Nile in the Delta (Rosetta and Damietta) and the canals that branched off them, which served to spread the flood waters, were deepened in order to be able to convey water at low flows (these canals were then called *seifi*). Water had to be lifted from the canals, so, by 1844, 52,800 *saquya* (water wheels) had been installed (de Sainte Marie 1989). However, this system proved to be too demanding in terms of labor and lifting power, prompting the solution to reduce the depth of the canals while raising water levels through the construction of cross-barrages. The construction of a first delta barrage began in 1833 but was not completed until 1862; several similar diversion barrages in Upper Egypt and on the Damietta branch were added later under British rule. This system

[1] One feddan = 0.42 ha, roughly one acre.

allowed the cultivated area to reach 1.3 Mfed by 1885. A barrage was constructed in Aswan in 1902 and a storage dam upstream of Khartoum in 1937 (de Sainte Marie 1989; MPWWR 1990), while steam pump stations were established in the first decade of the twentieth century. Considerable literature exists on the history of irrigation and water control (e.g. Linant de Bellefonds 1873; Barois 1911; Willcocks and Craig 1913; Hurst 1957) and the evolution of agrarian systems (e.g. Rivlin 1961; Richards 1982; Cuno 1992; Mikhai 2008; Ruf 1987), to which the reader can refer for further detail.

In his "Irrigation Policy" proposed in 1935, H. Sirry Bey (1935) unsurprisingly offered a picture of policy focused on issues of storage (raising the Aswan Dam, the planned Gebel Aulia Dam) and on the necessity of improving water supply to water-short or new areas through: the widening, deepening, opening, and connecting of irrigation canals, the conversion of *houd* in Upper Egypt to *seifi* (summer) irrigation, the building of pump stations and locks, and the strengthening or raising of river barrages. These measures aimed to make maximum use of the natural flow augmented by the releases from the Aswan and Sennar dams.[2] In addition, the expansion of cultivation was reasoned in terms of the available budget and labor force (expected to come from Upper Egypt) and improved drainage (redesigning the drains and installing 18 pump stations to drain land below the 2.5 m contour line). Likewise, the 1950 official report "Irrigation and drainage in Egypt" reviews the networks of main, branch, and distributary canals, headworks, regulators, siphons, culverts, and weirs, and discusses water requirements, irrigation rotations, and how to improve and extend water distribution. "Long-term storage" is shown to be necessary and is expected to be secured by dams in Sudan (Merowe) and upstream countries.

Despite successive raising works, the limited volume stored in Aswan Dam had long nurtured dreams of a high dam nearby. While options had been considered as early as 1946, it was after the 1952 revolution that Nasser rekindled the idea and preliminary studies were launched. When the UK, US, and World Bank employed loan conditionalities and tensions were raised following the nationalization of the Suez Canal in 1956, Egypt turned to the Soviet Union for assistance in building the dam (MPWWR 1990). The High Aswan Dam (HAD) was finally constructed between 1960 and 1968, with a capacity of 162 billion m^3 (Bm^3) (including 90 Bm^3 of live storage), completely eliminating the flood regime and allowing the wide-scale development of year-round cultivation. It promised to expand the irrigated area by 1.2 Mfed, converting the remaining basin irrigation (0.97 Mfed) to perennial irrigation and guaranteeing the irrigation of at least 0.7 Mfed of rice annually, as well as improving navigation, controlling floods, and generating hydropower (ibid.). The project's detractors pointed to the loss of natural fertilization through siltation, evaporation losses, the risk of malaria, and coastal erosion.

Various experiments were conducted in the 1960s to determine "water duties" and optimal rotations for the Valley and Delta, with a general system of 5 days

[2] The Sennar Dam is located in Sudan.

on/10 days off, a 7/7 system in summer, and a winter closure period in January. The 1975 water policy had a target of 16.76 Bm^3 of additional supply, mostly though the reuse of drainage water (12.2 Bm^3) and Upper Nile conservation projects, envisioning up to 2.5 Mfed of horizontal agricultural expansion. The 1980 policy reduced these figures, but updates in 1986 and 1990 aimed at an expansion of 1.6–1.7 Mfed with an "available water" volume of around 10 Bm^3 (Fahmy 1996).

Optimistic projections of available supply banked on the Jonglei Canal project in Sudan,[3] various rates of drainage water reuse and groundwater abstraction (in the Old Lands and oases), and greater efficiency. Yet they were "rather unrealistic and have not been based on solid scientific studies. Estimates given by experts are very subjective and are used to satisfy the mismatch between resources and requirements" (Fahmy 1996), without considering the interdependency between some of these measures. As noted by Hvidt (1995), despite the 1979–88 drought period causing Lake Nasser to receive 99 Bm^3 less water than expected and reducing the reservoir to an all-time low of 6.8 Bm^3 in 1988, the response still reflected a supply/managerial bias in line with entrenched planning practice. Indeed, the turn of the century would see the acme of an irrigation/water policy obsessed with expansion plans, including projects in the New Valley and targets reaching 3.4 Mfed.

Different dynamics were brought about by the liberalization of cropping patterns and prices in the late 1980s, the "counter-agrarian reform" unleashed by Mubarak in the early 1990s, and changes in water management due to the spread of individual pumps and technical problems faced by farmers, such as drainage. A shifting demand/supply ratio and pollution problems also began to materialize. Irrigation was becoming a more complex issue than the basic idea that had hitherto dominated water policy in Egypt: "that any excess water available should be devoted to the extension of the irrigated area" (El Quosy 2006).

9.3 Sustaining the Delta 'Irrigation Machine'

The Nile's water is diverted through eight river barrages to a 31,000 km network of public canals and 80,000 km of private *mesqas* (tertiary canals) regulated by more than 22,000 water control structures (MWRI 2002) that distribute water to plots, while 16,700 km of drains capture the return flow that is generated. Water distribution and drainage are aided by 583 large pump stations, not to mention the 3.6 million diesel pumps used by farmers. This 'machine' was not only very costly to set up but must constantly be maintained and improved in the face of particular challenges, such as local water shortages or soil salinization.

[3] This canal drains the Sudd wetland in Sudan to reduce 'losses' from evaporation.

9.3.1 Draining the Lands

It is often mistakenly believed that drainage problems arose with the perennial irrigation allowed by the HAD. Yet drainage problems were first recognized when the original Aswan Dam (1870) allowed some land to be irrigated in summer, meaning the year-round application of water. This prompted the opening of the first drains but not enough to prevent the collapse of the cotton crop in 1909 (van Achthoven et al. 2004). In the 1920s, H. Sirry Bey noted that, "the most convincing evidence of this [problem] is the deterioration noticed in the lands of Menufia and Kaliubia Governorates, which, up to the time of the Great War, were the richest lands of Egypt." In the 1930s, massive investment in pump stations allowed water to be drained from the lower half of the Delta where it could not be drained by gravity ("free flow"). This drainage "service" was provided to land owners at no cost. In 1949 drainage was made the responsibility of the Ministry of Public Works and the area serviced by open drains increased from 2.2 Mfed in 1952 to 6.9 Mfed in 1968, with a network of 17,000 km of drains (van Achthoven et al. 2004).

With the development of year-round cultivation, water tables rose across the Delta. Resulting waterlogging and salinity problems were initially neglected because of the lack of funds to cover the costs of secondary drainage system. By the mid-1970s, some 80% of the most productive land in the country had been affected (Richards 1980) and the issues were being addressed by a major World Bank-funded program of investment in drainage launched in 1970. The Egyptian Public Authority for Drainage Projects (EPADP) was established in 1973 (Abdel-Dayem et al. 2007; Ritzema 2009). Surface drains were excavated or dredged in order to maintain a water table depth of 1.5 m and an ambitious project was launched to provide all cultivated land with subsurface drainage (i.e. a network of pipes and collectors buried at a depth of 1.5 m) by the year 2012. Farmers were to pay 50% of the capital costs over 20 years, with a grace period of 4 years. Maintenance (un-clogging pipes by injecting pressurized air) is theoretically available for free at the farmers' request and lasts 25 years. In 1990, however, further costs were induced when it was necessary to replace concrete pipes with PVC/PE. In 2007, around 85% of the cultivated area had been equipped, at an actualized value of around US$1.5 billion. In 2017, EPADP had installed collectors over 6 Mfed, rehabilitated another 1.9 million (www.epadp.org.eg), with around half a million kilometers of draining pipes buried in the soil.

Experiments have shown that subsurface drainage has a substantial impact on yields of major crops, such as cotton and wheat, with increases ranging between 10% and 30% (Abdel-Dayem et al. 2007). Although well appreciated by farmers, the technology has its drawbacks with regard to the cultivation of rice, as it increases the water demand by around 30–40%. This problem has prompted the proposal of the so-called "controlled drainage" technique, where farmers may obstruct their subsurface collectors (that are accessible through manholes) in order to better retain water in their soil profile, but it proved hard to implement.

Draining the land also requires that secondary and outfall drains are free flowing until they join the sea. Due to the very flat topography of the Delta, it is necessary to close its various outlets to the coastal lakes or the sea and to artificially lower the water level in the drainage systems by pumping water over these closure points (maintaining a typical difference in level of 2.5 m). This battery of around 40 large-scale pump stations currently discharges an average annual volume of 13 Bm^3 to coastal lagoons and the sea (Molle et al. 2018).

9.3.2 The Cost of Maintaining the Machine

Following the construction of the HAD, the function of the canals changed from spreading the flood to supplying irrigation water to plots. The clear water released by the dam and the increased use of fertilizers promoted weed growth in irrigation and drainage channels. As a result, the focus of maintenance needs shifted from desilting channels to the control of aquatic weeds (MoI and USAID 1986).

In 1983, the annual cost of channel maintenance was 57 million Egyptian pounds (US$104 million at 2016 value) (World Bank 1986). With a World Bank loan injecting US$130 million over 8 years, the project aimed to introduce "modern channel maintenance practices," replacing the traditional system of desilting by excavation with a mix of weed mowing, herbicide treatment, and desilting. It built on similar projects developed since 1978 with Dutch assistance (biological weed control, training of EPAD staff), US$140 million from various USAID projects, and a CIDA-funded project. Between 1973 and 1999, the World Bank funded seven drainage projects, as well as Channel Maintenance Projects and a National Drainage Project. USAID also funded projects (in the 1990s, for example, the IMS project included "structural replacement" that built or upgraded 20,000 structures.

The "irrigation machine" also includes close to 600 large-scale pumping stations that deliver water to higher New Lands and drainage water into canals, or out to the sea and coastal lakes. Their construction and maintenance have been funded by various World Bank projects, most notably three successive "Irrigation Pumping Station Rehabilitation projects" (1983–1992/US$42M; 1993–1998/US$31M; 1999–2007/US$194M with KfW). The equipment is the responsibility of the Mechanical and Electrical Department. According to the World Bank (1999), the loans, "reduce the operation and maintenance costs of the pumping stations with related beneficial effects on the Government's recurrent budget." Irrigation totally depends on these pumping stations, which explains the attention and outlay continuously devoted to them. Achthoven et al. (2004) estimated that over the years US$3 billion has been spent on subsurface drains, remodeling and deepening open drains, and constructing and rehabilitating drainage pump stations, but a full accounting in present value has yet to be done.

9.3.3 Conjunctive Use of Water Against Shortages

The establishment of a system of year-round irrigation over 2.5 million ha in the 1970s was not achieved without problems. Due to the complexity and ramified nature of the hydraulic network in the Delta, certain areas faced water shortages. The simplest response to the insufficient discharge in some canals appeared to be the re-injection of water from the main drains into the main canals. A total of 28 pump stations were therefore planned and installed with the objective of supplementing the main canals with 12.2 Bm^3 of drainage water. Unfortunately, in the early 1990s, several of these stations had to be abandoned (some were never even used) because the low quality of untreated, or partially treated, municipal and industrial waste water that entered the drains was unfit for irrigation. By 2005 (APP 2005), only half of the 29 reuse pump stations that had been installed were functioning well or with only minor problems. Water reuse therefore remained limited at around 4 Bm^3. The plan to obviate the pollution problem involved "intermediate reuse" pump stations, as they came to be called. These would abstract drainage water from secondary drains and re-inject it into secondary canals before it joined the polluted main drains and the quality became degraded. In the 2010s, a large number of secondary drains were equipped with such reuse pump stations. The amount of drainage water reused in irrigation in 1996 was about 4.4 Bm^3, growing to 6.4 Bm^3 in 2014 (DRI 2015) with a target of 8.4 Bm^3 in 2017, according to the 2005 National Water Resource Plan (NWRP). The new NWRP (2017) estimates total drainage water reuse to have been 9.31 Bm^3 in 2015 and sets a "most probable" target of 16.26 Bm^3 by 2037, including 6.14 Bm^3 of treated wastewater.

However, this undeniably sound and effective way of increasing the available water for irrigation is reserved for the ministry and officially prohibited to farmers. Yet there is no attempt to enforce this probably obsolete regulation, as the ubiquitous individual pumps set up along the drains of the Delta have been, and still are, the chief response to water shortages. Elsewhere, topography allowing, irrigation canals are supplied by direct gravity diversion of drains (e.g. the Nashart drain in Kafr el Sheikh governorate). In the command area of the Meet Yazid Canal, for example, which serves an area of 85,000 ha in the central part of the Delta, a survey has found around 2500 pumps along main and secondary drains of the northern half (Molle et al. 2015). Intermediate reuse pump stations and individual pumps have each been found to provide (at times of peak demand) an additional supply equivalent to around 10% of the canal's maximum flow (IWMI and WMRI 2013).

Over the past 15 years, this combined use of irrigation and drainage water has coincided with a dramatic increase in the number of farmers resorting to groundwater to offset local shortages. Approximately 60% of the Delta is affected by this mushrooming of shallow wells tapping the superficial aquifer (that is replenished by seepage from waterways and irrigated fields), the salinity of groundwater being too high in the northern fringe of the Delta. As a response to shortages, individuals or groups of farmers generally invest in wells, and these have reached densities of over 1 well for 2 ha (El-Agha et al. 2017). This gradual, yet intense, development of conjunctive use is highly typical of irrigation systems experiencing growing water scarcity. However, the consequence of this process is increasing quantities of salts being transferred to agricultural plots through the use of saltier water from both the drains and the aquifer (in order of magnitude 1 dS/m and 0.7–2.5 dS/m respectively, against 0.42 dS/m for Nile water). This increases leaching requirements and can impact yield.

The mixing of fresh and drainage water is not only used to respond to local shortages. It has also been taken as the basis for expanding irrigation on the two sides of the Delta. The Salam Canal to the east receives water from the Damietta Branch of the Nile as well as from two main drains intersected on its way to the new lands developed in the Sinai. The intended 50/50 mix has not been achieved and only a third of fresh water was supplied for two thirds of drainage water (Mohamed 2013), keeping the salinity of irrigation water at around 1 g/l. To the west, the Umoum drain, which flows towards Alexandria, is also equipped with pump stations designed to transfer water to the Nubaria Canal that supplies the New Lands on the western edge of the Delta. But water salinity and pollution problems have so far worked against the full utilization of these schemes. In 1992–2000, more than US$3 billion was spent constructing wastewater treatment facilities. The cost of increasing the treatment capacity by 2 Mm^3/day by 2007 was estimated at US$2 billion (Bazaraa 2002).

9.4 Land Expansion Strategies

9.4.1 The Old 'New Lands'

The idea of expanding cultivation in the desert on a wide scale was first proposed in 1952 by Magdi Hassanein, as the Ministry of Agrarian Reform and Land Reclamation was being created alongside that of agriculture (Springborg 1979). In 1954, Hassanein was placed at the helm of the newly created Tahrir Province Organization and started advocating the idea of establishing a model collective farm, in contrast with the model of private ownership of small plots promoted by the Ministry of Agriculture and Irrigation and the Higher Committee for Agrarian Reform. Hassanein's vision was that, "such reclamation undertaking implied that land should remain under public property, and that the state should not divest it. On these new lands, Egyptian socialism will firm its victory through the possession by its people of the whole of this land" (in Gumuchian 1975). New lands were "not merely an agricultural project […] but an entire social experiment, the nucleus and vanguard of Egypt's new rural society" (Voll 1980). Newly reclaimed lands were divided into sectors of 50,000 fed on average, which were further divided into plots farmed by individuals under the supervision of cooperatives (Gumuchian 1975).

Socialist ideology was beefed up in 1964, as Khrushchev visited Egypt and announced that the Soviet Union would assist in the reclamation of 10,000 fed in the northern part of Tahrir Province (Fig. 9.1) and the establishment of a model State

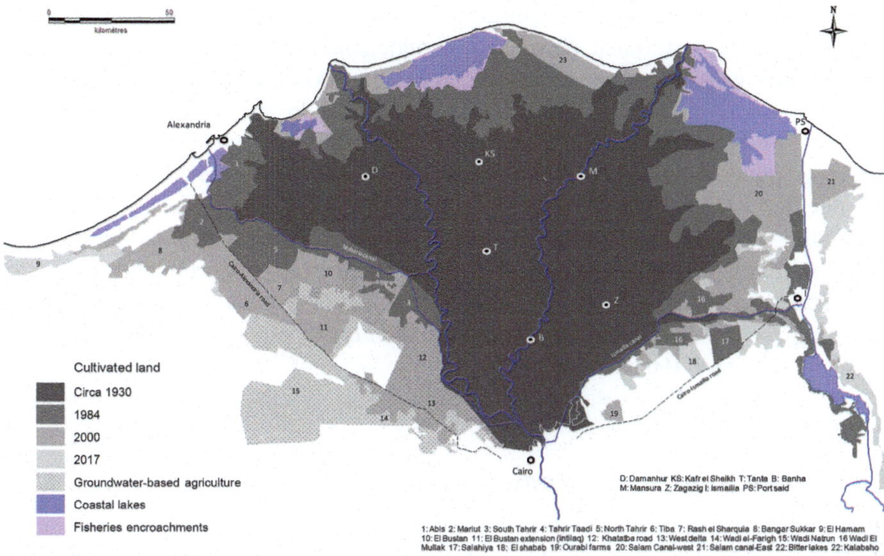

Fig. 9.1 Historical expansion of irrigated agriculture in the Nile Delta. Notes: This map shows gross reclaimed areas. It does not show (expanding) urban, or small non-agricultural areas, such as airports, army camps, residual sand dunes, *kom* (mounds), etc. In some cases, the area indicated as fully reclaimed may only be partially cultivated (as in the case of El Hammam or Kalabsho). We have not shown areas under conjunctive use of surface and groundwater because this now virtually includes the whole Delta with the exception of its northern fringe. The agricultural area around Cairo has decreased: the map shows the extension in 1984 (not 1930) that covers the actual limit, which is indicated by a solid white line. (Sources: Google Earth, Egyptian Survey maps)

Farm (Springborg 1979). State policy on land reclamation and associated land tenure patterns remained a hotly debated issue until the 1967 war, when Nasser himself attempted to satisfy all sides by stating that reclaimed land could be distributed to peasants, rented to individuals or companies, or farmed directly by the state, based on rational cost accounting and the specificity of each case. Reclamation had become a real bonanza for private construction companies as well as for "branches of the civil service and public sector with responsibility for reclamation and utilization of reclaimed land [… which] were liberally stocked with officers" (Springborg 1979) and whose private interests had become associated with the continuation of reclamation and state control over land.

In the 1960s, the Tahrir Province area and the Mariut sector were developed by the state, but settlement of small farmers also took place in Abis (south-east of Alexandria), as part of the Egyptian-American Rural Improvement Service Project (1952–1963) (Fig. 9.1). The bulk of the land developed during the 1960s was reclaimed by state companies, although some work was also done through private companies (Hopkins et al. 1988).

During the 1970s, land reclamation came to a halt (Meyer 1998) and political debates and decisions centered on the reorganization of the apparatus managing reclaimed land and on the redistribution of land. Many reforms were prompted by the disruption and financial difficulties associated with the 1973 war. Most of the new lands were partitioned between four independent companies established in 1973, which were given increased management autonomy, with an independent budget, in order to counteract excessive centralization and stifling bureaucracy. Peasants, however, remained integrated in the system of cooperatives, which set cropping patterns, provided seeds and equipment, and marketed production (Gumuchian 1975; Hopkins et al. 1988). In 1975, a decree organized the transfer of ownership of public land to the tenants who were farming it (Hanna and Osman 1995). A year later, a law was passed that widened the range of recipients of land to include new agricultural graduates, and, soon after, all graduates, as a way of compensating for the state's failure to keep its promise of providing every university graduate with a job (Hanna and Osman 1995).

9.4.2 *The New 'New Lands'*

In 1978, land reclamation was back on the agenda, with a target of 578,000 fed (242,000 ha). Most of this was to be to the west of the Delta (where new lands already totaled 200,000 fed in 1980: IFAD 1980) and supplied by the Nasr Canal, itself sourcing water from the Nubaria Canal (van Achthoven et al. 2004). In 1980, however, the World Bank (1980) stated that the performance of the reclaimed land was "sobering". It stressed the urgency of improving the selection and design of new land development projects. "Of the existing reclaimed land, less than 60 percent [was] under cultivation and possibly as little as 35 percent [was] being cultivated with profit. The principal difficulty had not been the unsuitability of the soils for crop production, rather it has been the deficiencies in technical design, such as inadequate drainage, and the 'company' approach to managing reclaimed lands." The Bank proposed a project that would introduce institutional change while making improvements to technical design, the small-farm settlement pattern, and agricultural services.

In 1981/1982, reclamation works resumed with the Salhiya Project (~about 23,000 fed) (Meyer 1998) and the El Shabab Project (about 33,500 fed) on the eastern side of the Delta (Fig. 9.1). On the western side, the World Bank funded a project to extend the Nasr Canal and expand cultivation in the area, which came to be known as Bangar Sukar. It was to be irrigated by gravity (basin), with the usual optimistic expectations that, "with good land leveling, training of farmers and adequate supervision, it is expected that the overall irrigation efficiency will be satisfactory" (World Bank 1980).

It was originally conceived that the settlers of this land would be experienced farmers with families. However, given Egypt's swelling population of unemployed graduates it was decided that they too could be allotted land. In 1978, 3000 fed were distributed to settlers and 5500 fed to graduates in Nahda state farm, near Alexandria—a process that extended to South Tahrir (Tahadi area) (Corey 1978). In the mid-1990s, a third category of settler emerged: tenants who had been evicted under land reform, *mutadarireen* (more on this later), while some land was also earmarked for the "Mubarak graduate villages" (Adriansen 2009).

However attractive it may have been, the idea of reclaiming desert land encountered many technical and social difficulties that were not well anticipated. A severe problem experienced in many locations was that of waterlogging: large quantities of water applied to the soil resulted in perched aquifers, which in turn led to soil salinization and a subsequent rise in the water table that uplifted and destroyed the concrete linings of the main canals (IFAD 1980). Other problems included wind, which covered crops with sand and was combated through planted rows of casuarina trees, sandy soil with low fertility and limited water-bearing capacity, and the presence of undesirable minerals, such as calcium carbonates, gypsum, boron, and selenium (Hanna and Osman 1995). Technical difficulties included dysfunctional collective pumping stations and pipe networks serving 400–800 fed (Corey 1978), a costly and intermittent electricity supply, pump station breakdowns, maintenance difficulties, and a lack of qualified technicians (DDC 1997). Most collective pump stations broke down and were replaced either by smaller stations (for four or eight farmers) or individual pumps. These were funded by special projects (IFAD, Abu Dhabi Fund, etc.) or the farmers themselves.

Social difficulties also bedeviled the settlements due to transportation difficulties, a lack of infrastructure, medical services and schools (Hanna and Osman 1995; Hopkins et al. 1988; Ibrahim and Ibrahim 2003), a lack of capital, credit, extension services, and farm machinery (Corey 1978), and abuses of power by the cooperatives' officially nominated board members (Hopkins et al. 1988). Faced with a substantial rate of graduates abandoning their land (or renting it out to other farmers, more or less illegally), and in order to ease the land market, in 2002, the parliament passed an "Investment Law" allowing graduates to sell their land after a transition period of 10 years (Adriansen 2009). In 2003, according to Adriansen's (2009) calculations, 343,000 fed of land had been granted to a total of 69,962 beneficiaries, 64% of whom were graduates. This should not occult the fact that 60% of the newly developed land was sold off or auctioned to private investors in parcels ranging from 200 to 50,000 fed (van Achthoven et al. 2004).

Neither can it be ignored that the history of the allocation of new land was rife with stories of corruption—high-ranking military or police officers, officials, or large companies receiving land for nominal sums. Many urbanites were able to buy cheap land, for example in Wadi Natrun or along the desert road, on condition that it would be used for farming or industrial purposes, and then develop real estate or resell it (Ibrahim and Ibrahim 2003). Despite the setbacks and difficulties, New Lands –to the extent that they are supplied with water- have progressively established themselves as relatively productive areas, with a gradual shift towards drip-irrigated fruit tree crops (Alary et al. 2017).

Since the turn of this century, land reclamation has been mainly based on the use of groundwater by investors on the western and eastern fringes of the Delta (Fig. 9.1). The best known area for corporate agricultural development is the so-called West Delta, located on the outskirts of Cairo, on both sides of the Cairo-Alexandria road. But these investors have now largely moved to other areas such as Wadi Natroun, Wadi el-Farigh but also East Oweinat et West Minyah (Acloque 2018).

9.4.3 The New Valley, Toshka, and Recent Expansion Plans

The chain of oases—Kharga, Dakhla, Farafra, and Baharia—to the west of the Nile Valley forms a line often referred to as the "New Valley." In 1958, Nasser aroused people's imaginations by proclaiming its potential for 3 Mfed of irrigated land. Thirty years later, the state had developed a few hundred deep wells and allocated 19,000 ha of land (only half of which was actually irrigated). The administration, and its new laws regarding the control of land and water, had largely displaced pre-existing community water and labor systems, creating a new peasantry dependant on the state. Traditional wells had dried up because of falling water tables and salinity built up for lack of drainage facilities (Faggi and Maury 1987).

The "Ground Water Pilot Scheme, New Valley, 1972" prepared by the FAO assessed the cultivation potential to be 400,000 fed for 200 years (Abou-Korin 2002). But the situation at the turn of the century was disappointing, as all wells had lost their artesian characteristic and 30% of the land that was cultivated in 1965 had been abandoned due to the shortage of water (ibid.). As a result, the area currently cultivated in the oases does not exceed 120,000 fed (Sims 2015).

In 1997, the oases were grouped together with Toshka and East Oweinat in the "South Valley Development Project." This was backed by the first version of a new 20-year NWRP ending in 2017 that envisioned the reclamation and irrigation of 3.4 Mfed, with 1.4 Mfed in the South Valley Project. Located to the west of Lake Nasser and believed to be a former course of the Nile River, the Toshka depression is also the spillway of the HAD and received large amounts of excess water in October 1996, rekindling old projects to divert water to the "New Valley." The project announced by Mubarak in 1997 consisted of a huge pump station that would abstract up to 5 Bm^3/year from the HAD and convey water through a 70 km-long canal to 540,000 fed of land to be reclaimed in the area (Wahish 1998). Toshka was presented as the "project of the new millennium," the "building of a new society" and a "new civilization." Accompanied by much fanfare and political propaganda, it was closely associated with President Mubarak's cult of personality (Deputy 2011).

Investors in the South Valley Development Project received a 20-year tax exemption and were exempted from import tariffs on capital equipment and machinery. They were able to buy the land for LE50/fed (~US$30/ha at current value) when the price of a feddan in the Delta could fetch over LE20,000 (Wahish 1998). Conspicuous among the (foreign) investors was Prince Al-Waleed Bin Talal, who acquired 100,000 fed of land. Diversely seen as a development plan destined to thwart extremism, as claimed by Mubarak, a presidential project under the army's control and to its economic benefit (Warner 2013), or a strategy to claim prior use of Nile water and strengthen its position at the basin level (Waterbury and Whittington 1998), the project faced many difficulties and never developed as planned. Although internal dissenting opinions were initially silenced, the lack of a feasibility study or cost-benefit analysis nurtured critiques that surfaced after Prime Minister Kamal el-Ganzouri, the project's key supporter, was replaced in 1999 and the project faced its first obstacles on the ground (Al-Ahram Weekly 2000; Sowers 2003). The project was gradually shelved. With the weakening of the regime, and in the aftermath of its fall in 2011, sharp criticism appeared freely in the media, asserting that the deal struck with Bin Talal had been "illegal" and "Mubarak's pyramids" a "mega-failure" (Egypt Independent 2014). Once in power, however, the Muslim Brotherhood—long-time critics of the scheme—unveiled a plan to settle one million people in Toshka.

Other flagship projects of the Mubarak era include East Oweinat (750,000 fed, close to the Sudanese frontier, which mainly seems to have benefited the army) and the Northern Sinai Agricultural Development Project (NSADP), which aimed to resettle 750,000 Egyptians on 92,000 ha of irrigated land west of Suez and 168,000 ha in the Sinai proper. The former area had been largely reclaimed from marshy land (and parts are now home to aquaculture), while the latter has been subject to very high salinity levels that could not be successfully leached. As a result, only part of the area is used, also for aquaculture. Other more distant areas into the Sinai have remained undeveloped for lack of water.

In a repetition of history, in 2014, President Sisi announced that the government was planning to reclaim 4 Mfed of desert land, a target that was soon reduced to a "first phase" dubbed the "1.5 million fed project" (Al Ahram weekly 2015). Ninety percent of the area, distributed over 12 governorates, would be irrigated with water from 3500 deep wells. Unsurprisingly, the project has so far been mired in issues including the availability of water, insufficient logistical capacity, the lack of a feasibility study, a lack of transparency, uncertainty over the type of crops, and encroachment on the targeted land.

Verhoeven (2015) suggests that the logic of land expansion, whether cloaked in Malthusian narratives or nexus positivist thinking, primarily serves to sustain the clientelist practices that are at the root of state building and reproduction, notably by providing opportunities for capital accumulation in land reclamation and agricultural development (Dixon 2013; Roccu 2013). Springborg (1979), Sadowski (1991), Sowers (2003), and Dixon (2013), among others, have described the alliances between business interests and the government and its bureaucracies. While land expansion has overall been very costly, politically oriented, and has attracted much criticism, New Lands have come to represent a substantial proportion of Egyptian agriculture—one that grows as Old Lands are gradually urbanized. According to the MWRI (2013), the total expanse of New Land, complete with infrastructure, was 2.27 Mfed in 2012, of which only 1.49 Mfed were cultivated (0.6 Mfed with groundwater), while 0.24 Mfed were said to be "in progress." But the reports also point to the "uncontrolled horizontal expansion" in the country. Estimated at 0.36 Mfed in 2010 (ibid.), this is irrigated by private initiative, mostly based on groundwater.

9.5 Irrigated Agriculture and the Policy Context

9.5.1 Farming in the Delta and the Valley

Egypt's 3.78 million ha of irrigated land can be roughly broken down as 60% (2.25 million ha) Old Lands and 40% (1.53 million ha) New Lands (CAPMAS 2018). Although 37 million people live in cities, 55 million still reside in rural areas (Faostat 2015). While the agricultural sector only accounts for 11% of the national GDP, its contribution to employment is 29% (Tellioglu and Konandreas 2017).

The main crop is wheat, with around 1.4 million ha planted each year—a key commodity with regard to family- and national-level food security—followed by maize, berseem (clover), vegetables, and rice. Sugar cane, sugar beet, and cotton form a second group, each covering an area of around 200,000 ha (Fig. 9.2). While the production of rice, wheat, vegetables, and fruit trees (in the New Lands) has been on the increase over the past 30 years, other crops, such as cotton and berseem have declined. Substitution within cropping patterns largely reflects the market and state subsidies, but the sheer dominance of wheat, rice, maize, and berseem gives a clear sign of the need to feed families and farm animals.

In 2000, of a total of 3.21 million farms, 68% were under 2 fed in size (22% of the total area) and 91% less than 5 fed (50.6% of the total area). Farms over 10 fed represented less than 3% of the total number but a third in terms of total area (Ayeb 2010). This crudely shows a skewed distribution of land, although very large farms run by investors are only found in New Lands.

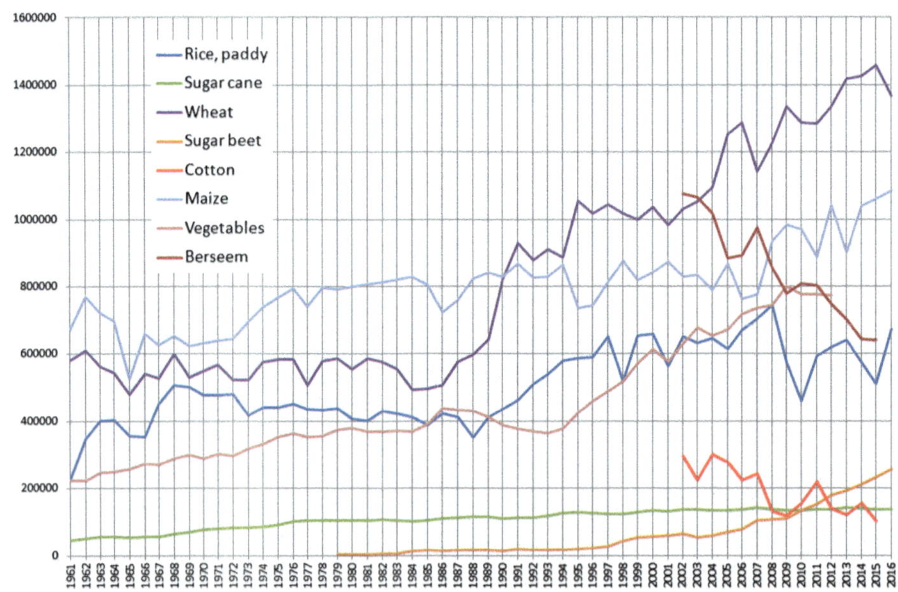

Fig. 9.2 Evolution of main irrigated crop acreages in Egypt, in ha. (Source: faostat & capmas)

Labor in irrigated agriculture has been found to be dependent on macro-level factors, such as the government's absorption of labor and outmigration, in particular to the Gulf countries. This saw a hike following the inflation of the 1970s and a drop after the Iraqi wars and had a big impact on real wages (Meyer 1996). Shrinking land and a growing rural population have meant there is sufficient labor, although the cost of paying for it has led farmers to adopt water-saving techniques, such as the direct seeding of rice.

9.5.2 Market Liberalization

Agricultural policy in Egypt was heavily influenced by American aid through the USAID-funded Agricultural Production and Credit Project (APCP) (1987–1995) and later the Agricultural Policy Reform Program (APRP) (1996–2002), as well as the World Bank (Agricultural Development Projects and Agricultural Modernization Project) during the same periods. In the 1980s and early 1990s, watchwords revolved around privatization and reform of the public sector (with the creation of "holding companies"), and the liberalization of agricultural markets and cropping patterns (El-Fellaly and Saleh 2004). In the mid-1990s, priorities shifted towards private investment and privatization in agribusiness, agricultural sector support services, and land- and water-resource investment and utilization.

Reforms of the agricultural sector in 1986 included the complete liberalization of producer prices for all products except cotton and sugar cane, the elimination of subsidies for input and credit, promotion of the private sector in processing and marketing, and the removal of barriers to imports and exports (Meyer 1996). In 1992, 1 year after Mubarak had agreed to an IMF and World Bank Economic Reform and Structural Adjustment Program (ERSAP), a crucial law removed constraints on land rental and terminated all lease contracts by 1997.

There is no consensus on what these reforms really produced. International Financial Institutions (IFIs) and the state boast huge improvements in productivity and diversification, while other observers question the statistics and point to evictions of poor, landless tenants (Bush 2007). Oddly, no comprehensive or convincing study of agrarian change has ever been conducted. Opposition to the land tenure reform generated conflict, but it did not spread widely, either from fear of violent repression or because the effects of reform were weathered out.

IFIs ceaselessly lament the persistent under-performance of agriculture and its poor management and structural weaknesses. Inevitably they recommend further liberalization of Egypt's trade regime—a call backed by the US and EU, who, at the same time, lavishly subsidize agriculture. What seems less controversial, however, is that the obsessive promotion of capital-intensive, export-oriented agriculture in the New Lands by both the government and its foreign supports favors capital and does very little to provide jobs or address issues of land and market access in the Old Lands, not to mention food security (Bush 2007). The official discourse, and that of the IFIs, has promoted the idea that "Privatization has become an urgent necessity in water management projects in Egypt […] Many experts view privatization of irrigation projects as a step that will lead to rationalizing the use of water and relieve the government of the huge costs of giant irrigation projects required in the coming few years" (OOSKAnews 2013).

Export-oriented and capital-intensive agriculture has tapped markets with high certification standards (supermarkets and export), and donors, notably USAID, have entertained the hope that small producers could benefit from such linkages through contract farming in different guises. Yet there are few emerging success stories beyond the time span of specific projects, and investors are instead experimenting with different forms of externalization of risk and cost by providing small farmers with training and/or land and water to produce crops, such as potato and strawberry, for their export market (Acloque 2017).

9.5.3 Conflicting Agricultural and Irrigation Policies

Egypt presents a typical case of an institutional setting where the responsibility for irrigation is divided between a Ministry of Agriculture and Land Reform (MALR) and a Ministry of Water Resources and Irrigation (MWRI). These administrations have different mandates and objectives that often translate into conflicting policies. The duty of the MALR is to expand agriculture and enhance its economic and social value, while the water administration tends to be more keenly aware of the limited availability of the resource, since overexploitation generates water shortages and management difficulties. A major policy fault line between the ministries is an understanding of how far it is possible to improve irrigation efficiency.

In its "Sustainable agricultural development strategy towards 2030" (ARDC 2009), the MALR proposes "to achieve a gradual improvement of the efficiency of irrigation systems to reach 80% in an area of 8 Mfed, and to reduce the areas planted with rice from 1.673 Mfed (2007) to 1.3 Mfed by 2030, in order to save an [overall] estimated 12.4 Bm^3 of water." "Savings" would in turn allow the expansion of irrigation over 3.1 Mfed, with a cropping intensity estimated at 199%, and an increase in economic water productivity of 119% by 2030. This proposal is dismissed by most water decision-makers in private but rarely openly criticized in the media, revealing the perception of the power balance between the two ministries. Indeed, water savings of 12.4 Bm^3 would almost exactly match the quantity of water pumped out to the sea, which is absurd considering the importance of this drainage for the removal of salts and pollutants from the Delta. Such debates, sometime framed with disconcerting arguments that partly reflect the competition for budgets and megaprojects.

Many other tensions exist between the two ministries. Issues include their roles, remits, and budgets in the reclamation of new lands and who should be responsible for farm-level pressurized distribution systems in the framework of Irrigation Improvement projects (see next section) between the MALR's cooperatives and Branch Canal Water User Associations. Turf battles extend to the question of data, where assessments of rice areas differ, and even to crop requirements: a joint working group has met for years without settling the issue because the two ministries have vested interests in establishing either low or high figures.

9.6 Current Water Issues and Challenges for the Irrigation Sector

9.6.1 Irrigation Modernization Policies: Technology to the Rescue?

Almost all irrigated land in Egypt (the few exceptions include Fayoum) uses machinery to lift water from waterways to the level of the fields. In ancient times, this was done by *shadoof*, *tanbur* (Archimedean screw) and the animal-powered *saquia* (water wheel). At the beginning of the eighteenth century, the saquia was enhanced so that the scooped water was collected and distributed at the level of the axis. In the early twentieth century, large-scale diesel pumps were introduced, allowing richer landowners to irrigate at higher elevations. In other locations, small farmers would form and share saquia rings, where water lifting and irrigation were determined by the location and capacity of the saquia. The advent of the HAD and the generalization of year-round agriculture generated the need for increased pumping capacity, and this was met by small diesel pumps (initially Indian models), which started to replace saquias in the 1980s. While initially these were shared (rented or borrowed), as can still be found in Upper Egypt, all farmers in the Delta now have one or more individual (fixed or mobile) diesel pumps. This has dramatically increased farmers' capacity to abstract water, and allows them to pump from any convenient location along canals or drains, although it has undermined the social arrangements of the saquia rings.

As a result of the ramified and inter-connected nature of the distribution network in the Delta, the typical 5 days on/5 days off summer rotation in secondary canals is hard to ensure. Unpredictability in the timing and quantity of available water results not only in a mismatch between supply and demand but also in risk-minimizing strategies on the part of the farmers, who over-irrigate as a means of storing water in the soil profile. In the 1970s, such problems of low efficiency and equity were analyzed by the Egypt Water Use and Management Project (EWUP 1977–1984) coordinated by the MWRI and researchers from Colorado State University. The two major and complementary recommendations were (1) to ensure *continuous flow* in secondary canals (through downstream control automatic gates and, later, baffle distributors at the head); (2) to establish *collective* pump stations at the tertiary (*mesqa*) level, serving buried piped distribution networks to deliver water to a number of valves and replace diffuse *individual* pumping from multiple points (canals and drains).

The IIP (Irrigation Improvement Project) was launched in 1984 with the support of USAID. It became a fully fledged program in 1989, and was expanded by the World Bank in 1995. The project was praised for having spearheaded the "modernization" of irrigation in Egypt; "the first step to bring the Egyptian irrigation system in line with the functional demands it will be facing by the turn of the 21st century" (Hvidt 1998). It was granted the status of a national project, with a dedicated "sector" within the ministry, and the long-term objective of being expanded nationwide. In 2006, it was renamed, and remains, the IIIMP (Integrated Irrigation Improvement and Management Project), with improved design criteria, including the use of electric pumps and the possibility of adding quaternary distribution pipes and farm-level hydrants (OFIDO and FIMP projects, implemented by the MALR). By 2007, the project had installed 2900 collective stations over an area of 200,000 fed (World Bank 2007), to which must be added 36,000 fed by IIP2 (funded by KfW), 67,000 fed improved by USAID before 1996, 66,000 fed improved by IIIMP until 2015, as well as areas improved with the national budget (around 50% of the cost is charged to farmers over 20 years), totaling around 0.4 Mfed. Yet Fahmy (1996) points out that the 1986 policy estimated that the IIP would cover around 2.5 Mfed, and that 2 Bm^3 of water would have been saved by the year 2000.

However, the implementation of the project has been bedeviled by a number of persistent problems. These include delays, high staff turnover and losses of trained personnel, inadequate career opportunities and salaries to attract new engineers (IRG 1998), escalating costs, faulty design and/or execution (e.g. pump inlet positioned too high, reversed canal slope) that "seriously undermined farmer confidence in the IIP and its abilities" (IRG 1998), difficulties faced by farmers in finding spare parts or the expertise to address technical problems.

The main shortcoming of the IIP projects has been the failure to establish continuous flow, "the key and lead technology of IIP" (IRG et al. 1998) and the main selling point to farmers. As a result, neither the pattern of water distribution to each branch canal nor the overall efficiency of the system has seen an improvement. Where mesqas have been filled in, farmers are now fully dependent on the collective pump station, but those located near the canal or drain have generally kept their individual pumps to abstract extra water. It is hard to make a definitive assessment of the IIP projects (see Molle et al. 2015): in the Meet Yazid area, 18% were found to be abandoned (either because the engine had been stolen or the group had been unable to manage conflicts or a technical breakdown), one third had shifted to electric engines, and many types of adjustment and transformation were observed. Farmers were found to be handling collective action in various informal ways and far removed from the formal, organizational mode that the project had originally imposed.

All in all, the farmers who see benefits in terms of reduced drudgery, equity in distribution, or pumping costs tend to be well situated in the upstream part of the branch canals. Environmental and social heterogeneity is very high and explains the contrasting situations observed. This, along with the fact that the 40% share of the investment to be paid by farmers was far from being fully recovered, suggests that investing in IIP programs is not a priority for the country and should be conducted on an 'on-demand' rather than subsidized basis (Molle et al. 2015). Tellingly, the new 2017 NWRP barely refers to IIP; nor does it set targets or discuss the future of the sector.

9.6.2 *From the System to the Farm: Elusive Efficiency Gains*

The question of whether improvement projects "save water" has been discussed since the start of the projects. While early studies claimed a reduction in the amount of water applied, years of monitoring and evaluation efforts were never able to substantiate this (Molle et al. 2015). In 1998, Merrey (1998) noted that "IIP has many other benefits, but the expectation that substantial water savings will occur is not realistic; and continued expansion of irrigation on the assumption that such savings will occur will lead to induced shortages in the Nile Valley [and Delta]."

There are several reasons for the failure to save water. First, continuous flow has not been achieved, meaning farmers still fully abstract whatever amount of water is delivered to their branch canal. Second, in the absence of a strict rotation within the branch canal, upstream farmers continue to abstract water disproportionately (in part as a protective strategy as supply remains unpredictable). Third, it is necessary to distinguish between applied and consumed volumes: where collective networks allow a more even and better distribution at the tertiary level, evapotranspiration, as well as yields, is likely to be enhanced. This would beneficially increase the depleted fraction and improve efficiency, but this small effect is impossible to evidence in a context where each season differs in terms of cropping patterns and available supply. Additionally, the corresponding reduction in return flows may affect the downstream farmers, who use drainage water, as well as increasing the salt concentration.

Regardless of the possible but unlikely effects of improvement projects on water use efficiency, the MALR, as part of its strategy for 2030 (ARDC 2009), is seeking to shift the average plot-level water efficiency from 50% to 80%. Aside from the huge investment required—which would exceed most farmers' financial capacities—and the fact that the technology is ill-suited to certain major crops, such as rice and berseem, this simplistic reasoning ignores the now well-established fact that plot-level efficiency cannot be extrapolated to the entire system. Ironically the Nile Delta was one of the key cases used in the 1990s to show that systems with substantial reuse of water require careful examination (Seckler 1996). The MARL's claim that savings could reach 12.4 Bm^3 and be used for land expansion (ARDC 2009) is flawed, as it fails to understand the macro-level water balance of the Delta and the importance of the drainage outflow for (1) keeping water levels in the drains of the northern part of the Delta low enough to allow cultivation and (2) flush out the salt load and pollutants. It also neglects the importance of growing rice or farming fish at least a year out of two to maintain an acceptable level of salinity in the northern fringe of the Delta, which is subjected to upward seepage of groundwater (this practice necessarily generates outflow).

Questions of irrigation efficiency and agricultural expansion into New Lands thus depend on a thorough understanding of the Delta balance. This vexed issue has recently been revisited by Molle et al. (2018), who have found that, despite continuous growth in the net area irrigated with Nile water, the outflow to the sea has remained relatively, and surprisingly, stable over the past 30 years, at around 12 Bm^3, with higher values clearly associated with excess hydrologic years and higher releases from the HAD. This stability is extremely puzzling, since neither the expansion of rice cultivation in the late 1980s and early 1990s nor the overall expansion of cultivated/harvested areas (net of urban encroachment) seems to have had a significant effect on overall return flows. Data such as cropping intensity, yields, or the use of groundwater are too inaccurate to fully solve this riddle. More importantly, a 3–4 Bm^3 reduction in the 12 Bm^3 outflow to the sea is an ambitious target in itself, as it would involve significant change to water distribution and use, but also because it is generally accepted that the lowest possible value for the Delta outflow is 8.5 Bm^3 (IRG et al. 1998).[4] If we take this volume as a constitutive fact of water management in the Delta, then the current efficiency of the Valley/Delta system is 93% of its estimated potential, and any improvement would account for only around 5% of the Nile's water resources (ibid.).

There are benefits in reducing return flows through improving efficiency at the plot level (although this is not true in all parts of the Delta, particularly where return flows are recycled) because their quality degraded. This reduction would mean that farmer "demand" in terms of gross abstraction would be reduced, easing distribution at the secondary and main distribution levels. The main potential for this lies in improving farm-level irrigation practices and reducing soil evaporation and unproductive losses through laser land leveling, the lining of *marwas* (quaternary ditches), distribution of water through piped networks, mulching, and raised beds (Karrou et al. 2011).[5] These measures have been and are being implemented by the MALR and farmers but have not been systematized.

System-level irrigation management is complex and has not been adequately described or understood beyond the formal rotation system. For example, Roest (1999) estimated spillage from canals to drains in the eastern Delta at 23% of total supply. Yet, 20 years later, this phenomenon is rare at the extremities of secondary and tertiary canals (IWMI and WMRI), showing a shift in the demand/supply ratio and adjustments in management. (Better) matching water supply and demand has been a returning (often donor-funded) objective[6] of the past 20 years (Ismail et al. 2001; El Quosy 2006; Mott MacDonald 2011). Several attempts at attuning main and secondary canal discharges to 'real' water requirements that would be based on actual cropping patterns observed in the field have failed. This is partly

[4] The 2005 NWRP had targeted an outflow to the sea of 9.5 Bm^3 by 2017. The new 2017 NWRP expected a value of 9.9 Bm^3 by 2037.

[5] See www.icarda.org/sites/default/files/u158/Science%20Impact%20Raised-Bed_final.pdf

[6] For example, the MALR/USAID Agricultural Policy Reform Program (APRP) issued a policy brief announcing, "MALR and MWRI work together to deliver water to farmers at the right time and in sufficient quantities."

due to an illusory idea that "demand" can be precisely computed based on theoretical parameters and partly due to foot-dragging by local managers who fail to imagine a shift from management based on water levels to one based on discharges, given the nature of the hydraulic infrastructure and the uncertainty in supply from higher levels.

Another recurring theme with regard to the regulation of agricultural demand is water pricing. According to a former minister, "with the increasing scarcity of water from which the country is suffering [...] it has become indispensable that farmers pay for irrigation services in an attempt to rationalize the use of irrigation water and establish sustainable drainage techniques" (OOSKAnews 2013). Contrary to this suggestion, however, successive reports have consistently concluded that, "water pricing under the present and expected near-term conditions may not be feasible in conserving water in the Old Lands" and advanced several arguments: the transaction costs would be tremendous and may exceed the revenue to the government (El Assiouty 1984); the price of water cannot exceed the cost of its delivery because, according to Islamic law, it is essentially free; it is impossible to relate the price to "demand" or actual volumetric supply; water pricing should only be seen as a cost-recovery mechanism through a flat tax (Perry 1996).

9.6.3 Institutional Development and Water Users Associations

Egypt has rich experience when it comes to developing water user groups in the field of agriculture. This includes a diversity of geographical situations (groups around tubewells in the oases, large-scale irrigation schemes in the Valley, Delta, and New Lands), water management arrangements (collective pumps for four to eight farmers in the New Lands, or at the tertiary canal level in the Old Lands), and various scales (from the tertiary [mesqa] and the secondary [branch-canal] to district level [around ten branch canals] (Fig. 9.3). Over the past 30 years, several projects have attempted to organize farmers, improve the interface/coordination between farmers and irrigation managers, or develop district-level "water boards" to ensure the participation of all stakeholders (see Molle and Rap 2014).

As early as the 1970s, the EWUP project recommended that farmers' participation should be sought in the field of both irrigation distribution and maintenance, and called for the establishment of a specially trained cadre of professionals (Irrigation Advisory Service: IAS). A direct consequence of the collective pump stations proposed by IIP projects was that collective action was needed to operate and maintain the pump, organize water distribution, and pay the energy costs. This called for the establishment of mesqa-level Water Users Associations (WUAs).

However, WUAs initially had no legal status to act as independent bodies. This changed in 1994 with the modification of the 1984 Law 12, wherein WUAs were defined as legal organizations at the mesqa level in the improved irrigation systems (IIP) in the Old Lands, while Water Users Unions (WUUs) were made legal entities for the New Lands. The Bylaws of Law 213/1994 (Decree No 14900 of 1995)

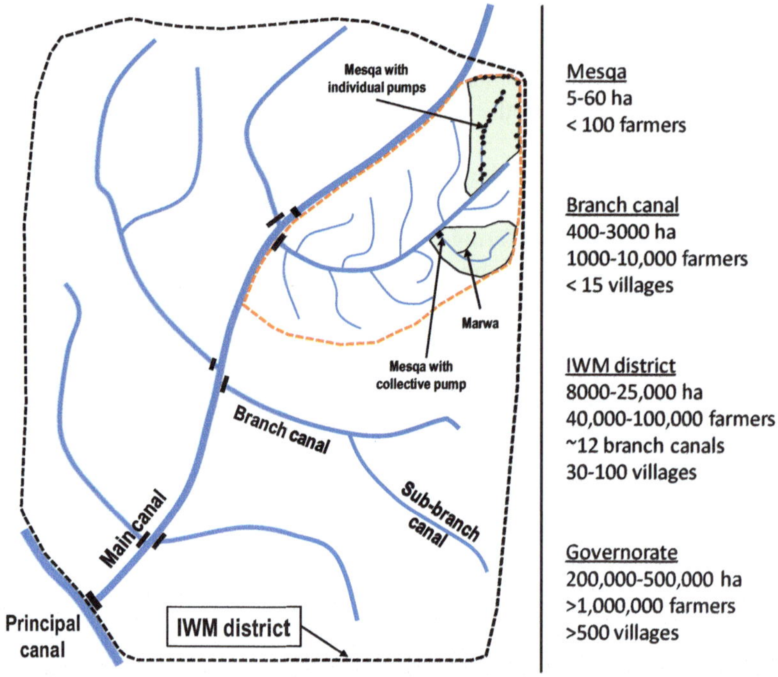

Fig. 9.3 Management levels in Egyptian irrigation (Delta). (Molle et al. 2015)

detailed the rights and duties of the WUAs and WUUs, and the recovery of capital costs of improved irrigation facilities (ICG 2009).

In 1995, the Dutch-funded Fayoum Water Management Project (FWMP) established the first user organizations at the branch canal level, known as "Local Water Boards" (Abdel-Aziz 2003). These were responsible for the operation and maintenance of irrigation intake structures of all mesqas and of secondary drainage infrastructures in their command areas, as well as weed control and domestic water use derived from canals and drains. All water users, whether farmers, residents, or industry, were obliged to become members of the Water Board. Between 1994 and 2009, Water Boards were tested at the district (*markaz*) level in the Fayoum governorate. They were conceived as small "water parliaments" consisting of civil society stakeholders. The project was expanded to other regions, until 900 Water Boards had been set up across the country.

Under USAID's Agricultural Policy Reform Program (APRP) (1996–2003), strong support for various kinds of decentralization and Irrigation Management Transfer translated into several policy initiatives and changes. The MWRI introduced secondary-level Branch Canal Water User Associations (generically called BCWUAs) and "Integrated Districts." Nine initial BCWUAs were formed by ministerial decree (IRG 2002b). In 2006, the IIIMP project took a much broader approach than the IIP by also considering the establishment of BCWUAs and a "From Mesqa to District" approach (APP 2007; World Bank 2005).

However, branch canal-level experiments have faced several setbacks (Rap et al. forthcoming). First, BCWUA board members were often appointed by the ISAs rather than being freely elected—a top-down approach that privileged traditional village leaders and power holders close to the government (Gouda 2016). Second, mechanisms to involve the BCWUAs in the establishment of maintenance priorities and monitoring the work of contractors, although welcome by farmers, often proved to be short-lived. Third, while BCWUAs could play a useful role in cost recovery and financial administration, the law does not recognize them as legal entities, meaning they have no authority to collect service fees or manage a budget. Fourth, BCWUAs were unable to enforce rotations without the approval of government engineers, since they lacked the legitimacy and resources to intervene, resolve disputes, or apply penalties in cases of infringement. Fifth, although farmers called for greater consultation and interaction with the district engineer (El-Zanaty & Associates 2001), some irrigation sector engineers were not convinced of the benefits of the increased involvement of the BCWUAs. They saw them as threatening to their prestige, power, legitimacy, and jobs (Hvidt 1998), and also recognized that they are unable to commit to a given supply pattern due to their dependence on higher level supply. The BCWUAs' lack of legal empowerment is seen to result from the ministry's reluctance to divest its discretionary power, and that of MPs to lose their vote-winning role in mediating occasional water crises.

General assessments of participatory water management reveal worrying trends (APP 2007; Molle and Rap 2014). Barakat concluded that user-group participation in water management is extremely low, that none of the actors properly understand their rights, and that WUOs and MWRI field staff do not see themselves as partners. Likewise, based on the monitoring and evaluation of 150 WUOs over several years, Bron (APP 2007) found user participation in water management to be very low, even when organized into water users' associations, and that "no water users' organization in Egypt [had] reached a level of institutional strength that [could] be considered sustainable."

9.6.4 Institutional Problems Within Ministries and the Limits of Reform

A diagnostic of institutional problems within the MWRI carried out 20 years ago (IIMI 1995a, b; Merrey 1998) identified problems that are still considered systemic and "sticky." These include considerable duplication of functions (in part due to donors' insistence on the creation of autonomous entities to manage their projects, producing, for example, the EPADP[7] as well as the Irrigation Improvement Sector), insufficient cooperation and information sharing between, and even within,

[7] Other semi-autonomous authorities include, among others, the High Aswan Dam Authority, the Egyptian Shore Protection Authority, a holding company responsible for North Sinai and the South Valley, and another for the Western Delta.

ministries, low salaries leading to "seepage" of the best professionals to the private sector or Gulf countries, a highly centralized structure with power concentrated in the hands of a few, and a lack of transparency.

A further major problem has been a multiplication of, and lack of consistency between, operational units (directorates). Vertically linked to their respective sectors, these deal with irrigation, drainage, or mechanical/electrical issues (as well as groundwater, once it was promoted to a separate sector in 1999). Irrigation and drainage administrations have separate buildings, equipment, and work schedules (Merrey 1998).

This has prompted the MWRI, with considerable persuasion from donors, to adopt "a policy to integrate all water management functions at the district level to support decentralized management" (IRG 2002b). The goal was to merge into one integrated district the various sectors (irrigation, drainage, mechanical) that are defined with different boundaries (neither of them corresponding to administrative districts). This was intended to: (1) reduce the number of staff and put all of them under the authority of a single district engineer, (2) do away with the intermediate layer of the inspectorate, (3) integrate the various functions of water management for coordinated planning and management. The pilot IWMDs of Zifta and Ibrahimia in the Delta were recognized in 2001 by Ministerial Decree No. 506, and, by 2007, 27 districts had been established.

The implementation of the IWMDs faced several obstacles and issues, including (IRG 2002a): (1) setting the new boundaries (often taken as those of the irrigation district), (2) selecting the officers (with conflict arising between the three departments over who would head the IWMD), (3) allocating budget and determining operational procedures (with finances coming from different departments), (4) a lack of water monitoring programs (needed for improved management but requiring funding for equipment), (5) a reluctance to delegate authority and decision-making from the general directorate to the IWMD, (6) limited cooperation from the Drainage and Mechanical equipment sectors, which maintained or shifted their best equipment and staff within the directorate (above the district) where such rationalization has not been attempted.

The USAID-funded LIFE-IWRM Project (from 2004 to 2012, in two phases) provided technical assistance to the MWRI to implement decentralized and participatory IWRM over an area of 485,000 ha (15% of Egypt's irrigated area) (El Atfy et al. 2007). The project's stated achievements include the establishment of 45 districts in 8 irrigation directorates, with the formation of 622 BCWUAs (covering 2 Mfed and 1.3 million users), trained to participate effectively in decision-making.

BCWUAs should be involved in the annual planning, prioritization, and selection of maintenance and minor works, with inspection of the canals and drains to be carried out jointly with the IWMD staff. They should also participate in the monitoring of progress and quality control during the execution of works (Barakat 2009). BCWUAs are expected to monitor, measure, and record the water levels at key control points to detect and report anomalies and shortages. The participation of all BCWUAs in the management system of the IWMDs was found to have a positive influence on the quality and the equity of water distribution among in the IWMDs (El Atfy et al. 2007).

In 2005, the MWRI set up an Institutional Reform Unit (IRU), which issued a first Institutional Reform Vision and Strategy document that included the formation of water users' associations at branch canal, district, and directorate levels, the transfer of O&M management and financial responsibilities to WUOs, and the restructuring of MWRI local administration into public bodies known as Regional Water Management Authorities. The strategy proved to be overambitious. A review carried out in 2010 (Salem 2011) again identified problems with the politicization of staff recruitment, the brain drain generated by low salaries and incentives, and the permanence of a structure of sector-based silos where communication is almost exclusively vertical.

There have long been discussions about introducing a degree of privatization into water management (APP 2001), mostly at the instigation of donors. One issue is whether special authorities should be dismantled when their role comes to an end. For example, when the EPADP's drainage program is complete, its virtual monopoly of drainage must give way to a more service-oriented and demand-driven mode of operation. Drainage *services* would be provided to private drainage users—investors or water boards—for new development, rehabilitation or maintenance, possibly by operators other than, or in competition with, EPADP (van Achtoven et al. 2004).

Overall, institutional reform within the ministry has not been able to displace entrenched cultural and bureaucratic models, whose hyper-centralization, believed to date back to Ottoman times (Salem 2011), is associated with problems of transparency and knowledge-sharing.[8] Efforts to rationalize and integrate irrigation and drainage, and reduce the number of and inconsistency between operational units, were discontinued in 2014, when the minister decided to halt the IWMD reform. Sector leaders had understood that reform would result in loss of power, discretion, and prestige, and consequently were resistant to it (ibid). Even the participatory definition of maintenance priorities with farmers has been sidelined, as these are often dictated "by powerful figures who used to belong to the ruling party or the Local Popular Councils or other affiliates of power corridors" (Salem 2011).

Inter-ministerial committees and other bodies set up to facilitate coordination or integration are numerous, but, "in many cases they are either not functional or leave little trace due to unclear mandates, lack of permanent supporting structures, and ineffective feedback mechanisms" (Luzi 2010). They are often established at the insistence of donors, who, although sometimes serving as catalysts, produce an overkill of institutional initiatives and programs that may promote fatigue and inertia.[9]

[8] The National Water Resources Plans (2005 and 2017) are not available on any official website; some theoretically public annual reports can only be accessed through personal relationships, even within the ministry; the ministry website is poor, out of date, etc.

[9] Between 2003 and 2005, for example, Egypt was involved in the NWRP with the Dutch, the Water Policy Reform Program (WPRP) supported by USAID, the Integrated Water Resources Management (IWRM) Action Plan supported by the World Bank, the National Environment Action

9.6.5 The Water-Food-Poverty Nexus and Policy Linkages

At least since Nasser's time, food self-sufficiency has always featured as an unquestionable policy objective. This is understandable when one considers both the high population growth and the political vulnerability associated with a dependence on food imports and world market food prices. However, by the early 1980s, Egypt was importing around half its food requirements (MoI and USAID 1986), and this is still more or less the case.

The three "sacred cows" of water and agricultural policy—namely, food security, job creation, and limited per-capita land endowment in the Old Lands—are constantly used as an indisputable rationale for the expansion of irrigation. This is well illustrated by an MWRI policy document, which stated that "unless a very ambitious and serious plan to expand land resources, rationalize the use of limited water resources, and increase the efficiency of using both land and water in agricultural production to its maximum, food security in basic crops can never be achieved" (MWRI 2002).

Yet the rhetoric of the expansion of irrigation for food security purposes (Acloque 2017) is squarely contradicted by the promotion of the New Lands as a frontier of modernization devoted to high-value, export-oriented agriculture. It may be dictated by a need to maximize the return on private investment or improve the balance of payments, but it will do little to improve food security in the Old Lands (Ayeb 2008) and will even worsen it where the New Lands divert water from them (such as the Toshka project).

Since Egypt is the world's largest wheat importer, special attention and governmental support is extended to its production. Leaders are mindful of the food riots of 1977, when Sadat tried to impose an IMF deal in Egypt, and of those in 2008, when the Government of Egypt (GoE) attempted to reduce subsidies after the international wheat price had tripled (Ayeb 2008; Bush 2010). More than half of Egypt's population lives below the poverty line and the price of bread is a key parameter of social stability.

Rice is also important for food security, and its profitability has made it a favorite crop of farmers, despite its high water demand and the government's attempts to discourage it. In August 2016, the cabinet issued a decree aimed at limiting the extension of rice to 1.1 Mfed to avert water shortages (Daily News Egypt 2016), but it was later overruled by the prime minister under pressure from MPs. In 2017, the government defended the idea in parliament and secured an agreement for a quota of 0.7 Mfed for 2018: a tall order when the balance of 2017 was 1.87 Mfed against an allowable area of 1.08 Mfed. It is unlikely that command-and-control measures will have much effect, although, in 2009, the ministry claimed to have reduced rice acreage from 2.2 to 1.2 Mfed by cracking down on violations (Al Ahram Hebdo 2013).

Plan (NEAP) supported by UNDP, the European Union Water Sector Reform Program (EUWSRP), the EU Water Initiative (EUWI), the EURO-MED Partnership in Egypt, and so on.

Even with optimistic projections in terms of land expansion and productivity, any potential gains will be largely nullified by population growth, and food security is set to remain a moving target. Allan (2003) has proposed viewing cereal (foodstuff) imports as transfers of embedded "virtual water" as a means to virtually increase the country's water supply. But the idea of relying on global trading systems to provide food and solve water deficits proved to be a "no-go area in public discourse" (ibid.), something that can be understood in light of the destabilizing impact of global food price hikes and of Egypt's political dependence on the US.

9.6.6 Planning Ahead: The New National Water Resources Plan 2037

In 2017, the MWRI issued its new NWRP for the next 20 years. The plan includes a series of notable discursive shifts. Most importantly, it stresses that, however necessary they may be, supply- and demand-oriented approaches have *limited scope*, and that, "Government agencies, the private sector, and civil organizations—as well as individual farmers and users of domestic water—will have to find ways to adapt to increasing scarcity" (NWRP 2017). It projected a 10% reduction in the amount of water available for crops. Combined with a predicted rise in crop evapotranspiration, this means, "that farmers will have to choose crops or cultivation systems that consume less water, reduce cropping intensity, or accept suboptimal production" (ibid.).

The unrealistic land expansion targets of the 2005 NWRP now give way to a scenario where, "the total agricultural area including New Lands will remain fairly stable: gains in cultivated area due to land reclamation projects will be offset by loss of arable land due to urbanization, construction of roads, etc.". The NWRP (2017) gently dismisses MARL's 2030 strategy and its expansion target of 3.1 Mfed allegedly made possible by an increase in the average plot-level irrigation efficiency from 50% to 80%. But the fast-track measures to tackle Egypt's future water needs delineated by a committee set up by prime ministerial decree in 2016, and which again include a strategy "to save 10 Bm^3 by 2030," show the lingering political attractiveness of grand but unrealistic promises.

A further noteworthy change is the attempt to decentralize water management towards the governorates by assigning them specific amounts of water through a bulk allocation system. With the IWMD reform shelved, a continuing brain drain, and no sign of the political will to enact significant reform, the MWRI remains vulnerable to shocks, which could, for example, come from a series of dry years combined with reservoir development in the upstream part of the Nile Basin.

9.6.7 Can the Nile Take Its 'Gift' Back?

Herodotus' proverbial 'Gift of the Nile', Egypt is making use of all the water that flows into the HAD. As we have shown earlier, very little of it can be "saved." In official discourse and water balances, this inflow is unfailingly taken at 55.5 Bm^3—the "share of Egypt" as specified in a 1959 treaty between Sudan and Egypt. A long time "hegemon" in the Nile Basin (Warner et al. 2017), Egypt has seen its position weaken over the past five decades, both for internal reasons and as a result of the growth of upstream neighbors. The most notable of these, Ethiopia, is home to the headwaters of the Nile that contribute 85% of the runoff accruing to Sudan and Egypt, and is laying not unreasonable claim to the waters of the Nile. The ongoing construction of the Grand Ethiopian Renaissance Dam (GERD) is a clear challenge to the current distribution of the river's resources, although the most crucial issue lies in how fast GERD—with its 74 Bm^3 of storage—will be filled (Wheeler et al. 2016). Once filled, the dam is expected to have a very limited impact in the short term on the inflow to the HAD, and its 169 Bm^3 of storage capacity, because it is a hydropower dam and diversion within Ethiopia for irrigation or otherwise will take time to materialize on a substantial scale. Irrigation expansion in Sudan probably dwarfs the potential of upstream countries, as vast swathes of flat land are available along the river course. International investors, including Egyptian companies, have spotted this bonanza and are currently developing large-scale irrigation facilities, tapping surface or groundwater. The expansion of irrigation in Sudan probably poses a much bigger risk to Egypt, and one which is already gradually materializing, though not as spectacularly as that of GERD (Yassin 2014; Hussein 2015).

Egypt's response to this challenge has wavered between confrontational and conciliatory tones, while remaining highly emotional. In a televised speech, the deposed President Morsi emphasized that, "The lives of the Egyptians are connected around [the Nile] as one great people. If it diminishes by one drop then our blood is the alternative" (BBC 2013); more recently, the minister of water resources and irrigation was reported as saying, "We won't accept cuts to our share of Nile waters, even a cup less" (Al-Masry Al-Youm 2017). While the country is publicly buttressing the defense of its "historical share of Nile waters," there are signs that it is already preparing for temporary reductions in supply, for example by building a mega seawater desalination plant (Tawfeek 2018) and working to reduce rice acreage.

Egypt's stance is further weakened by its policy contradictions. On the one hand, it claims it would collapse should one drop be withdrawn from its "share"; on the other, it periodically announces plans to expand irrigation by several Mfed, thereby undermining its own claim[10] to be facing a deficit of 30 Bm^3. Yet almost daily media coverage of the threats to Egypt's water resources also provide fertile ground for grand projects, such as the mega desalination plant, or flawed, multi-billion-dollar

---

[10] See, for example, a recent statement by the minister of water resources and irrigation (Egypt Today 2018).

schemes to "improve efficiency." There has even been the occasional quixotic mega-project proposal, like the idea to distribute the HAD waters through huge pipes bordering the valley and avoid "losses," urged by scientists wanting Egypt to adopt a "third hydraulic revolution" to follow Mohamad Ali and Nasser (Al Ahram Hebdo 2013).

The supply/demand ratio is shifting, whether due to land expansion, new cropping patterns, or reduced water supply. We must note, however, that continued urban expansion over agricultural land works to decrease demand. As early as 1980, the annual loss in Egypt was estimated at 45,000 fed/year (Richards 1980), while the Ministry of Agriculture recently released a study showing that, between 1974 and 2007, 32,000 acres were lost each year on average, increasing to 41,000 fed between 2007 and 2010 (Al Ahram 2017; Ahram Online 2017). Since the revolution, several formal and informal sources report that this figure has multiplied by three.

9.7 Conclusion

Egypt's irrigation is a fascinating history of the gradual transformation of the Nile natural environment towards a cultivated area of around 3.78 million ha, a cropping intensity close to 2, and high land productivity. Water managers have learned to distribute around 60 Bm^3 of water through a maze of 31,000 km of canals and operate hundreds of large-scale pump stations to lift water from canals and drains. Farmers have responded to various constraints on input factors by intensifying, diversifying, and mechanizing their production.

Since the 1960s, the relatively familiar picture of irrigated agriculture in the Old Lands has been paralleled by expansion into New Lands: the socio-technical landscape of the New Lands east and west of the Nile Delta, and further afield in the desert, have introduced new capital-intensive ways of farming. Although New Land development has come under criticism for various technical, economic, social, and political reasons, it has now established itself as a relatively high-performing sector, with a trend towards fruit tree cultivation. This expansion partly makes up for the dramatic loss to urbanization of arable land in the Delta and Valley, but it also signals a policy shift in favor of export-oriented, capital-intensive agriculture. "The idea that has dominated so much of the water policy in Egypt [...] that any excess water available should be devoted to the extension of the irrigated area" (El Quosy 2006) has also served bureaucratic, financial, and political interests. It has been pushed to its limits, creating competition with Old Lands over Nile water, groundwater depletion, and economic inefficiencies (part of the equipped area is left idle for lack of water or soil salinization).

The Egyptian water administration has been repeatedly described as a centralized and vertical structure, with poor communication across internal branches/sectors. Reform programs devoted only to operational levels have a limited impact (Jacobs 2005), and those that are targeting institutional change—such as establishing Integrated Districts or BCWUAs—are eventually abandoned or put on the backburner once donor-funded programs come to an end. As Richards (1980) puts it, "The engineers of the bureaucracy naturally tended to think of the fundamentally social problems of Egyptian agriculture as essentially technical ones" and relatively little is known about what is really going on in the countryside, in terms of the labor and land markets, for example, and even water management itself. Research remains insufficient, if judged by the crucial importance of what is at stake.

The story of irrigation policy is intertwined with the wider political economy of the country. Accessing (irrigated) land has long been a privileged way to accumulate wealth and capital and also a means of relieving the pressure on the peasantry. Issues of food security, social stability, and geopolitics undergird the management of water resources. The increasing uncertainty generated by climate variability and water resource development in the upper Nile Basin, the overextension of the irrigation network, the high efficiency of water use at system level, and the institutional rigidities combine to foreshadow potential turbulence and herald the end of water abundance. In this, they echo the recent NWRP and its emphasis on the need to work within acknowledged limits.

References

Abdel-Aziz, Y. (2003). Decentralisation and water user participation. Water Demand Management Forum. IDRC.

Abdel-Dayem, S., Abdel-Gawad, S., & Fahmy, H. (2007). Drainage in Egypt: A story of determination, continuity, and success. *Irrigation and Drainage, 56*, S101–S111.

Abou-Korin, A. A. (2002). Sustainability requirements of desert development in Egypt: The case of New Valley Region. *Journal of Engineering Science.*

Acloque Desmulier, D. (2017). Recomposition des acteurs et des espaces agricoles Egyptiens: enjeux politiques, socio-économiques et alimentaires. *Revue d'Économie Régionale & Urbaine, 3*, 457–482.

Acloque Desmulier, D. (2018). Personal communication by email.

Adriansen, H. K. (2009). Land reclamation in Egypt: A study of life in the new lands. *Geoforum, 440*(4), 664–674.

Ahram Online. (2017). *Urban encroachment, lost development, and Egypt's next president.* Sunday, 15 January.

Al Ahram. (2017). *Experts: Ministry of irrigation deceived "people and the government"*. http://agri.ahram.org.eg/News/55542.aspx

Al Ahram Hebdo. (2013). *Irrigation: Nouveau débat sur les eaux du Nil.* 13–19 février 2013 – Numéro 961.

Al Ahram weekly. (2015). *Reclaiming land.* June 25, 2015.

Al-Ahram Weekly. (2000). Toshka in the crossfire, 27 Jan. – 2 Feb. 2000, Issue No. 466.

Alary, V., et al. (2017). Desert land reclamation programs and family land dynamics in the Western Desert of the Nile Delta (Egypt), 1960–2010. *World Development, 104*(2018), 140–153.

Allan, J. A. (2003). Virtual water – The water, food, and trade Nexus. Useful concept or misleading metaphor? *Water International, 28*(1), 106–113.

Al-Masry Al-Youm. (2017). *We won't accept cuts to our share of Nile waters, even a cup less, Egypt*, December 2, 2017.

APP (Advisory Panel Project on Water Management). (2001). *Workshop on privatization/sector reform within MWRI* (Roundtable discussion). 31st May 2001, Cairo, Egypt.

APP (Advisory Panel project on water Management). (2007). *Proceedings of the workshop on Water Boards: From Policy to Strategy*, 13–14 January 2007, Cairo.
APP (Egyptian-Dutch Advisory Panel on Water Management). (2005). *Workshop on intermediate reuse of drainage water*. Cairo, Egypt, June, 23–2005.
ARDC (Agricultural Research and Development Council). (2009). *Sustainable agricultural development strategy towards 2030*. Cairo: Ministry of Agriculture and Land Reclamation, Egypt.
Ayeb, H. (2008). Crise alimentaire en Égypte: compétition sur les ressources, souveraineté alimentaire et rôle de l'État. *Hérodote, 131*(4), 58–72.
Ayeb, H. (2010). *La crise de la société rurale en Egypte. La fin du Fellah ?* Paris: Karthala.
Barakat, E. (2009). *IWRM I branch canal water user association assessment*. Report No. 5. EPIQ. July 2009.
Barois, J. (1911). *Les irrigations en Egypte* (2nd ed.). Paris: Béranger.
Bazaraa, A. S. (2002). *Reuse of agricultural drainage water: Revision of the reuse mixing policy*. Cairo: Advisory Panel Project on Water Management.
BBC. (2013). *Egyptian warning over Ethiopia Nile dam*. 10/06/2013. www.bbc.com/news/world-africa-22850124.
Bush, R. (2007). Politics, power and poverty: Twenty years of agricultural reform and market liberalisation in Egypt. *Third World Quarterly, 28*(8), 1599–1615.
Bush, R. (2010). Food riots: Poverty, power and protest. *Journal of Agrarian Change, 10*(1), 119–129.
CAPMAS. (2018). http://www.capmas.gov.eg/
Corey, G. L. (1978). Trip report. To assist mission team with defining scope of New Lands Development Project and in a site selection for the project. USAID.
Cuno, K. M. (1992). *The pasha's peasants. Land, society, and economy in Lower Egypt, 1740–1858*. Cambridge: Cambridge Press.
Daily News Egypt. (2016). *Cabinet curbs water-consuming plants' cultivation to avert water shortage*. 02 – 08 – 2016.
de Sainte Marie, C. (1989). État et paysans dans les systèmes hydrauliques de la vallée du Nil (Egypte). *Études Rurales, 115–116*, 59–91.
Deputy, E. (2011). *Mubarak & Toshka: Let's do it for the Kids (2011)*. APSA 2011 Annual Meeting Paper. https://ssrn.com/abstract=1901986
Desert Development Center (DDC). (1997). *Desert irrigation efficiency*. Final Report to International Development Research Center of Canada (IDRC).
Dixon, M. (2013). *The making of the agri-food corporate system in Egypt*. PhD thesis, Cornell University.
DRI. (2015). *Yearbook 2013/2014*. Cairo: Ministry of water and irrigation.
Egypt Independent. (2014). *On Toshka New Valley's mega-failure*. Thursday 6 November 2014.
Egypt Today. (2018). Egypt plans to face water scarcity, allots LE 900B, *Egypt Today*, Feb. 25, 2018.
El Assiouty I. (1984). *Pricing of irrigation water in Egypt*. Cairo: Supreme Council of Universities.
El Atfy, H.I., Viala, E., Fredericks, J., & Svendsen, M. (2007). *Improving the performance of irrigation and drainage systems in Egypt through decentralization and participation*. Paper presented at the ICID Conference, October 2007, Sacramento.
El Quosy, D.D. (2006). *A study on matching irrigation supply and demand (MISD)*. Report prepared for Eco ConServ-Environmental Solutions. Cairo – July 2006.
El-Agha, D. E., Closas, A., & Molle, F. (2017). Under the radar: Dynamics of groundwater use in the Nile delta. *Hydrogeology Journal, 25*, 1621–1631.
El-Fellaly, S. H., Saleh, E. M. A. (2004). *Egypt's experience with regard to water demand management in agriculture*. Eighth International Water Technology Conference, IWTC8 2004, Alexandria.

El-Zanaty & Associates. (2001). *Knowledge, attitudes and practices of Egyptian farmers towards water resources*. National Survey 2001. EPIQ Water Policy Reform Program (Report No. 54). USAID.

Faggi, P., & Maury, G.R. (1987). Étatisation de l'eau dans le projet «nouvelle vallée» en Égypte. In *L'homme et l'eau en Méditerranée et au Proche-Orient. IV. L'eau dans l'agriculture* (pp. 103–110).

Fahmy, H. (1996). Comparative analysis of Egyptian water policies. *Water International, 21*(1), 35–45.

Faostat. (2015). www.fao.org/faostat/

Gouda, D. M. (2016). *Social capital and local water management in Egypt*. Cairo: AUC Press.

Gumuchian, H. (1975). La République Arabe d'Egypte à la conquête de ses déserts. *Revue de Géographie Alpine, 63*(2), 225–252.

Hanna, F. & Osman, M. A. G. (1995). *Agricultural land resources and the future of land reclamation and development in Egypt*. Options Méditerranéennes, Sér. B/n°9.

Hopkins, N., et al. (1988). *Participation and community in Egyptian New Lands: The case of South Tahrir, Cairo papers in social sciences*. Cairo: AUC.

Hurst, H. E. (1957). *The Nile. A general account of the river and the utilization of its waters* (Revised ed.). London: Constable Publishers.

Hussein, W. (2015). *Will Saudi agricultural investments in Sudan leave Egypt high and dry*? Al Monitor – The pulse of the Middle-East.

Hvidt, M. (1995). *Current efforts to improve irrigation performance in Egypt*. Paper presented at The third Nordic conference on Middle Eastern Studies: Ethnic encounter and culture change, Joensuu, Finland, 19–22 June 1995.

Hvidt, M. (1998). Implementing new irrigation technology in upper Egypt: Political and bureaucratic considerations. In P. Mollinga (Ed.), *Water control in Egypt's canal irrigation – A discussion of institutional issues at different levels* (pp. 23–40).

Ibrahim, F., & Ibrahim, B. (2003). *Egypt: An economic geography*. London/New York: I.B. Tauris.

ICG (International Consultants Egypt). (2009). *Institutional framework and decision-making practices for water management in Egypt*. INECO project report.

IFAD. (1980). *Report and recommendation of the president to the executive Board on a proposed loan to the Arab Republic of Egypt for The West Beheira Settlement Project*.

IIMI. (1995a). *An action plan for strengthening irrigation management in Egypt*. USAID.

IIMI. (1995b). *Water resources, irrigation operations and institutional issues: An analysis of the Ministry of Public Works and Water Resources, Government of Egypt*. IIMI-MPWR Study, 'Strengthening Irrigation Management in Egypt'. Report Number 2. Cairo.

IRG (International Resources Group), Winrock International and Nile Consultants. (2002a). *Integrated water management district: MWRI plan for pilot implementation* (APRP-Water Policy Activity, Report No. 62). Cairo: USAID.

IRG (International Resources Group), Winrock International and Nile Consultants. (2002b). *Assessment of the impacts of the water policy reform program* (APRP-Water Policy Activity, Report No. 58). Cairo: USAID.

IRG (International Resources Group); Winrock International; Nile Consultants. (1998). *National policy for drainage water reuse*. Report No. 8. APRP – Water Policy Reform Program.

IRG. (1998). *Water user association formation outside the irrigation improvement program area* (APRP-Water Policy Activity, Report No. 9). Cairo: USAID.

Ismail, R. et al. (2001). *Matching irrigation supply and demand. Solving water delivery problems in a liberalized agricultural economy. Policy Brief*. MALR/USAID Agricultural Policy Reform Program (APRP).

IWMI and WMRI (2013) *An exploratory survey of water management in the Meet Yazid Canal command area of the Nile Delta. Water and salt management in the Nile delta*. Project Report No. 1. IWMI, WMRI: Cairo.

JACOBS. (2005). *Egypt: Water sector reform programme, identification report*. London: JACOBS Consultancy.

Karrou, M., Oweis, T., Benli, B., & Swelam, A. (Eds.). (2011). *Improving water and land productivities in irrigated systems. community-based optimization of the management of scarce water resources in agriculture in CWANA* (Report no 10). Aleppo: ICARDA.

Linant de Bellefonds, L. A. (1873). *Mémoires sur les principaux travaux d'utilité publique, exécutés en Egypte depuis la plus haute antiquité jusqu'à nos jours*.

Luzi, S. (2010). Driving forces and patterns of water policy making in Egypt. *Water Policy, 12*, 92–113.

Merrey, D. J. (1998). Governance and institutional arrangements for managing water resources in Egypt. In P. Mollinga (Ed.), *Water control in Egypt's canal irrigation – A discussion of institutional issues at different levels* (pp. 1–22). Wageningen: Wageningen University.

Meyer, G. (1996). The impact of liberalization and privatization on the agricultural sector in Egypt. In H. Hopfinger (Ed.), *Economic liberalization and privatization in socialist Arabic countries: Algeria, Egypt, Syria and Yemen as examples* (pp. 121–143). Gotha: Perthes.

Meyer, G. (1998). Economic changes in the newly reclaimed land: From state farming to smallholding and private agricultural enterprises. In S. Hopkins & K. Westergaard (Eds.), *Direction of change in rural Egypt* (pp. 334–356). Cairo: American University of Cairo Press.

Mikhai, A.M. (2008). *The nature of Ottoman Egypt irrigation, environment, and bureaucracy in the long eighteenth century*.

Ministry of Irrigation (Egypt), USAID. (1986). *Strategies for irrigation development in Egypt. Water management synthesis II, project agricultural and irrigation engineering*.

Mohamed, A. I. (2013). Irrigation water quality evaluation in El-Salam canal project. *International Journal of Engineering and Applied Sciences, 3*(1), 2305–8269.

Molle, F., & Rap, E. (2014). *Brief retrospective on water user organizations in Egypt. Water and salt Management in the Nile delta* (Project Report No. 3). Cairo: IWMI, WMRI.

Molle, F., Rap, E., Al-Agha, D. E., Ismail, A., Abou El Hassan, W., & Freeg, M. (2015). *Irrigation improvement projects in the Nile Delta: Promises, challenges, surprises* (Water and Salt Management in the Nile Delta Project Report no. 4). Cairo: IWMI, WMRI.

Molle, F., Gafaar, I., Ezzat Al-Agha, D., & Rap, E. (2018). The Nile delta's water and salt balances and implications for management. *Agricultural Water Management, 197*(15), 110–121.

Mott MacDonald. (2011). *Matching water supply and demand*. Technical Report 57. National Water Resources Plan Coordination Project (NWRP-CP).

MPWWR (Ministry of Public Works and Water Resources). (1990). *The Nile and history of irrigation in Egypt* (444 p). Cairo: MPWWR.

MWRI. (2002). *Adopted measures to face major challenges in the Egyptian water sector*. Cairo: MWRI.

MWRI. (2013). *NWRP measures. National Water Resources Plan, Coordination Project (NWRP-CP)*. Technical Report 44. Ministry of Water Resources and Irrigation.

NWRP. (2017). *The national water resources plan NWRP 2017-2030-2037 Egypt*. Cairo: The Ministry of Irrifation and Water Management.

OOSKAnews. (2013). Egypt allows private investment in irrigation. Weekly Water Report, Middle East and Africa. *OOSKAnews*. 05/21/2013.

Perry, C. J. (1996). *Alternative approaches to cost sharing for water service to agriculture in Egypt* (Research Report 2). Colombo: International Irrigation Management Institute (IIMI).

Richards, A. (1980). The agricultural crisis in Egypt. *The Journal of Development Studies, 16*(3), 303–321.

Richards, A. (1982). *Egypt's agricultural development, 1800–1980: Technical and social change*. Boulder: Westview Press.

Ritzema, H.P. (2009). *Drain for gain. Making water management worth its salt*. PhD thesis. UNESCO-IHE Institute for Water Education.

Rivlin, H. A. B. (1961). *The agricultural policy of Muhammad Ali in Egypt*. Cambridge, MA: Harvard University Press.

Roccu. (2013). *The political economy of the Egyptian revolution Mubarak, economic reforms and failed hegemony*. Basingstoke: Palgrave Macmillan.

Roest, C. W. J. (1999). *Regional water distribution in the Nile delta of Egypt.*

Ruf, T. (1987). L'histoire de la maîtrise de l'eau en Egypte. In: *Aménagements hydro-agricoles et systèmes de production*. Montpellier: CIRAD.DSA, (6), 361–374.

Sadowski, Y. (1991). *Political vegetables? Businessman and bureaucrat in the development of Egyptian agriculture*. Washington, DC: Brookings Institution.

Salem, O. (2011). *Review of the institutional reform strategy of the ministry of water resources and irrigation*. Second Draft.

Seckler, D. (1996). *The new era of water resources management* (Research Report 1). Colombo: International Irrigation Management Institute (IIMI).

Sims, D. (2015). *Egypt's desert dreams: development or disaster?* Cairo: American University in Cairo Press.

Sirry Bey, H. (1935). *Irrigation policy*. Bulak/Cairo: Government Press.

Sowers, J. L. (2003). *Allocation and accountability: State-business relations and environmental politics in Egypt*. PhD thesis, Princeton University.

Springborg, R. (1979). Patrimonialism and policy making in Egypt: Nasser and Sadat and the tenure policy for reclaimed lands. *Middle Eastern Studies, 15*(1), 49–69.

Tawfeek, S. (2018). Sisi vows to protect Egypt's share of Nile water. *Egypt Independant* January 9, 2018. www.egyptindependent.com/sisi-vows-protect-egypts-share-nile-water/

Tellioglu, I., & Konandreas, P. (2017). *Agricultural policies, trade and sustainable development in Egypt. Geneva: International Centre for Trade and Sustainable Development (ICTSD)*. Rome: United Nations Food and Agriculture Organization (FAO).

van Achthoven, T., Merabet, Z., Shalaby, K.S., & van Steenbergen, F. (2004). *Balancing productivity and environmental pressure in Egypt. Toward an Interdisciplinary and Integrated Approach to Agricultural Drainage* (Agriculture & Rural Development Working Paper 13). World Bank.

Verhoeven, H. (2015). The nexus as a political commodity: Agricultural development, water policy and elite rivalry in Egypt. *International Journal of Water Resources Development, 31*(3), 360–374.

Voll, S. P. (1980). Egyptian land reclamation since the revolution. *Middle East Journal, 34*(2, Spring), 127–148.

Wahish, N. (1998). Toshka turns millennial green. *Al-Ahram Weekly* On-line, 27 Aug. – 2 Sep. 1998, Issue No.392.

Warner, J. (2013). The Toshka mirage in the Egyptian desert – River diversion as political diversion. *Environmental Science and Policy, 30*(2013), 102–112.

Warner, J., Mirumachi, N., Farnum, R. L., Grandi, M., Menga, F., & Zeitoun, M. (2017). Transboundary 'hydro-hegemony': 10 years later. *WIREs Water, 4*, e01242.

Waterbury, J., & Whittington, D. (1998). Playing chicken on the Nile? The implications of micro-dam development in the Ethiopian highlands and Egypt's New Valley Project. *Natural Resources Forum, 22*(3), 155–164.

Wheeler, K. G., et al. (2016). Cooperative filling approaches for the Grand Ethiopian Renaissance Dam. *Water International, 41*(4), 611–634.

Willcocks, J., & Craig, J. I. (1913). *Egyptian irrigation* (3rd ed.). London: E&FN Spon.

World Bank. (1980). Report and recommendation of the president of the international development association to the executive directors on a proposed credit to the Arab republic of Egypt for a new land development project (West Nubariya), November 25, 1980.

World Bank. (1986). Arab Republic of Egypt: Channel maintenance project. Staff appraisal report.

World Bank. (1999). *Second pumping stations rehabilitation project*. Implementation completion report. Arab Republic of Egypt.

World Bank. (2005). Project appraisal document on a proposed loan in the amount of US$120 million to the Arab Republic of Egypt for an Integrated Irrigation Improvement and Management Project (IIIMP).
World Bank. (2007). *Implementation completion and results report.* IIP Project. World Bank.
Yassin, N. (2014). Why is Egypt's hydro-political concern more intense on Sudan? *The Reporter* – English Edition, May 31, 2014.

Chapter 10
Irrigation Policies in the Mediterranean: Trends and Challenges

François Molle and Carles Sanchis-Ibor

Abstract Irrigation is a central feature of agriculture in Mediterranean countries. During the twentieth century, and more specifically after World War II, most states have invested massively in large-scale dams, interbasin transfers, and public irrigation schemes. More recently, farmers have capitalized on increasingly cheap pumping devices to tap groundwater, whether in conjunction with surface water or not, and the introduction of pressurized networks has come to constitute a major technological change in the region. This chapter first reviews the changing relationships between irrigation and the wider water sector as scarcity builds up, and ponders on the respective prospects for both supply- and demand-management options. It then takes stocks on recent technical and institutional changes in the irrigation sector before turning to the importance of economic dimensions and context. Finally irrigation is analyzed in terms of policy and political process, and of its relationship with other sectors, while the final section explores the challenges posed by water overexploitation, environmental degradation and climate change.

Keywords Irrigation policy · Water management · Water governance · Participatory management · Groundwater water-energy nexus · Modernization · Mediterranean basin

---

F. Molle (✉)
UMR-Geau, Institute of Research for Development (IRD), University of Montpellier, Montpellier, France
e-mail: francois.molle@ird.fr

C. Sanchis-Ibor
Valencian Centre for Irrigation Studies (CVER), Universitat Politècnica de València, València, Spain

F. Molle et al. (eds.), *Irrigation in the Mediterranean*, Global Issues in Water Policy 22, https://doi.org/10.1007/978-3-030-03698-0_10

10.1 Introduction

Irrigation is a central feature of agriculture in Mediterranean countries.[1] Dry environments or relatively low precipitations that largely fall outside of the summer periods have long pushed communities to develop small-scale infrastructures to capture, store and divert water to their fields. During the twentieth century, and more specifically after World War II, most states have invested massively in large-scale dams, interbasin transfers, and public irrigation schemes. More recently, farmers have capitalized on increasingly cheap pumping devices to tap groundwater, whether in conjunction with surface water or not, and the introduction of pressurized networks has come to constitute a major technological change in the region (whether dominated by sprinkler like in France or northern Italy, or drip, like Israel and Spain).

The expansion of irrigation to an extent that hovers around 23 million ha in Mediterranean countries has gradually faced a host of financial, economic, technical, environmental and institutional problems. The large-scale manipulation of water has altered the hydrologic cycle, affecting springs, wetlands, and flows between streams and aquifers, but also generated problems of contamination and soil salinisation. Infrastructure have proved to be costly to maintain and upgrade and shifting these costs onto users to be problematic. Relationships between the state and irrigators have been largely imbalanced (although Spain and a few places offer some exceptions) and irrigation has become but one user within a larger water management framework that must accommodate dwindling and increasingly variable resources, increases in temperature and evapotranspiration, sectoral competition with a priority to urban and tourist uses, and the recognition of water as an environmental resource. Last, irrigated agricultural production finds itself reshaped by the globalization and vertical integration of food production.

In this concluding chapter, we build on the preceding country-level analyses to offer a general overview of the trends and challenges faced by irrigated agriculture in the Mediterranean.

10.2 Irrigation as the Main Water User

10.2.1 Irrigation as a Wasteful Practice?

Worldwide, irrigation is responsible for 70% of total water withdrawals (Molden et al. 2011) and around 92% of water consumption (Allan et al. 2015). In the Mediterranean, as seen in the introduction to this volume, these percentages are even higher due to the semi-arid to arid climate that prevails in a large part of the

[1] Unless otherwise indicated, the information synthesized in this chapter is described in more detail in respective country chapters, where the reader will also find sources and references.

region. Since irrigation and environmental needs receive in general a lower priority than domestic and industrial uses, they generally bear the brunt of the hydrological variability and of water shortages. But irrigation also comes under criticism for being a wasteful practice. "At present, it is fairly common to find that more than half the amount of water withdrawn from the resource does not even reach the fields being irrigated", note Hamdy and Lacirignola (1997) in a typical statement put forth to suggest that here lies the key to solving water scarcity if –only– irrigation were able to improve its overall efficiency.

The pervasiveness of the flawed extrapolation of plot-level efficiencies to wider scales by cross-multiplication is puzzling. The fallacy has been sustained by international organizations like UNEP's Blue Plan, which derived from it unrealistic water-saving targets at the country level (Blinda 2009). In the Nile Delta, a place ironically used in the 1990s to develop the understanding that systems with substantial reuse of water need careful examination (Seckler 1996), the Ministry of Agriculture entertained the idea that savings could reach 12.4 Bm^3 that could be used for land expansion (a value that is coincidentally roughly the amount of water pumped out to the sea). In Morocco, official documents and statements refer to water saving targets associated with drip irrigation anywhere between 0.85 and 3 Bm^3, without substantiation. Similar expectations have been found in Spain and in Turkey, in the Konya basin, for example. Such claims are made in river basins that are more often than not 'closed', that is, where outflows are roughly limited to uncontrollable sporadic floods. While some northern Mediterranean rivers (e.g. Ebro, Rhone, Po, Seyan), or the Moulouya in northern Morocco still carry substantial volumes of water to the sea, most other rivers don't, and are closing or closed (e.g. Oum Er-Rbia, Tensift and Souss-Massa in Morocco, Júcar and Segura in Spain, Chelif in Algeria, Nile, Orontes and Jordan rivers, etc.).

Basin closure process, whereby the demand induced by water resource development invariably comes to exceed the resource available, is a generic process observed worldwide and well described in the literature.[2] In countries where viable agriculture is highly dependent on irrigation, farmers (and the state) have gradually tapped whatever surface or groundwater can be abstracted and this includes –crucially– return flows from existing irrigation use, whenever this return flow is not degraded by its flowing to a sink like a saline aquifer. As basins close they are increasingly akin to a zero-sum game where evapotranspiration cannot be increased (either by intensification or expansion) without reducing it somewhere else in the same proportion (unless this increase is based on additional groundwater depletion). This hydrologic reality is often either conveniently glossed over and/or non-understood by decision-makers and managers in the region.

[2] See Keller et al. (1998), Molden et al. (2001), and Molle et al. (2010).

10.2.2 Irrigation and the 'Double Squeeze'

Irrigated areas have soared during the second half of the twentieth century globally and in the Mediterranean. This has been the result of large public investments in collective schemes and, more recently, of the boom in groundwater-based irrigation (more on this later). As a result, irrigation water demand has ballooned and in many places outpaced the amount of water mobilized and controlled. At the same time municipal and industrial water demands (the latter to a lower extent because of improvements in technology) have grown steadily. Since this demand is in all countries considered to be a priority (whether by law or de facto), the hydrologic variability is fully passed on to the environmental and agricultural demand. In EU, the former has now been recognized and prioritized over the latter, but the opposite is true in most other countries. The competition from cities is illustrated by Tunisia, Southern Italy or Spain, where irrigation allotments often had to be capped or reduced. The vulnerability of this huge demand is now compounded by both a higher climatic variability and a downward trend in terms of annual water supply. Consequently irrigation water demand is now facing a 'double squeeze' that reflects both declining supply and the growth of other priority uses, resulting in a higher frequency of years where demand from agriculture and the environment cannot be met (Fig. 10.1). The shortfall either results in curtailing irrigated agriculture and/or in depleting groundwater stocks.

The environmental demand shown on the figure reflects the growing awareness, and sometimes legal recognition, of a 'demand' that has long been ignored or overlooked. In France or in Spain, for example, because of limited knowledge about natural resources systems, the lack of environmental concern in the past, and the

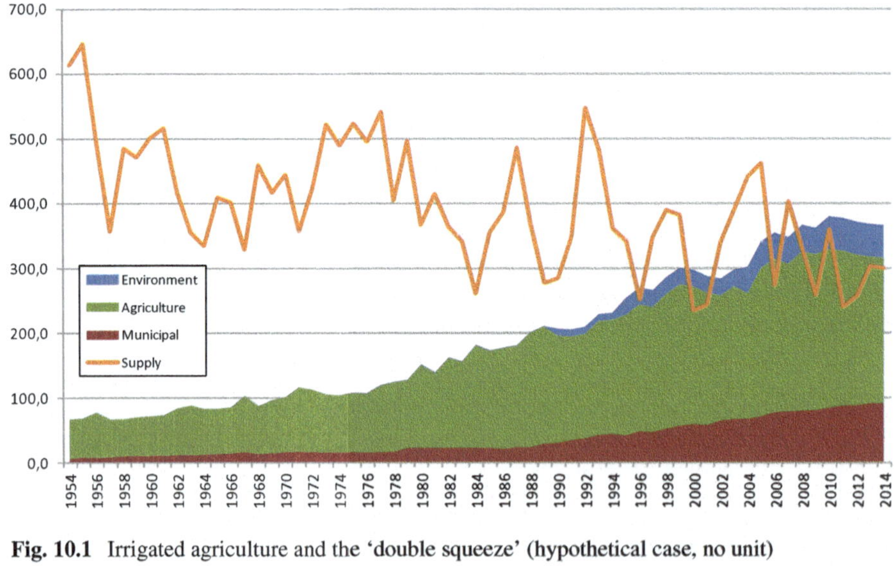

Fig. 10.1 Irrigated agriculture and the 'double squeeze' (hypothetical case, no unit)

importance of water for the local economy and politics, abstraction authorizations delivered frequently exceeded the available resource.

A mitigating factor of the 'double squeeze' is the reduction of irrigated areas due to urbanization. In Tunisia unplanned urban expansion, combined with the development of industrial zones, is encroaching on public irrigated areas (e.g. *Grand Tunis*) despite regulations to protect this type of land (of which 28% is affected by this phenomenon). In Egypt, the yearly loss of arable land to urbanization was estimated at 10,000 ha but this rate has rocketed up after the 2011 revolution. In Marrakech, or in the *Conca d'Oro* of Palermo where a magnificent citrus garden was destroyed between 1950 and 1980, villas and tourist facilities are slowly replacing agriculture. In Spain, citizens are reacting against this urban sprawl processes and the defenders of the Granada, Zaragoza, Murcia and Valencia historical gardens have joined in an umbrella organization (*Intervegas*) to demand legal measures for land protection and agricultural conservation, particularly gravity irrigation practices.[3] Other conservation and restoration actions have been developed throughout Europe (Leibundgut and Kohn 2014).

Another mitigating factor is the fact that a part of the water going to cities is not consumed and returns to the river/aquifer system. By reusing this water (whether treated or not) agriculture can recover a large part of the water that had been diverted away to cities. Nowhere is this clearer than in Israel where agriculture is largely decoupled from (fresh)water, as treated wastewater is now its main resource. This decoupled model is also being put in practice in some tourist areas of Spain, where WUAs are seasonally exchanging water rights with the urban sector, using wastewater for irrigation against a financial compensation.

10.2.3 The Groundwater Revolution

Groundwater has for centuries been tapped through *qanats* (also called *khettaras* in Morocco, or *foggara* in Algeria) and wells, but technology constrained the amount of water that could be abstracted. In the past 30 years or so, cheaper pumping devices have fuelled a boom in groundwater use that occurred in three different types of situations. First, wells have gradually appeared in most large-scale public irrigation schemes to make for the growing shortfall of water. The intensity of conjunctive use is a good indicator of the (in)adequacy of irrigation water supply. In schemes like Tadla in Morocco, individual wells have been massively dug since the 1980s (see Hammani and Kuper 2008), when the Oum Er-Rbia basin started to be overexploited. A similar situation was observed in the Gharb scheme in Syria's

[3] In the case of the Valencia 'market garden'(*horta*), whose degradation was defined by Courtot (1994) as foreshadowing "the end of a myth", a recent regional law (2018), promoted by a popular mobilization, has protected all the historical lands of the "Horta" from urban development, and an Agricultural Development Plan is to be launched before the end of 2018 to strengthen agricultural activities.

Orontes Valley, in Central Anatolia, and Spanish or Italian public schemes. Even in the *Consorzio di Bonifica di Piacenza* in Emilia, in the Po Valley, 40% of irrigation water is sourced from aquifers. In the southern Capitanata Scheme, this percentage varies between 50% and up to 100%, depending on the year and the water situation. In other places this phenomenon is more recent, like for example in the Nile Delta where intensive well drilling has been observed in the past 10–15 years (El-Agha et al. 2017). In some places (Egypt or Turkey) the state has even drilled its own wells to supplement its public irrigation canals. Second, tubewells have provided supplemental irrigation water to crops in appeared in areas with rainfed agriculture, typically in Spain, France or Turkey, but also in Morocco or Syria, providing smallholders with an opportunity to intensify (Allan 2007). Last, tubewells have allowed irrigated agriculture to expand into arid or desert areas, like in Morocco, Algeria or Egypt, in particular around oases that had long enjoyed artesian wells and springs.

In all Mediterranean countries, states have attempted to establish registers, permits, or rights for the use of groundwater. But whether in Italy, Spain, or Southern countries this proved insufficient to prevent illegal and/or excessive water abstraction. Typical state control instruments such as permits, well spacing, depth limits, local bans, meters, pricing, have been as little successful as elsewhere (see Molle and Closas 2017) and illegal wells have flourished throughout the region. EU countries have started to define quotas at the aquifer level and to involve mandatory water users' associations (Spain) or OUGC (France) in collective management through the definition of quantitative thresholds and measures to be taken depending on the hydrologic situation. But in Spain only 392 of the 730 groundwater bodies are now in good state and in the Mediterranean basin altogether most of the overexploited basins, with typical drops in the water table of 1 m/year, have little hope to see their level stabilized, let alone to recover. Impacts are particularly severe in coastal areas, where sea-level rise and aquifer drawdown combine to promote salinity intrusion. This can lead to the destruction of irrigated agriculture, as observed in Italy, Morocco or Tunisia. Other impacts observed concern wetlands (e.g. the Tablas de Daimiel, Spain; Marrakech palm grove; Azraq Oasis in Jordan).

Regulation is made extremely difficult by a high number of disperse and often unknown wells, lack of data, and sometimes legal or religious considerations. But the reasons for a general laissez-faire attitude are largely political (Kuper et al. 2016; Molle et al. 2017a): groundwater is often the resource that small individual farmers are able to tap locally to make a livelihood, or to intensify or safeguard their crops. Tough action by the state is poised to generate social unrest and strong reactions. Local tensions and conflicts around groundwater in Morocco (Souss Massa) or Tunisia (Bsissi), or environmental degradation resulting from overdraft in France (Beauce) and Spain (La Mancha), show how states can be forced into action to initiate deliberative processes with users. In some other countries (e.g. Egypt, Algeria, Morocco), groundwater use is also associated with personal or corporate interests close to decision-makers that also work against regulation.

The region is also home to a few interesting experiences with collective groundwater-based irrigation schemes. Tunisia (GDAs) and Morocco (in the Souss basin), but also Spain under Franco (Closas 2018), have built groundwater-based public irrigation networks. Pump stations on a single well but used and managed by a collective are frequent in Turkey, where Groundwater Irrigation Cooperatives have been established since the late 1960s, and Tunisia, with some cases also in Egypt. This has not been seen as a viable option by other governments, although farmers themselves have taken the initiatives to invest collectively in wells for their irrigation, notably in Spain and the Nile Delta.

10.2.4 Bringing More Water (and Sustaining Unsustainablity)

When irrigation activities are threatened by a lack of water, whether because supply is reduced by drought events or competition from other sectors, or demand increased by state-driven basin 'overbuiding' (Molle 2008) and individual initiatives, governments usually look for supply-augmentation options. Bringing more water in, at public expense, is always politically easier than leaving angry farmers dealing with shortages, or imposing on them harsh measures to curtail their use.

Inter-basin water transfers for irrigation have been popular in Spain (0.35 Bm^3 from Tajo to Segura, but frustrated in the case of the Ebro transfer), in Greece, in France (SCP delivering water to the coast), in Italy (0.8 Bm^3 transferred across Southern Apennines mountains), in Tunisia (0.4 Bm^3 from northern mountain ranges to the south), Algeria (numerous projects) and in Morocco, which is planning the transfer of 0.8 Bm^3 from north to south.

Such transfers are often presented as a way of 'redressing nature imbalances', as a question of hydro-solidarity (Spain, Morocco or Tunisia), but they are costly (tunneling and pumping are generally needed) and increasingly resented or combated by donor-basin regions (e.g. in Spain). The Tajo-Segura Water Transfer in Spain and the Rocade Canal, in Morocco, illustrate how such projects are designed based on high volume targets but only end up delivering a smaller portion of water (35% for the former, 60% for the latter), thus generating water scarcity, shifts towards groundwater, or even a demand for additional water transfer infrastructure…

Wastewater has also been identified as a complementary "resource" to be treated and reused in agriculture (Qadir et al. 2007). United Nations' 2017 report entitled "Wastewater…The Untapped Resource" has recently put the issue center stage. Tunisia has been a frontrunner, with the reuse of Tunis water initiated in the 1960s. Turkey's area irrigated with wastewater was believed to be about 200,000 ha in 2004. Israel turned to wastewater reuse in the 1990s and as of today 86% of the treated wastewater is reused, mostly in agriculture. Spain came to it in the 2000s (with around 450 Mm^3/year of water, while further expansion plans have been delayed by financial difficulties). Some countries like France, Morocco, Algeria or Egypt have not yet developed ambitious strategies to tap this resource. In the latter three countries, wastewater is actually already used in many places, even though it is not treated.

Desalination is also a way to augment supply and plants are gradually appearing around the Mediterranean, notably in Israel, Algeria and Spain. But high production costs confine this resource to domestic supply. Exceptions include areas with high-value horticulture crops for export around Murcia/Almeria (Spain), and the plant under construction in the Massa basin (southern Morocco) through a PPP arrangement where water will be sold to investors and make for dropping aquifers.

Last, supply augmentation options also include on-farm storage or small communal reservoirs (like in the "1000 days 1000 reservoirs" program in Turkey) (see Le visage et al. 2018). In France on-farm storage is the main response to irrigation needs in summer, after pumping from small streams came under criticism for drying up rivers.

Responding to shortages by increasing supply without regulating uses has the well known effect of encouraging users to expand, in a vicious circle poised to generate more shortages.

10.2.5 Water Scarcity and Demand Management

In the late 1990s early 2000s, the disqualification of supply-augmentation policies generated an emphasis on demand management as an alternative (Chohin-Kuper et al. 2014). Its insufficient potential to redress severe imbalances has transformed this emphasis into a call for not antagonizing both approaches and carrying them out in parallel; when supply management does not come back strategically couched in discourses about green development, sustainability, and the like (Crow-Miller et al. 2017).

Demand management combines interventions to reduce 'losses' in urban or irrigation distribution networks, measures to reduce demand by users (retrofitted appliances, drip irrigation, water pricing, awareness raising campaigns, etc.) and to reallocate water. In the Mediterranean the UNEP Blue Plan and GWP have been spearheading the policy emphasis on demand development (Blinda 2009; GWP 2012). Although these measures are in general desirable, they have been found wanting in many situations and their potential now appears as being much more limited than often enthusiastically believed, especially in the field of irrigation.

There are two major reasons for this state of affair: first, reducing demand often implies a form of constraint and likely discontent, which makes this policy politically unattractive; second demand policies are implemented in water scarce basins where little water is 'lost'. In what follows we will in particular briefly examine the record of interventions aimed at raising efficiency, irrigation water pricing, and participatory managerial approaches.

10.3 Irrigation Management and Technical Change

10.3.1 Irrigation's Varying Context and Purposes

During the past decades, agriculture has lost economic and political weight in Mediterranean countries, as its contribution to national GDPs and, to a lesser extent, it importance in the provision of jobs have steadily declined (Figs. 10.2 and 10.3).

Markets liberalization and global competition, the development of food corporations, climate vagaries, labor-saving technologies, and farmers' ageing have slimmed down the whole sector agricultural in the region. In this context of decline, irrigation has become the most common strategy to offset climate uncertainty and risk, secure or increase farms' productivity and competitiveness, and in some cases, the only opportunity for many smallholdings to maintain their activity. In Turkey, the benefit of irrigation in terms of yield increase is 273% for cotton, 147% for fruit trees and 155% for citrus, while the largest increase (a sixfold increase in yield) is achieved for maize (DSI, 2017). In Spain, the mean net margin per hectare in irrigation is 4,4 higher than in rain-fed areas (Gómez-Limón 2014). Similar rates have been observed in the rest of Mediterranean countries.

Irrigation is paramount for both agribusiness and family farming, and plays a major role in retaining population in the rural areas of European shores and in combating rural poverty in other countries. The socio-economic importance of irrigation is still used as an argument to justify expansion policies and to forge a consensus between the farming sector and policy-makers. In Turkey, an alliance of stakeholders -the state bureaucracies, WUAs, chambers of agriculture, engineers and other concerned NGOs- has sustained the recent expansion of the irrigated lands, and firmly supports the target of 8.5 million ha to be developed. Egypt announced plans

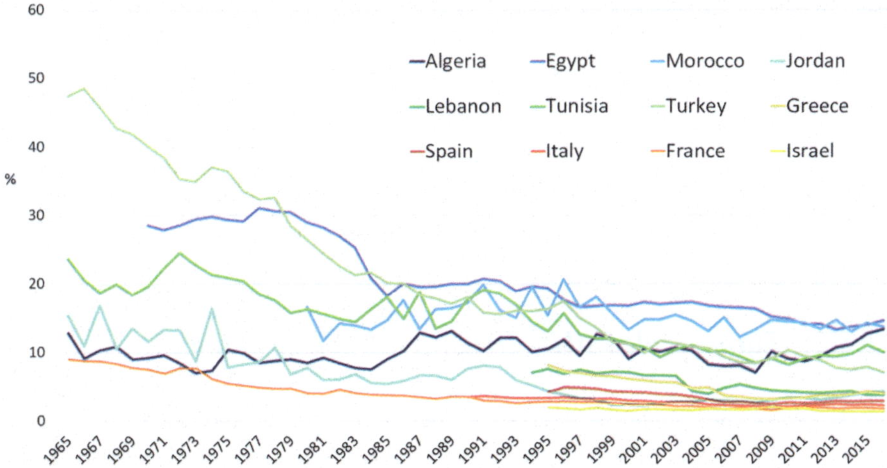

Fig. 10.2 Agriculture, value added (% of GDP) in selected Mediterranean countries. (Source: World Bank 2018)

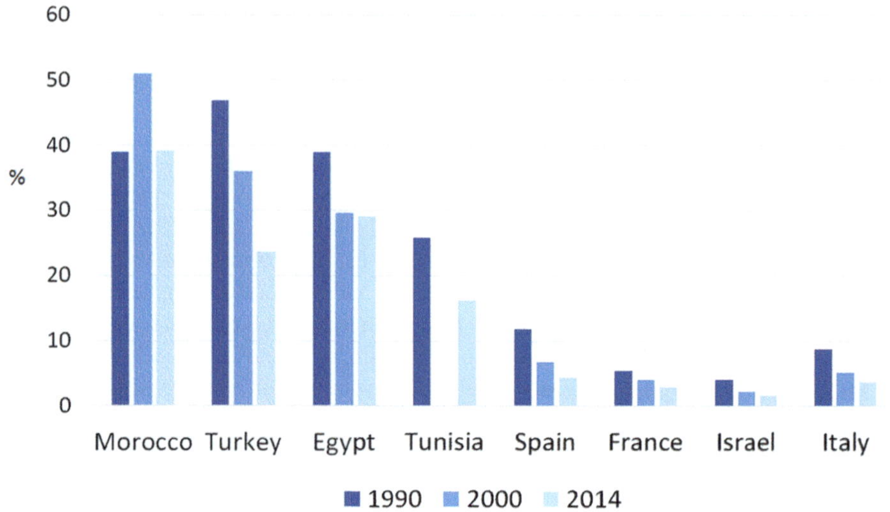

Fig. 10.3 Agrarian employment in selected Mediterranean countries (% of the total workforce). (Source: FAO)

to expand agriculture in the desert on 400,000 ha although past experience with such plan is not encouraging. In Tunisia, Morocco and Spain lands irrigated with groundwater have been expanding during the last two decades, with the support (or passivity) of the administration. In Spain irrigation areas have expanded from 3.3 to 3.7 Mha between 2000 and 2017, and the administration and the FENACORE[4] defend this growth because of the alleged decrease in the water used for irrigation during the same period (from 17.6 to 14.6 Bm^3, according to the Ministry of Agriculture), attributed to the generalization of pressurized networks and other water-saving technologies.

Nevertheless, France and Italy, after several decades of growth, followed a different trend during the twenty-first century. In France, the area equipped for irrigation decreased by 12% between the two last agricultural census (2002 and 2012), as the socio-political support for this kind of investment weakened. In Italy irrigated lands have been decreasing since 2000, but non-irrigated agricultural areas decreased even more (also in France and Spain), suggesting that irrigation contributes to sustaining agriculture, slowing down the general decline of the sector.

10.3.2 Large-Scale Public Irrigation Under Question

Iconic large-scale public irrigation schemes have been developed by the states, mostly during the 1960s up to the 1980s. This has notably been the case in Spain during Franco's regime, in central and southern Italy (Reclamation and Irrigation

[4] Federación Nacional de Comunidades de Regantes de España (Spanish National Federation of Water User Associations).

Consortia), Morocco (where King Hassan II's 'hydraulic policy' sought to develop one million hectares of irrigated land), in Algeria during the Socialist period (in coastal plains and in the Sahara), in northern Tunisia, Syria (Orontes river basin), not to mention the continued expansion of the Nile delta/valley system.

This 'hydraulic mission'has been undertaken by, but also strengthened, powerful bureaucracies such as the Ministry of Water Resources and Irrigation in Egypt, the Ministry of Agriculture in Tunisia, the National Institute of Colonization (NIC) and the "*Dirección General del Agua*" in Spain (created during Franco's regime), Mekorot in Israel (funded in 1937), or the DSI in Turkey. The DSI was modeled after and supported by the US bureau of Reclamation, and its first director (S. Demirel) even became president of the country. It is the only administration in the region to still have a substantial irrigation expansion and construction program.

The hydraulic mission time correspond to times when there was a considerable demand for rural development and large-scale public projects (Molle et al. 2009) and 'development states' could count on the responsiveness of development banks and transnational consulting companies. This mission conducted in the name of national sovereignty, food self-sufficiency or poverty alleviation gave rise to dams, large-scale irrigation schemes and other hydraulic infrastructures. They also served the objective of nation-building and of consolidating symbolic and political capital for ruling regimes (Franco in Spain, Asad in Syria, Nasser in Egypt, Hassan II in Morocco, etc.)(ibid.).

Because of management problems, low technical efficiency, contested economic profitability, recurring financial burden in the form of periodic rehabilitation, environmental impacts (soil salinisation, waterlogging, dry springs, rivers and lakes, etc.), they have therefore frequently been the object of successive waves of reforms addressing technical (modernization), financial (water pricing), or institutional (participatory management; privatization) aspects. These reforms are discussed in the following sections.

Though these public investments incurred substantial outlays, it was found that many are surprisingly underutilized, with low cropping intensities.[5] In Tunisia, the overall cropping intensity in irrigated areas is estimated at 60% of the potential due to several constraints (water shortages, soil salinisation resulting from a lack of drainage facilities, preference for using groundwater, but also reluctance by some farmers to engage in capital-intensive agriculture when input and output markets are too uncertain). In Algeria, although the equipped area reached 223,377 ha in 2012, the area effectively irrigated was as lows as 68,690 ha, that is, 31%... In Turkey, a considerable portion of land within an irrigation scheme is also not irrigated due to urbanization, adherence to rainfed agriculture or other socio-economic aspects (Özerol et al. 2012), with an average irrigation ratio of about 65% countrywide. In Egypt there is no clear statistics for the New Lands, but a substantial part of it is abandoned or only partly irrigated due mainly to a lack of irrigation water.

In Italy, since 1970 the irrigated land declined with regard to irrigable land, and is currently only two thirds of the latter. The two main reasons are low water availability

[5] The ratio of the area harvested every year to the area equipped with irrigation facilities.

and low profitability of irrigated crops, driven by changes in the Common Agricultural Policy that reduced the profitability of fruit and vegetable but also increasing regional competition. In France, low rates of use of irrigation facilities (68% in 2010) reflect changes in rainfall and crop subsidies, urban expansion (e.g. SCP company) or a planned water demand that never fully materialized (the Bas-Rhône Languedoc company). In Catalonia, the failure of the ambitious Segarra-Garrigues Canal (12,000 ha irrigated out of the 95,000 ha planned), in contrast with other successful past experiences in Spain, suggests that the time for these large-scale projects has passed.

Although Morocco is subject to the same problems, it illustrates that insufficient availability of surface water can be partly compensated by resorting to groundwater as a complement. Irrigation is largely maintained but sources up to half of its water from aquifers in a non-sustainable way.

10.3.3 The Rallying Call for Modernization

In the face of water shortages, the promise of a ‘technical fix’ provided by micro-irrigation and interventions like canal lining has been extremely attractive for politicians and managers (Venot et al. 2017). Israel has invested in drip-irrigation since the 1960s and the technology now amounts to 80% of the irrigated area (the remaining being under sprinkler). Spain and Morocco, in particular, have designed ambitious multi-billion euro programs to ‘modernize’ irrigation and ‘save’ water (1.16 Bm^3 at an estimated cost of €7.3 billion for Spain, and 0.8 Bm^3 at the cost of €3.7 billion for Morocco), and Egypt’s Ministry of Agriculture has also floated proposals for massive technological shifts, as part of a 640 billion pound[6] agricultural strategy until 2030.

In Algeria, Tunisia, Italy or Turkey water-saving technologies have also been encouraged and subsidized by the state. In Turkey grants support 75% of the costs for collective pressurized irrigation applications and 50% of within-parcel modern pressurized irrigation investments but the use of water saving technologies is far under target because of increased costs due to land fragmentation. In Italy the investments required by these technologies have been mainly supported by Rural Development Policies, covering up to 60% of the costs. The same applies to Tunisia. In Morocco, subsidies for drip irrigation from the ‘Green Morocco Plan’ now reach 80–100% according to the type of farm. Spain modernization programs (which supports between 40% and 100% of the investments) and farmers’ own strategies have now resulted in 44% of the total irrigated land being under drip. This also includes sprinkler systems that may be better suited for certain crops and terrains (e.g. the modernization of networks in Aragon, Spain and France, where sprinkler systems dominate).

[6]Around $120 billion, in 2010 values. This is only half of the investment envisaged, the other half being expected to come from the private sector.

It is important to note, however, that the spread of sprinkler and drip irrigation owes little to water scarcity, as generally believed. In France and Italy increasing labor costs and the ease of use are the main factors. In Spain, Ortega-Reig et al. (2017a, b) have identified a double discourse, whereby while the administration and the government boards of the WUAs justify the investment for water-saving purposes, farmers attribute their interest in drip irrigation to labor saving/comfort and productivity. In numerous regions, the adoption of drip goes with an expansion of, or a shift to, high-value crops, mainly fruit trees and vegetables, which demand a better application of water. Drip also allows for fertigation and altogether improved agricultural practices generate substantial increases in yields and better uniformity of produce (an important standard of supermarkets' demand, notably in the fruit and vegetable sectors of EU countries). But this also means that such intensification must enjoy a secure source of water compatible with frequent irrigations. This is why the shift to micro-irrigation is largely associated with (individual) groundwater-based supply where, in addition, pumping from well may also directly provide the pressure needed to serve the distribution systems (frequently, though, farmers will build an intermediary on-farm storage from which they will pump again to distribute water).

What drip irrigation really does to water has, in recent years, been the object of much debate and controversy in places such as California, China, Australia, Spain or Morocco. First, it has been observed that for a variety of reasons farmers (in general smallholders) did not always reduce the amount of water applied to their plot. Second, even where this is the case it appears that the amount of water effectively consumed at plot level is basically unchanged, and even sometimes increased due to the large increases in transpiration allowed by better and more frequent application of water (Perry and Steduto 2017). Third, the shift to drip irrigation is often accompanied by: (a) a shift to more water-consuming crops, (b) a densification of tree plantations, (c) an extension of the cultivated area made possible by reduced per hectare application rates (the so-called rebound effect) (Ward and Pulido-Velazquez 2008; Playán and Mateos 2006; Willardson et al. 1994). The combination of these effects can be observed in Morocco (Molle and Tanouti 2017), although there are also other Mediterranean cases where existing plantations are left untouched (Soto-García et al. 2013; Sanchis-Ibor et al. 2017a). It is apparent that drip and agricultural intensification in general do the opposite of what decision-makers expect or announce: they do improve productivity and reduce labor costs and drudgery, but do not reduce water *consumption*. For some reasons, Turkey, Italy, Tunisia and even Israel have so far largely avoided this debate, despite its relevance for regions such as the Konya closed river basin in Turkey, for which the literatures is confined to plot-level technical issues.

Not all the cases of irrigation expansion are associated with a *shift* to drip irrigation. In the vineyards of southern France, for example, irrigation has tripled since 2000, stimulated by subsidies and wine market requirements, despite the reduction in the total vineyard area. Similar processes have been observed in old rain-fed areas in Spain, such as the olive grove areas of Andalusia, where drip is only part of a strategy to increase production, with negative effects on groundwater overexploitation and energy consumption, or in Valencia, where the Basin Authority conditioned the expansion of citrus irrigation into rain-fed areas upon the installation of pressurized distribution network (Sese-Minguez et al. 2017). Drip is, in such cases, the vector for irrigation expansion and, despite being often considered as a demand-side measure, the vehicle to perpetuate old irrigation expansion policies, particularly in closed basins.

More generally, modernization or rehabilitation of existing infrastructure are made necessary by their degradation, the intent to better and volumetrically control supply, or to deliver water under pressure instead of by gravity. Morocco, for example, has undertaken to transform part of its large-scale gravity schemes (with a total of 200,000 ha) into collective piped networks with a pressure allowing the use of drip irrigation at the farm level. In Spain, the second pillar of the EU Common Agricultural Policy (rural development) still prioritizes the policies of irrigation modernization in the Regional Development Plans (RDP), even though the budget is going to be reduced. In Spain, or in France, telecontrol has been widely used to monitor and operate irrigation structures in real-time from conveyance canals and in-line reservoirs to water delivery networks and drip or sprinkler irrigation systems. Consequently, labor requirements have been reduced and working conditions have improved in farms and irrigation districts.

Modernization also relates to improvements at the farm level. Aside from the spread of micro-irrigation discussed above, field moisture monitoring systems are being adopted in countries such as France, Spain or Israel. They provide growers with real-time computerized field monitoring capacity. Remote sensing or drones are also proving to be useful in the identification of spatial heterogeneities at the (large) farm level and the development of precision agriculture. In sum, while better control of water quantity and quality, and of nutrients, is enhancing land and water productivity, and generally reducing the volumes of water applied, better technical control also has to go with a better control and predictability/reliability of supply, as well as with improved technical services and market linkages (Playán and Mateos 2006).

When one zooms out to the scale of the irrigation system, aquifer or river basin it is frequent that no water is 'saved' since return flows have already been appropriated. This is notably the case in systems over quaternary aquifers where seepage either replenishes the aquifer (often intensively tapped through wells) or goes back to the river system. 'Savings' estimated by cross-multiplication on efficiency coefficients before and after adoption of drip (Blinda 2009), are illusory. Even more so when these 'paper water' savings are transferred to other areas, as happened with the water 'saved' through the modernization of traditional irrigation systems in the Júcar basin planned to be transferred to the Vinalopó basin.

Beyond this efficiency debate, the introduction of pressurized irrigation might be the most significant technical change in the Mediterranean region since the Medieval Arab Agricultural Revolution. Microirrigation is also changing norms and practices in WUAs, with traditional farmers' knowledge being replaced by engineering principles and centralized irrigation water management and decision-making, as observed in Spain (Ortega-Reig et al. 2017a, b). In Morocco, Benouniche et al. (2014) have identified processes of "bricolage", by which farmers manipulate the drip kits and perform their own technical solutions to adapt the technology to their needs. Modernization may incur a loss of cultural heritage in some ancient hydraulic systems. This has raised some public concern, which is slowing down or stopping the advance of drip in some traditional waterscapes. In Spain, in some historical irrigation systems farmers resist to the adoption of microirrigation, defending the natural flow of water as a matter of identity and a contestation of the powerful notion of modernization.

10.3.4 Irrigation and User Participation

In 1846, Catalan lawyer Jaubert de Passà published his "*Recherches sur les arrosages chez les peuples anciens*", a unique four volume global compendium on irrigation. Rooted in his experience with communal irrigation in the French Pyreneans, he describes at length the virtues of self-management in ancient communal irrigation (Ingold 2008; Mollard 2004). Mediterranean countries all have a rich history of small- to medium-scale irrigation systems, largely developed endogenously, which have often endured up to these days.

In Italy, nearly 500 Reclamation and Irrigation Consortia (*Consorzi di Bonifica e Irrigazione*), associations of landowners involved in collective water and land management, manage most of the Italian irrigation. The consortia of Northern Italy mainly derive from spontaneous associations of landowners, whether small (in the alpine valleys) or large (in the Po valley), while those from the Center or the South are generally quite large and were established in the twentieth century. They have management and maintenance duties and a public juridical personality to share the costs and collect related payments. In France, approximately 2000 irrigation user associations (ASAs) manage a total of over 300,000 ha of irrigated networks of all sizes, some dating back to 500 years ago. ASAs have considerable decision-making autonomy in terms of the operating rules governing the relationships between members and the operational management of the area but are subject to a strict regulatory framework and supervision from the state (notably with regard to financial management). In Israel, Water associations are regional cooperatives whose members are *kibbutzim* and *moshavim* (settlement communities). These water associations are also platforms for political activity and negotiations with the public officials on issues such as quotas or water rates. In Spain, numerous irrigation communities (*comunidades de regantes*) have their origin in the Islamic period (*jmaa*) and were recognized after the *Reconquista*, before being regulated by the modern state in the 1866 Water Law. Today, more than 7200 communities manage collective irrigation systems, encompassing the 80% of the irrigated lands. Most of them belong to FENACORE, their national federation. FENACORE acts as a powerful lobby and was also the promoter of the Euro-Mediterranean Irrigators Community (EIC), integrating associations from Italy, Greece, France, Portugal, Spain, Tunisia, Morocco, Egypt and Germany.

In other countries, despite an equally rich communal irrigation background, water user associations (WUAs) established in public schemes by the administration have been bedeviled by a number of difficulties and constraints (Ghazouani et al. 2012). In Morocco, the lack of willingness to empower associations is apparent and the law on WUAs lists servitudes and duties, while no autonomy, financial or otherwise, is granted. In Tunisia, 1253 GDAs are responsible to a varying extent for a total surface area of 213,000 ha of public irrigated areas. The technical and financial management of irrigation networks by GDAs and the performance of these groups in terms of service quality vary widely: problems such as water shortages, insufficient financing for major repairs needed, elite capture or political meddling, have triggered several successive reforms but not eliminated problems. In Egypt, comprehensive experimentation with WUAs at different levels –from the tertiary to the district level- as part of different donor-funded projects has not succeeded in durably institutionalizing a degree of participatory management. In Turkey, in the early 1990s DSI started to transfer O&M responsibilities to a variety of collectives according to a model of irrigation management transfer (IMT) inspired by the Mexican example and largely motivated by the intent to externalize costs onto users.

All in all, these experiences have not been very successful as the states (or their water bureaucracies) have been reluctant to devolve or share power, financial autonomy has been an elusive goal, and a top-down process of formation has rarely elicited viable collective action.

10.3.5 The Water-Energy Nexus

Because water must frequently be lifted and/or pressurized (in the case of piped networks or farm-level sprinkler or micro-irrigation), many irrigation schemes come with substantial energy needs. When water first needs to be treated or desalinated, these energy requirements –and associated costs- quickly balloon. This also applies to interbasin transfers that come with pumping costs (e.g. Algeria or Tunisia), one main obstacle to the planned North-South Moroccan transfer.

Hardy and Garrido (2010) observe that the modernization of irrigation in Spain has brought with it an increase in farm consumption of electricity, against a backdrop of rising energy prices. This general increase is caused by new irrigated areas, a shift from surface water to pressurized systems, although numerous groundwater-based farms decrease energy consumption when they adopt drip irrigation (Jackson et al. 2010; Sanchis-Ibor et al. 2017a). Likewise in Tunisia, various factors contribute to raising the energy bill: the deepening of wells and boreholes; adoption of energy-intensive irrigation techniques and increased regional transfers of water.

The rise in the electrical bill caused by modernization has sometimes prompted a financial support of the State to the agricultural sector. In Morocco, agricultural policies (notably the conversion to drip irrigation) have been designed with little consideration of their impact on the country's €10 billion energy bill (96% of Morocco's energy is imported), 13% of which on account of agriculture. In 2011, total subsidies involved in energy consumption by agriculture amounted to €0.75 billion. In Italy, some Regional Administrations financially contribute to covering WUAs' O&M costs in order to offset the structural handicaps involving higher energy costs for lifting and distribution. In Spain, the national federation of WUAs organized massive public demonstrations in 2014 to demand, unsuccessfully, this type of subsidies for the irrigation sector.

Egypt's irrigated land probably forms the biggest pump station on earth. Both irrigation and drainage water is lifted, whether through 1 of the 600 large-scale pump stations or through close to 4 million individual pumps. Energy subsidies make up 70% of total state subsidies and therefore represent an outstanding financial burden.

Because of the importance of the groundwater economy for livelihoods and agriculture, policies to use solar energy to reduce groundwater pumping costs for users have been considered. Countries like Morocco and Egypt encourage and even subsidize the expansion of solar-based pumping, and some Spanish regions have recently included such subsidies in their modernization programs. However, Jordan has rightly anticipated the associated negative impacts on groundwater use and the Ministry of Water has so far brushed the idea aside. The recent report by FAO (2018) titled 'The benefits and risks of solar powered irrigation' – duly emphasizes that this option "if not adequately managed and regulated – bear the risk of supporting unsustainable water use" but nevertheless promotes it in contexts where regulation is precisely problematic.

10.4 Irrigation and Its Economic Context

10.4.1 Investments and Financial Sustainability

As discussed above, the financial sustainability of public irrigation schemes and the recurring burden for state coffers have long been a key policy concern. Infrastructure deterioration and the need for maintenance of hydraulic structures and pump stations, desilting, dredging, weeding, etc. have taken their toll on state budgets. Solutions have included 'sharing' management and costs with users through participatory irrigation management (PIM) or more assertive Irrigation Management Transfers (IMT), most particularly, as far as the Mediterranean is concerned, in Turkey or EU countries.

The World Bank supported the idea of privatization in irrigation on the model of the utility privatization drive of the 1990s. But the recognition that large-scale irrigation systems are different from urban networks or power grids has percolated very slowly at best. It became apparent that the potential of irrigation for generating profit to a hypothetic private company taking over the management of a scheme was very low, as transpired for example in the feasibility study of the privatization of the *Office de Mise en Valeur Agricole* in Morocco. Farmers' ability-to-pay in such schemes is largely linked to the remuneration dictated by markets and this value is low for most major crops.

Emphasis has therefore been shifted to Public-Private Partnerships (PPPs) as a means of both involving the private sector in the financing and management of irrigation and favoring a type of agriculture that "can pay for a service" (World Bank 2007). That cash crops are frequently associated with individual groundwater-based irrigation suggests that they could expand to public schemes and generate a higher willingness- and ability-to-pay for water. But this makes light of the facts that horticulture expansion has limits (it represents only 7% of cultivated areas worldwide), and that capital-intensive agriculture is risky for farmers without capital (in particular in the face of pest pressure and market vagaries).

Commercial, climatic and political risks have been identified as hindering the adoption of PPPs, as could be seen in the case of the West Delta project in Egypt. In the Guerdane project, Morocco, a private company built and operates a pipe to transfer water from a dam to the investors. Forty-eight percent of the project costs have been shouldered by the state. Another project is starting in the Coastal Chaouia region, south of Casablanca, where the delivery service has been entrusted to a private company; and another project is being completed in the Massa region (south of Agadir), where a desalination plant will be built to irrigate market crops and citrus for export.

Social and economic risks have been observed in Spain, where the re-collectivization of irrigation management after privatization has been analyzed in the Jucar River basin (Sanchis-Ibor et al. 2017b). Drip irrigation has acted as a Trojan horse, facilitating the penetration of private companies into irrigation management, displacing farmers that had collectively (and successfully) managed irrigation systems during the last millennia, rising O&M costs and creating distrust and social disarray.

10.4.2 Irrigation Water Use and Economic Tools

Irrigation water pricing is a prominent tool of the EU Water Framework Directive. Intensely promoted in the 1990s, the idea was shifted to the backburner in the late 2000s,[7] partly due to the recognition that the field of irrigation, especially large-scale irrigation, presented key differences with the urban water or energy sectors, where pricing tools have well documented applications (see Cornish et al. 2004; Molle and Berkoff 2007, for a full analysis of the reasons for the failure to achieve irrigation water savings through pricing).

In Spain, water pricing is based on public and private tariffs that are essentially what users pay for the delivery of water. Public prices include resource management fees for the regulation of surface or groundwater conducted by the State, on top of

[7] Comparison between the 1993 and the 2003 World Bank's water Policy Papers clearly reflects these changes.

costs of amortization of the investments they have made, and O&M running costs. Tariffs and fees vary widely among Spanish irrigators as a function of the resource origin, its size, the method of water application, etc. They have never been used by the administration as a means of reducing demand, although this idea was brought to the National Water Council in 2006 but was widely refused by regional governments and water user associations' representatives. In Italy, Dono et al. (2012) note that for selected schemes applying a volumetric pricing would not appreciably increase overall efficiency and would especially redistribute income among farmers. Operating costs also reflect the size of the irrigated land more than the volume of water supplied, and inefficiencies are largely due to the oversizing of infrastructures and would therefore not be reduced by an increase in the price of water.

In Israel, water for agriculture has historically been heavily subsidized as a means of encouraging and supporting settlements. Despite strong lobbying, in 2006 farmers agreed to a gradual rise in the price of irrigation water that would reflect the average cost of water supply. At the same time wastewater remained subsidized so as to encourage farmers to shift to treated wastewater.

In Tunisia, the overall increase in both prices and the cost-recovery rate between 1990 and 2002 served to cover a considerable portion of the rising water system operation and maintenance costs. Prices were frozen as of 2002 because of the strong resistance among irrigation users, which had a considerable impact on the financial balance of the GDAs, weakening their ability to fund servicing and maintenance work. "Political interventions" regarding pricing, bypassing regulatory bodies, viewing minimizing water charges as a way of improving farmers' revenues, has often had a negative impact on the sector. A similar situation can be found in Morocco, where tariffs were raised during the 1997–2006 period, also primarily for financial reasons. But raising prices sufficiently to cover O&M costs has proven to be problematic. For Doukkali (2005) "the inability to complete water-pricing reforms clearly suggests how political costs have overshadowed the real socio-economic and resource costs".

Financial sustainability also features highly in Turkey, where WUAs depend on the fees they collect to survive financially and to pay their personnel. The proportion of fees that the associations can collect from their users has increased from 40% to 60–100% after the IMT reforms, but many associations are financially non viable, especially if they have pumping costs due to the privatization of electricity services (Le Visage 2015). In Egypt, there is no attempt to charge farmers for water and only users dependant on a same collective pump station collect money for O&M costs.

In sum, the main challenge with water pricing is the financial sustainability of both WUAs and infrastructures. In Spain, France, Italy, Turkey, Egypt (for collective pumps) and Tunisia WUAs can legally recover costs, or part of them, from users. Full cost recovery is rare and includes cross-subsidizing from other sectors, and in no case covers all environmental costs. High prices rarely promote water savings because the marginal value of water in terms of production is far higher than its cost to the farmer, especially in water-scarce settings; they may push farmers to shift to groundwater; economic studies invariably find that price levels that would achieve water savings would severely dent incomes and are therefore politically unfeasible (see Molle and Berkoff 2007; Chohin-Kuper et al. 2014). The same applies to individual groundwater pumping.

Volumetric pricing, wherever technically achievable, and where supply is sufficiently controlled so that an on-demand system can be established, can be used with increasing block tariffs to both allow and discourage the use of water beyond a certain established quota (e.g. Israel and some schemes in Spain, France or Italy). In specific cases where water costs are very high (deep groundwater pumping or desalinated water) use efficiency becomes paramount but is already a feature of the high value commercial agriculture which can endure in such conditions (e.g. Almeria, Spain; Massa basin, Morocco). Water tariffs have also been used by some Spanish WUAs to incentivize the adoption of drip irrigation, penalizing farmers who are reluctant to abandon gravity practices.

While direct administration of water prices can at best achieve recovery of O&M costs, and sometimes of additional 'resource management' and even part of investments costs, water use is sometimes affected by energy prices. Energy prices are amenable to regulation but, in general, they are subsidized for agricultural water use rather than used as a lever.

Other potentially interesting economic tools have been noted. They include Temporary Right Transfers in Spain, and temporary buy out of rights in France or in Spain, where farmers are compensated for not using the water they are entitled to. This mechanism has been criticized because of the risk of creating potential "water-owners" in regions where agriculture is decreasing. In Spain, despite the initial enthusiasm for this mechanism, water authorities are now closely reviewing (and adjusting) WUAs' water rights, reducing the scope for such exchanges.

10.4.3 Corporatization of Irrigation?

A notable trend is observed with regard to the development and promotion of capital-intensive agriculture. "Plasticulture", with intensive use of greenhouses, plastic mulch, lines of drippers, etc. and largely devoted to vegetables, or intensive fruit-tree plantations (e.g. citrus, vine, guava, stone-fruit trees,…) has been developed in different contexts by both local investors (e.g. Israel; Almeria, Spain; Jordan Valley), but also national or foreign corporate capital (e.g. Egypt's New Lands, Morocco's oases, Algeria' Sahara, etc.) (Dixon 2017). Plasticulture has now expanded to most countries, creating agro-technological clusters in Almeria (Spain), Souss-Massa (Morocco), Albenga (northern Italy) and Southern Sicily, Demre and Kumluca and Southern Anatolia in general (Turkey); Tartous coast (Syria); Halba (northern Lebanon); or Israel.

This type of agriculture has been discursively promoted as 'modern' and 'efficient' in countries like Egypt and Morocco, where performance in terms of water productivity, efficiency but also contributions to export and the balance of payment are underlined. But emphasis on 'efficiency' performance often results, purposefully or not, in sidelining the other two dimensions of IWRM -equity and environment-

and in belying the conventional consensus that the three 'pillars' of IWRM have to be considered and reconciled. Corporate agriculture often benefits from preferential, state-facilitated, and frequently state-subsidized access to land and water resources. In Guerdane, southern Morocco, water has been piped out of the main valley to 10,000 of fruit trees mainly belonging to outside investors. Everywhere, deep groundwater pumping impacts water tables to the detriment of small farmers who do not have the means to deepen their wells.

10.4.4 Changing Markets and Incentive Structures

The spread of liberalism in the last decades of the twentieth century has increased the internationalization of Mediterranean agricultures and has also remodeled the public support to the irrigation sector. These processes have resulted in different impacts among the Mediterranean countries, depending on the type of agricultural production, farm structure, capital availability, and other regional and sectoral conditions. There has been much debate, often ideologically biased, about the results of these processes. The only evidence is that there have been winners and losers, and that losers are mainly the weaker smallholders.

In Israel this turn towards international markets has been largely successfully managed. With the opening to the international markets, the drought periods in the 1980s and, later, a reduction in the subsidies for water as well as in the political clout of the agricultural lobby in the Parliament, farmers had to adapt and innovate, developing an export-oriented high value agriculture irrigated with drip (Kislev 2013).

The nations of the EU also progressively increased their exposure to world markets since the McSharry reform of the CAP (1992), and particularly after the Mid-Term Review of 2003, which decoupled subsidies from production and almost totally reduced the import levies to fulfill WTO standards. Despite a still significant level of protection (Producer Support Estimate indicator is still 21%, against 39% in the 1980 decade),[8] the decreasing number of farms and the slight increase in the average size of agricultural holdings, numerous European farmers are unable to compete satisfactorily. In Spain, Italy and France, the viability of irrigated agriculture is threatened by the low remuneration of productions in the national and international markets (Atance 2013).

Reliance of migrant workers is high in Italy and Spain, and also in Portugal and Greece, because of the need for reducing costs and the lack of interest from locals, despite high unemployment rates (see Corrado et al. 2017). These contingents of workers are underpaid, suffer from precarious labor conditions, and in many cases do not have residency permits, which makes them socially vulnerable. This phenomenon, frequently made invisible, constitutes a clear form of social dumping (Laurent 2013) that also takes place in other Mediterranean countries. Nearly half

[8] 17% for Israel and 27.9% for Turkey, according to the OECD (2017) estimations for 2015.

of all Syrian refugees arrived in Turkey work in farms for very low wages (Erturk 2017) and Israel, after the second Intifada, has replaced Palestinian workforce by other migrants –mainly through the Thailand-Israel Cooperation on the Placement of Workers–, whose precarious conditions have been denounced by several NGOs (HRW 2015).

Changes in the supply chain of fruits and vegetables, where small operators tend to be replaced by large companies vertically integrated with retailers, have affected family agriculture (Petriccione and dell'Aquila 2011), while the growing competition from southern countries has weakened their position in national markets (Malorgio and Hertzberg 2007; Nomisma-Unaproa 2016). Farmers' associations frequently organize demonstrations against the importation of foreign agricultural products, publicly complain about price volatility, and demand policies that provide a more equitable distribution of the added value in the commercialization chain (García Álvarez-Coque and Martínez 2014).

Technology has facilitated recent improvements in productivity (sprinklers and drip), but has also caused a dependence on agricultural inputs, particularly energy, that is detrimental in a context of low agricultural prices. However, we lack of quantitative studies analyzing the economic impact of such technologies at the landholding scale, because researchers have concentrated their efforts on the analysis of water costs or water productivity, often not considering all the factors involved in the farming budget. This is a relevant issue, because in France and Spain, many farmers consider irrigation as pivotal for maintaining competitiveness and in Spain elderly farmers frequently admit they would have abandon agriculture without such 'modernization'.

The swings of the CAP during the last decades, and particularly, the changes in the Common Market Organizations, have induced substantial changes in irrigated landscapes. Maize subsidies played a major role in the expansion of irrigated maize in France, and the 2005 reform of the Common Market Organization for sugar has brought to zero the demand for sugar beet in many irrigated areas of the center and south of Italy. Similarly, the abolition of milk quotas has reduced milk production, and has indirectly caused a decrease in the demand for corn, affecting several Italian irrigated districts.

In Egypt, during the decades of 1980 and 1990, under the influence of donors and development banks, priorities shifted towards price liberalization and private investment in agriculture. There is no consensus on what these reforms have eventually produced but the capital-intensive export-led agriculture in the New Lands neither resulted in significant job opportunities nor in increased food security. Linking small farmers to markets through contract farming has also been little successful.

In Egypt and Morocco, but also more generally, a dual model has been consolidated, with a 'modern' capitalized high-tech sector, and a traditional smallholders sector supposed to be transitioning towards a 'modern' agriculture (Faysse et al. 2014; Akesbi 2014). The future of both sectors is linked to the evolution of factor prices, market conditions, and public support to irrigation, including expansion and modernization. Market risks are well illustrated by seasonal dramatic drops in prices, for example of tomato and citrus in Morocco, or various vegetables in Spain.

10.5 Policy-Making and Institutional Dimensions

10.5.1 The Policy Process

Many irrigation reforms have been conducted worldwide in the 1980s and 1990s (Mollinga and Bolding 2004). After the turn of the century, the topic has somehow become less fashionable, needed or urgent, depending on the viewpoint considered. Earlier reforms were largely motivated by the need of financially strapped states to divest part of their recurring rehabilitation and O&M costs onto users, combined with the frustration of donors in front of the perceived low performance of the sector and repeated rehabilitation needs. They were also– part of a more general "rolling-back the state" neo-liberal context and a period in which both economic tools and "participation" came to be seen as the major solutions to the ailments of the sector (Mollinga and Bolding 2004; Molle and Berkoff 2007). More recent global policy emphasis has shifted to privatization, PPPs, and supporting large-scale private investments.

Institutional change in the irrigation/water sector has been found to be seriously under-researched (Mollinga and Bolding 2004). This applies to issues such as the role and strategies of hydraulic bureaucracies, the constant reshuffling of ministries and restructuration of the administrations in charge of agriculture, irrigation, water, the environment, public works, etc., participatory irrigation reforms, the role and transfer of 'policy models', or the pervasive and glaring gap between formal policies and laws on paper and their implementation or enforcement on the ground. Most countries do "tick all the boxes" and have incorporated in their legislation participatory management, economic tools, river-basin planning or organizations, user-pay or polluter-pay principles but these remain largely cosmetic and/or little effective in most countries.

While this volume did not either undertake original research on such issues, the fact that a large part of the literature is confined to the description of formal regulations, implicitly assuming their overarching importance, blinds us to the multi-faceted dimensions of policy-making, and in particular to the politics of reform and policy-making.

The role of 'hydraulic bureaucracies' in the evolution of the irrigation sector is under-documented, although it is well established that bureaucratic interests are an essential part of it (Molle et al. 2009). We observe a resilience of the 'hydraulic bureaucracies' in the Mediterranean, whether they are under the Ministry of Agriculture (Spain since 2008, Tunisia) or a water or energy ministry (Egypt, Lebanon), but even where these have been put under or pooled with the environmental administration (Morocco, Spain between 1996 and 2008, Turkey between 1991 and 2011). For example, in Morocco Tanouti (2017) shows how the hydraulic bureaucracy has preserved its interests despite being put under an administration in charge of the environment. However, while countries like Morocco and Turkey still have a substantial dam/irrigation building program, in most countries the lack of adequate sites and scarce funding have clearly curbed the historical 'hydraulic mission'. Large-scale public irrigation, with the exception of the GAP in Turkey, is now mostly a thing of the past.

Little is known also about the role of the private sector. Many studies on Egypt show the link between irrigated land expansion and the clientage practices that are at the root of state building and reproduction, notably through providing opportunities for capital accumulation in land reclamation and agricultural development (Dixon 2013; Roccu 2013). In Morocco, national or international companies investing in irrigated agriculture also carry substantial political power. In Algeria, some powerful people are also able to invest in desert agriculture.

10.5.2 Policy Contradictions

Lack of "alignment" or sectoral contradictions are pervasive in the fields of agriculture in general and irrigation in particular (Özerol and Bressers 2015). They are linked to antagonistic objectives pursued by the different ministries despite frequent claims that 'coordination' and 'integration' are achieved through various inter-ministerial committees or high-level agricultural, or environmental, councils.

The most ubiquitous contradiction is that pitting against one another the water and agriculture administrations. This right hand ignoring what the left hand does is particular glaring in Morocco and in Egypt. In the former, the Plan Maroc Vert (PMV) powerfully thrust by the Ministry of agriculture promotes the expansion and intensification of irrigated agriculture, even in areas where water tables are already dropping by 1 m a year due to overexploitation. The disregard for both water resources and the environment illustrates an on-going regain of political power of the agricultural sector which can perhaps be partly ascribed to both the importance of the sector in the social stability of the country and the private interests that have invested in it. In Egypt, the Ministry of Water Resources and Irrigation and the Ministry of Land Reform and Agriculture have disputes about various issues, including on what is the potential for water saving, the latter promising to "modernize" agriculture and save 12 Bm^3, despite the former making explicit the impossibility of such target.

Policy contradictions are also apparent with regard to subsidies that promote certain crops that increase water use. This has been the case, for example, with EU subsidies to maize (France, Spain) although since 2003 subsidies are decoupled from farmers' production decisions through a system of direct payments. A recent report of the European Court of Auditors (ECA 2014) has concluded that, on the one hand "delays in the implementation of the WFD have, as a matter of fact, hindered the integration of water policy objectives into the CAP" while, on the other, "monitoring and evaluation systems both directly related to the CAP [...] did not provide the information necessary to fully inform policymaking as regards pressures on water coming from agricultural activities".

Integration between irrigation/water issues and environmental dimensions is also very limited in most countries (Jordan and Lenschow 2010). Examples include environmental impacts of stream and aquifer overexploitation on river base flows (e.g. Spain or Greece), or springs and wetlands (Azraq oasis, Jordan; Marrakech Palm grove, Morocco; Ichkeul lake, Tunisia; Konya Lake, Turkey; Tablas de Daimiel, Spain; Lake Karla or Argolis, Greece, etc.), and coastal areas (saline water intrusion in all countries).

A review of countries in the Arab world has identified a "lack of integration between sectoral water-related policies which leads to fragmented programs and inefficient utilization of technical capacities and financial resources", together with an absence of social and economic dimensions in developed water policies and insufficient awareness on environmental issues (GWP Med 2007). This is a pervasive problem that has no simple rational solution because lack of integration also reflects turf battles between ministers and ministries in the attempt to gain a bigger slice of the state budget and maximize bureaucratic or political power.

In Turkey, "water and environment affairs are managed under different ministries [which…] has caused some difficulties in executing water and environment works because of duplications of their duties and responsibilities" (Delipınar and Karpuzcu 2017). DSI remains the dominant administration in the Ministry of water, with little integration with the Ministries of agriculture and of the environment (Özerol et al. 2012). In Egypt, these three sectors also come under three distinct ministries, marked with hyper-centralization, vertical integration, and problems with knowledge-sharing and cross-sectoral coordination. In Morocco, the Economic, Social and Environmental Council recently issued a disquieting report emphasizing the absence of an operational regulatory body in the water sector (the inter-ministerial Committee being non-operational for years), failures to properly implement the 1995 Water Law and insufficient coordination between departments concerned with water.

One solution to inter-ministerial competition that is often proposed is to merge water and agriculture in the same ministry (as in Tunisia or Turkey, before 2011), or water and environment (as in France or Italy), or the three together (as in Spain). This does instill some consistency in overall water management but does not provide any check and balance and irrigated agriculture may still be supported beyond the availability of the water resource. The example of France shows that environment and development (*aménagement*) can be put in the same ministry but that resolving tradeoffs internally requires ad hoc mechanisms and balancing acts between stakes and sectors (Lascoumes 2014).

10.5.3 Political Dimensions

Chapters in this volume cover some institutional and policy changes and briefly hint at a number of political factors that interplay with irrigation policy elaboration and implementation. Already mentioned are the interests of the state in announcing irrigation projects and water resource development projects that are presented as unquestionably desirable or national priorities. A recent illustration is President

Sisi's "1.5 million feddan project" that seeks to expand groundwater-based irrigation in Egyptian deserts and has been accompanied with much hype, as had been earlier versions of the same ideas since Nasser's time. In Turkey, state authorities assert that an additional two million jobs will be generated once the targeted areas have been developed for irrigation by the year 2023 and benefits are presented as unquestionable while negative impacts, or the fact that one third of irrigation facilities are currently idle, are glossed over (Özerol et al. 2012).

An important political element of irrigation/water policies are electoral concerns from either the state or MPs and other politicians. In the past MPs have been instrumental, for example, in opposing water pricing policies (e.g. Egypt) or hikes in tariffs (e.g. Tunisia, Jordan). Energy subsidies are also maintained for similar reasons and their potential for triggering social unrest in case they are discontinued. Crucially, political considerations have worked against regulation of groundwater, which largely explains the current worrying overall situation of overdraft. Access to groundwater is not only the compensation and the alternative for farmers in public schemes who do not receive adequate supply (e.g. Algeria or Morocco), but also a "livelihood escape valve" in that it provides an opportunity for raising rural incomes and reducing poverty (Kuper et al. 2016). In the volatile post-2011 context in the MENA region, controlling illegal wells and other violations, such as urban encroachment on agricultural land (e.g. Egypt) or water theft (e.g. Morocco), has become close to impossible.

Politics also speak to the relative political power of the agricultural sector, which may reflect the interests of capital-intensive private farming, the social and employment importance of agricultural activities, or (in Europe) the need to maintain a minimal activity in rural areas, whether for "landscape management" or sustaining public services. The sectoral weight and political clout of agriculture, combined with the weight of hydraulic bureaucracies and/or the development drive of the ministry of water/agriculture, have in general been quite high in the Mediterranean, although they have certainly declined in the last three decades. This is well illustrated by the case of Israel, where the crises revealed by the 1989–1991 and 1999–2001 droughts laid bare the fact that over-exploitation of water was the outcome of longstanding policy to deliver subsidized water to colonies to encourage and support settlements, and of an impasse between the agricultural lobby, which prevented an increase of water rates for agriculture and a decrease in supply, and the Treasury, which blocked desalination. In 2006 the Water Commissioner, originally under the Minister of Agriculture, was replaced by a Water Authority. The Authority is responsible for the management, operation and development of the water sector and is placed under the Ministry of Energy and Water, reflecting the shift in water policy away from irrigation and a weakening of the agricultural lobby. Such a change can also be witnessed in Jordan, where policies now seek to reduce the agricultural use of groundwater in ways that were not possible 20 years ago (Al-Naber and Molle 2017).

The political clout of the agricultural and/or irrigated sector remains high, in a sheer contrast with the contribution of the sector to national GDPs. Agricultural lobbies remain strong in Spain and France, for example, and also in countries where capital-intensive export-driven agriculture is benefiting powerful actors and is promoted by the states (e.g. Turkey, Egypt or Morocco). In Spain, the Spanish government and the agricultural trade unions have fought maintaining the status quo in the internal distribution of the European subsidies, so that the irrigation farmers are receiving an economic support significantly higher than the rain-fed producers.

10.6 Challenges Ahead

10.6.1 Quantitative Easing/Squeezes

In the Mediterranean, many river basins are closed and many aquifers over-exploited, with the annual volume of water consumed sometimes exceeding the average annual available water. Because such systems present a very high degree of water recycling, it is only possible to restore a degree of sustainability through a reduction in consumption (evapotranspiration), which explains –as discussed earlier- the limits to what can be achieved through demand management. In such a situation there is a need for stricter volumetric management of water resources. This has implications in terms of data collection, monitoring and evaluation, and research, all of which are at present much under-developed with regard to needs (Chohin-Kuper et al. 2014).

The inexorable squeeze of agricultural supply discussed in Sect. 10.2.2 has generated calls for volumetric management by the EU, with a stricter integration of irrigators, whether individual or collective, in their respective river basins as a means of enhancing control over water quality, minimum flows or floods. In France, it is too soon to evaluate how OUGC (*Organisme Unique de Gestion Collective*) will handle the annual/monthly amounts of water they are granted and asked to apportion among members, especially in times of shortage when this volume is reduced. Spain also shows cases of enhanced volumetric management, like for example in the eastern Mancha.

In non-EU countries, data collection is insufficient and confuses decision-making (MED-EUWI 2007). In Morocco the official number given for private groundwater-based irrigation (441,430 ha), the most dynamic variable of the water sector, is the same since 2004… In Egypt there are often differences on water or crop data between official reports and their overall consistency appears questionable when one attempts to establish the water balance of the Nile delta (Molle et al. 2018). In Jordan official statistics point to a stable groundwater abstraction in the Highlands along the past 30 years, while Agricultural statistics indicate that the cultivated area has doubled (Molle et al. 2017b). Yet, analysts often overlook that collecting data such as land or water use is extremely demanding and costly, much beyond the capacity of most agencies, not to mention occasional data 'massaging' when they point to the wrong direction and are politically unpalatable.

The other side of the quantitative squeeze is the 'quantitative easing' that is likely to remain an attractive policy option from a political point of view. On-farm storage in France is now given priority and has relatively limited impacts (filling takes place in winter and use in summer). But large-scale transfers or desalination, notwithstanding the potential of treated wastewater, are still on the drawing board, especially in Morocco, Algeria and Turkey. They have considerable implications in social and environmental terms, capital investment, and energy consumption among other aspects.

10.6.2 Where Is the Environment?

As mentioned above, the massive mobilization of water resources, the fast expansion of irrigation over rain-fed lands, and the intensification of agricultural practices have jeopardized water resources –in terms of quality and quantity-, and altered numerous ecosystems. The roster of landscapes degraded or destroyed by irrigation expansion and intensive agriculture include all types of aquatic ecosystem and all Mediterranean regions.

Wetlands, frequently located at the end of the hydrographic systems and extremely sensitive to changes in the water budget, tend to receive the accumulated impacts of agricultural overexploitation and pollution. This is the cause of the recent crisis of the Mar Menor in Spain, the Ereğli and Eşmekaya reedbeds in Turkey, or the Sidi Moussa-Walidia complex in Morocco, where over-fertigation and pesticides are causing groundwater pollution, while water withdrawals for irrigation reduced the wetland by 21% between 1957 and 1991. The Dead Sea and several other inland lakes (e.g. in Tukey) are slowly drying up. In most cases the creation of protected areas is proving to be insufficient to preserve endangered ecosystems and their associated biota. Other soil- or water-quality related challenges include the pollution of rivers and aquifers by nitrates (causing public health problems and incurring treatments costs), soil salinisation (e.g. Tunisia), or waterlogging due to excess irrigation and to lack of drainage infrastructures (e.g. in Turkey).

Environmental preservation therefore stands up as a major challenge at a time when socio-political conditions push governments to give priority to production and the provision of livelihoods rather than to considering water as an ecological asset. In thc EU, the control now imposed by the WFD and the priority given to the conservation of aquatic ecosystems constitute an enormous challenge for countries where a large water demand for irrigation has already created significant water scarcity and water quality problems (Albiac 2006). The WFD is demanding, but at the same time a necessary lever to move administrations that have historically faced difficulties with the enforcement of their legislation on groundwater control or ecological flows.

10.6.3 Climate Change

According to the IPCC scenarios, the Mediterranean basin is considered as a region particularly vulnerable to climate change (CC). Climate models outline a pronounced rise in mean temperatures and a greater recurrence of high temperature events. A significant decrease in precipitation due to increased anticyclonic circulation and the northward shift of the Jet stream has also been highlighted. The recently observed trends coincide with some worrisome temperature scenarios, but downscaling rainfall evolution is more difficult due to the complex orographic structure of the Mediterranean basin (Giorgy and Lionello 2008; IPCC 2014). ET increase seems unquestionable, but the behavior of rainfall in the Mediterranean cells show more variegated regional patterns.

In Turkey, the average summer temperature of the twenty-first century is 1.5° higher than that of the 1960–1970 period, and Spain and France have also undergone a 1–1.5° rise in mean temperatures since 1970 and 1959 respectively (González-Hidalgo et al. 2016; Ribes et al. 2016). The pattern of rainfall change is regionally irregular. In Turkey, precipitation remains stable since 1980, while Italy has suffered a moderated decrease in the Thyrrenian regions (−5/8%), and a growth in the Adriatic (4%). In Spain, rainfall on the Atlantic side has decreased by 25% but Mediterranean precipitations have remained stable (Miró et al. 2009). Changes in rainfall patterns, reforestation processes, and optimistic overestimation of water resources at the end of the twentieth century have resulted in a downward revision of the water basin plans in Morocco and Spain, with decreases between 30% and 45% in the available water resources in the last 30 years (El Gueddari and Arrifi 2009; Avellà et al. 2014).

The 'double squeeze' affecting irrigation discussed earlier has prompted research on adaptive strategies. Beyond fine-tuning of irrigation to reduce evaporation losses and increase productivity, technical fixes –most notably drip irrigation and new supply augmentation options including wastewater reuse, desalination, and interbasin transfers– are favored by experts and administrations. CC has largely allowed a rebranding and greening of the old capital-intensive infrastructural hydraulic mission. Although Turkey's national adaptation strategy to CC places "expansion of rain-fed agricultural areas to irrigation" on the top of its list of adaptation actions, and Italy also included this measure in its national strategy, these policies seem to be based on flawed conceptions of irrigation efficiency and to be oblivious of the continued double squeeze underway, that rules out the extension of irrigation (IPCC 2014; Olesen et al. 2011). We have discussed the reasons for such a preference and the risks attached to delaying the recognition of the natural limits of water use.

Today, while there is sufficient scientific evidence to support that CC is one of the most relevant challenges for the irrigation sector in the Mediterranean region, for many farmers CC ranks lower among their concerns (see Richard-Ferroudji et al. 2013 for France, and Ortega-Reig et al. 2018 for Spain), as they not only already have to cope with water scarcity but also because they primarily feel vulnerable to market and socio-political uncertainty.

10.7 Conclusions

From a common base of ancient small-scale communal irrigation systems, all Mediterranean countries have developed large-scale public irrigation schemes from the 1960s up until the 1980s. Physical and financial sustainability issues have motivated reforms in the 1980s and 1990s that included PIM, IMT and economic tools but these reforms largely failed. Although problems of deferred maintenance are still pervasive, interest of donors and priorities of governments have significantly shifted to supporting capital-intensive agriculture, technological (drip-irrigation) or supply-oriented (transfers, wastewater treatment, desalination) fixes, PPPs, and the development of groundwater, while issues of water supply and sanitation, water quality, and sectoral water allocation have become more prominent.

Loosely planned/controlled water resource development by both the state and individuals resulted in the closure of many river basins and the overexploitation of most major aquifers, with the annual volume of water consumed at the basin level sometimes exceeding the average annual available water. On paper, rational basin-level planning and other mainstream policy principles have generally been adopted and incorporated into legislation but implementation on the ground is hindered by a lack of both material means (for monitoring, data collection, sanctioning, etc.) and political will: because water-short systems generally present a very high degree of water recycling, it is only possible to restore a degree of sustainability through a reduction in consumption (evapotranspiration) by certain users, which explains the associated political difficulties.

Much less public money is invested in irrigation, despite food security concerns in some countries, but support to the sector is manifest through subsidies which signal a still significant political clout of agriculture in most countries. Capital-intensive export-oriented irrigated agriculture –including plasticulture- is promoted for its economic virtues, sometimes to the detriment of small-scale 'traditional' agriculture, but its social relevance is weak and its water footprint locally high. Market integration up to fast foods, supermarkets and export channels, with strict requirements in terms of quality and caliber, also favors the former over the latter. While some small farmers or local investors succeed in tapping niche markets the majority is sticking to a (relatively stable and extensive) low-benefit farming system for lack of know-how, information and capital to face the much higher risk of cash-crop farming.

Irrigation technology shows advance in remote control, micro-irrigation, fertigation, on-farm storage, precision farming with the use of drones, GIS and remote sensing. Largely subsidized water-saving technologies enhance productivity and reduce labor needs but rarely deliver in terms of water economy (*consumption*, often confounded with application). These techniques come with added capital and maintenance costs and are ill-adapted to fragmented, small-sized de-capitalized farms with gravity irrigation. The low level of infrastructure use, for a diversity of factors ranging from water shortage, poor profitability of agriculture, and risk to urban encroachment, salinisation, or optimistic planning signal the various threats and difficulties faced by the public irrigation sector. In contrast, the vibrant groundwater economy almost obscures the fact that it is based on declining stocks and therefore largely unsustainable.

Also noteworthy is the dire lack of reliable/updated data and the critical lack of in-depth research on land and water use dynamics, as well as policy-making, at a point where overuse of water calls for a more quantitative data-intensive management. But beyond data, irrigation systems are too often perceived as merely technical devices, obfuscating their systemic nature and linkages with economics and politics at various levels (Lankford et al. 2016). With a 'double squeeze' exerted by environmental constraints and growing domestic use, combined with declining and more erratic precipitations, irrigation will struggle to keep the same level of abstraction, although wastewater reuse is already showing its potential in assuaging this pressure, re-allocating to agriculture the water that had been taken away by cities.

References

Akesbi, N. (2014). *Le Maghreb face aux nouveaux enjeux mondiaux.* Les investissements verts dans l'agriculture au Maroc. Notes de l'IFRI.

Albiac, J. (2006). *The case of the water framework directive and irrigation in Mediterranean agriculture.* Human Development Occasional Papers (1992–2007) HDOCPA-2006-34, Human Development Report Office (HDRO), United Nations Development Programme (UNDP).

Allan, T. (2007). Rural economic transitions groundwater use in the Middle East and its environmental consequences. In M. Giordano & K. G. Villholth (Eds.), *The agricultural groundwater revolution: Opportunities and threats to development.* Wallingford: CABI.

Allan, T., Keulertz, M., & Woertz, E. (2015). The water–food–energy nexus: An introduction to nexus concepts and some conceptual and operational problems. *International Journal of Water Resources Development.* https://doi.org/10.1080/07900627.2015.1029118.

Al-Naber, M., & Molle, F. (2017). Controlling groundwater overabstraction: Policies vs. local practices in Jordan Highlands. *Water Policy, 19*(4), 692–708.

Atance Muñiz, I. (2013). *Sostenibilidad económica de la agricultura española. La sostenibilidad de la agricultura española.* Almería: Cajamar Caja Rural.

Avella, L., Garcia-Molla, M., & Sanchis-Ibor, C. (2014). *Problemática del sistema de explotación del Júcar en el contexto de los planes hidrológicos de cuenca (1998–2008)*, XII Congreso Nacional del Medio Ambiente, http://www.conama2014.conama.org/conama2014/download/files/conama2014/CT%202014/1896711595.pdf

Benouniche, M., Zwarteveen, M., & Kuper, M. (2014). Bricolage as innovation: Opening the black box of drip irrigation systems. *Irrigation and Drainage, 63*(5), 651–658.

Blinda, M. (2009). *Mediterranean strategy for the sustainable development: Water use efficiency.* Sophia Antipolis: Plan Bleu Regional Activity Centre.

Chohin-Kuper, A., Garzón Delvaux, P. A., & Strosser, P. (2014). *Approche économique de la gestion de la demande en eau en Méditerranée: Instruments économiques, Plan Bleu, Valbonne.* (Les Cahiers du Plan Bleu 15).

Closas, A. (2018). Groundwater, the state, and the creation of irrigation communities in Llanos del Caudillo, Spain. *Water Alternatives, 11*(1), 19–39.

Cornish, G., Bosworth, B., Perry, C., & Burke, J. (2004). *Water charging in irrigated agriculture: An analysis of international experience* (FAO Waters Reports 28). Rome: FAO.

Corrado, A., de Castro, C., & Perotta, D. (Eds.). (2017). *Migration and agriculture: Mobility and change in the Mediterranean area.* London: Routledge.

Courtot, R. (1994). La Huerta de Valencia: la fin d'un mythe? *Bulletin de l'Association de Géographes Français, 71*(2), 181–186.

Crow-Miller, B., Webber, M., & Molle, F. (2017). The (re)turn to infrastructure for water management? *Water Alternatives, 10*(2), 195–207.

Delipınar, Ş., & Karpuzcu, M. (2017). Policy, legislative and institutional assessments for integrated river basin management in Turkey. *Environmental Science & Policy, 72*, 20–29.

Dixon, M. (2013). *The making of the agri-food corporate system in Egypt*. PhD thesis, Cornell University.

Dixon, M. (2017). Plastics and Agriculture in the Desert Frontier. *Comparative Studies of South Asia, Africa and the Middle East, 37*(1), 86–102.

Dono, G., Giraldo, L., & Severini, S. (2012). The cost of irrigation water delivery: An attempt to reconcile the concepts of cost and efficiency. *Water Resources Management, 26*, 1865–1877.

Doukkali, M. R. (2005). Water institutional reforms in Morocco. *Water Policy, 7*, 71–88.

ECA. (2014). *Integration of EU water policy objectives with the CAP: A partial success*. Special report, European Court of Auditors, https://www.eca.europa.eu/Lists/ECADocuments/SR14_04/SR14_04_EN.pdf

El-Agha, D. E., Closas, A., & Molle, F. (2017). Under the radar: Dynamics of groundwater use in the Nile delta. *Hydrogeology Journal, 25*, 1621–1631.

El Gueddari, A. B. S., & Arrifi, M. (2009). L'agriculture irriguée au Maroc face à la rareté des ressources en eau? In *L'Etat des ressources en eau au Maghreb en 2009*. Rabat: UNESCO.

Erturk, S. A. (2017). Refugees in the agricultural sector: Some notes on Syrians in Hatay province, Turkey. In A. Corrado, C. de Castro, & D. Perotta (Eds.), *Migration and agriculture: Mobility and change in the Mediterranean area* (pp. 168–180). London/New York: Routledge.

FAO. (2018). *The benefits and risks of solar powered irrigation – a global overview*. Rome: FAO.

Faysse, N., Rinaudo, J. D., Bento, S., Richard-Ferroudji, A., Errahj, M., Varanda, M., Imache, A., Dionnet, M., Rollin, D., Garin, P., Kuper, M., Maton, L., & Montginoul, M. (2014). Participatory analysis for adaptation to climate change in Mediterranean agricultural systems: Possible choices in process design. *Regional Environmental Change, 14*, 57–70.

García Alvarez-Coque, J. M., & Martínez Gómez, V. D. (2014). *La regulación de los mercados en la nueva PAC y el equilibrio de la cadena de valor* (pp. 59–82). Almería: Reflexiones en torno a la PAC Cajamar Caja Rural.

Ghazouani, W., Molle, F., & Rap, E. (2012). *Water users associations in the Near-East Northern Africa Region. IFAD interventions and overall dynamics*. Submitted to IFAD, October 2012. IWMI.

Giorgy, F., & Lionello, P. (2008). Climate change projections for the Mediterranean region. *Global and Planetary Change, 63*(2–3), 90–104.

Gómez-Limón, J. A. (2014). La agricultura de regadío en el contexto de la nueva PAC. In I. Bardají (Ed.), *Reflexiones en torno a la PAC* (pp. 109–140). Almería: Cajamar Caja Rural.

Gonzalez-Hidalgo, J. C., Peña-Angulo, D., Brunetti, M., & Cortesi, N. (2016). Recent trend in temperature evolution in Spanish mainland (1951–2010): From warming to hiatus. *International Journal of Climatology, 36*(6), 2405–2416.

GWP. (2012). *Water demand management: The Mediterranean experience*. UNEP (Blue Plan) and GWP.

GWP Med (Global Water Partnership Mediterranean). (2007). *Guidelines for legislative and institutional reforms needed for the implementation of IWRM at national level in ESCWA region*. Athens.

Hamdy, A., & Lacirignola, C. (1997). *Use of water in the Mediterranean: Sectorial distribution and prospects* (pp. 17–49). Options Méditerranéennes : Série A. Séminaires Méditerranéens.

Hammani, A., & Kuper, M. (2008). Caractérisation des pompages des eaux souterraines dans le Tadla, Maroc. In M. Kuper & A. Zaïri (Eds.), *Economies d'eau en systèmes irrigués au Maghreb* (p. 2008). Montpellier: Cirad.

Hardy, L., & Garrido, A. (2010). *Análisis y evaluación de las relaciones entre el agua y la energía* (Papeles de Agua Virtual 6). Madrid: Fundación Botín.

HRW. (2015). *Israel: Serious abuse of Thai migrant workers, human rights watch.* https://www.hrw.org/news/2015/01/21/israel-serious-abuse-thai-migrant-workers

Ingold, A. (2008). Les sociétés d'irrigation: bien commun et action collective. *Entreprises et histoire, 1*(50), 19–35.

IPCC. (2014). *Climate Change 2014: Synthesis report.* Contribution of working groups I, II and III to the fifth assessment report of the intergovernmental panel on climate change [Core Writing Team, R. K. Pachauri & L. A. Meyer (Eds.)]. Geneva: IPCC, 151 pp.

Jackson, T. M., Khan, S., & Hafeez, M. (2010). A comparative analysis of water application and energy consumption at the irrigated field level. *Agricultural Water Management, 97*(10), 1477–1485.

Jordan, A., & Lenschow, A. (2010). Environmental policy integration: A state of the art review. *Environmental Policy and Governance, 20*(3), 147–158.

Keller, J., Keller, A., & Davids, G. (1998). River basin development phases and implications of closure. *Journal of Applied Irrigation Science, 33*(2), 145–164.

Kislev, Y. (2013). Water in agriculture. In N. Becker (Ed.), *Water policy in Israel: Context, issues and options* (pp. 51–64). Springer.

Kuper, M., Faysse, N., Hammani, A., Hartani, T., Marlet, S., Hamamouche, M. F., & Ameur, F. (2016). Liberation or anarchy? The Janus nature of groundwater use on North Africa's new irrigation frontiers. In A. Jakeman, O. Barreteau, R. J. Hunt, J.-D. Rinaudo, & A. Ross (Eds.), *Integrated groundwater management: Concepts, approaches and challenges* (pp. 583–615). Cham: Springer International Publishing.

Lankford, B. A., Makin, I., Matthews, N., Noble, A., McCornick, P. G., & Shah, T. (2016). A compact to revitalise large-scale irrigation systems: A 'theory of change' approach. *Water Alternatives, 9*(1), 1–32.

Lascoumes, P. (2014). *Le développement durable: Une nouvelle affaire d'État.* Paris: PUF.

Laurent, C. (2013). The ambiguities of French Mediterranean agriculture: Images of the multifunctional agriculture to mask social dumping? In D. Ortiz-Miranda, A. Moragues-Faus, & E. Arnalte-Alegre (Eds.), *Agriculture in Mediterranean Europe: Between old and new paradigms* (pp. 149–171). Bingley: Emerald Group Publishing Limited.

Le Visage, S. (2015). *Dynamiques de l'irrigation par les eaux souterraines: quels usages, organisations et utilisations autour de la ressource dans la région d'Izmir – Turquie*? MSc thesis, ISTOM, France.

Le Visage, S., Kuper, M., Venot, J.-P., Yercan, M., & Atış, E. (2018). Pursuing the state's hydraulic mission in a context of private groundwater use in the Izmir Province, Turkey. *Water Alternatives, 11*(2), 421–438.

Leibundgut, C., & Kohn, I. (2014). European traditional irrigation in transition part II: Traditional irrigation in our time—decline, rediscovery and restoration perspectives. *Irrigation and Drainage, 63*, 273–293.

Malorgio, G., & Hertzberg, A. (2007). *Competitiveness of the Southern Mediterranean Countries in the Italian agri-food market, I Mediterranean Conference of Agro-Food Social Scientists.* 103rd EAAE Seminar 'Adding Value to the Agro-Food Supply Chain in the Future Euromediterranean Space'.

MED-EUWI. (2007). *Mediterranean groundwater report.* Technical report on groundwater management in the Mediterranean and the Water Framework Directive.

Miró, J., Estrela, M. J., Pastor, F., & Millan, M. (2009). Análisis comparativo de tendencias en la precipitación, por distintos inputs, entre los dominios hidrológicos del Segura y del Júcar (1958–2008). *Investigaciones Geográficas, 49*, 129–157.

Molden, D., Sakthivadivel, R., & Samad, M. (2001). Accounting for changes in water use and the need for institutional adaptation. In C. L. Abernethy (Ed.), *Intersectoral management of river basins:* Proceedings of an international workshop on integrated water management in water-stressed river basins in developing countries: Strategies for poverty alleviation and agricultural growth, IWMI and Inwent, pp. 73–87.

Molden, D., Vithanage, M., de Fraiture, C., Faures, J. M., Finlayson, M., Gordon, L., Molle, F., Peden, D., & Stentiford, D. (2011). Water availability and its use in agriculture. In P. Wilderer (Ed.), *Treatise on water science* (Vol. 4, pp. 707–732). Oxford: Academic.

Mollard, E. (2004). "Recherches sur les arrosages chez les peuples anciens" de F.J. Jaubert de Passa (1846): une histoire de la gouvernance avant l'avènement de la technocratie. In: R. Thierry & A. Rivière-Honegger (Eds.), *La gestion sociale de l'eau, concepts, méthodes et applications* (Territoires en Mutations, 12, pp. 17–31). Montpellier: Université Paul Valéry.

Molle, F. (2008). Why enough is never enough: The societal determinants of river basin closure. *International Journal of Water Resource Development, 24*(2), 217–226.

Molle, F., & Berkoff, J. (Eds.). (2007). *Irrigation water pricing: The gap between theory and practice*. Wallingford: CABI.

Molle, F., & Closas, A. (2017). *Groundwater governance: A synthesis*. Report submitted to USAID (vol. 6). IWMI. http://gw-mena.iwmi.org/outputs/

Molle, F., Wester, P., & Mollinga, P. P. (2009). Hydraulic bureaucracies: Flows of water, flows of power. *Water Alternatives, 2*(3), 328–349. www.water-alternatives.org

Molle, F., Wester, P., & Hirsch, P. (2010). River basin closure: Processes, implications and responses. *Agricultural Water Management, 97*, 569–577.

Molle, F., Closas, A., & Zubari, W. (2017a). Governing groundwater in the Middle East and North Africa Region. In K. Villholth, J. van der Gun, E. Lopez-Gunn, K. Conti, & A. B. Garrido (Eds.), *Advances in groundwater governance*. Leiden: CRC Press/Balkema.

Molle, F., El-Karablieh, E., Al Naber, M., Closas, A., & Salman, A. (2017b). *Groundwater governance in Jordan: The case of Azraq*. A Policy White Paper. Groundwater Governance in the Arab World. IWMI/USAID.

Molle, F., Gaafar, I., Ezzat El-Agha, D., & Rap, E. (2018). The Nile delta's water and salt balances and implications for management. *Agricultural Water Management, 197*, 110–121.

Mollinga, P., & Bolding, A. (Eds.). (2004). *The politics of irrigation reform. Contested policy formulation and implementation in Asia, Africa and Latin America*. Aldershot: Ashgate.

Nomisma-Unaproa. (2016). *2° Rapporto Nomisma-Unaproa sulla Competitività del settore ortofrutticolo nazionale*. Roma: Agra Editrice Srl.

OECD. (2017). *Agricultural policy monitoring and evaluation 2017*. Paris: OECD Publishing. https://doi.org/10.1787/agr_pol-2017-en.

Olesen, J. E., Trnka, M., Kersebaum, K. C., Skjelvåg, A. O., Seguin, B., Peltonen-Saino, P., Rossi, F., Kozyra, J., & Micale, F. (2011). Impacts and adaptation of European crop production systems to climate change. *European Journal of Agronomy, 34*(2011), 96–112.

Ortega-Reig, M., Sanchis-Ibor, C., & García-Mollá, M. (2017a). Drip irrigation in Eastern Spain. Diverging goals in a converging process. In J. P. Venot, M. Kuper, & M. Zwarteveen (Eds.), *Drip irrigation for agriculture: Untold stories of efficiency, innovation and development*. Abingdon: Earthscan from Routledge.

Ortega-Reig, M., Sanchis-Ibor, C., Palau-Salvador, G., García-Mollá, M., & Avellà-Reus, L. (2017b). Institutional and management implications of drip irrigation introduction in collective irrigation systems in Spain. *Agricultural Water Management, 187*, 164–172. https://doi.org/10.1016/j.agwat.2017.03.009.

Ortega-Reig, M., Sanchis-Ibor, C., García-Mollá, M., Pulido-Velázquez, M., Girard, C., Marcos, P., Ruiz-Rodríguez, M., & Garcia-Prats, A. (2018). *Adaptación de la agricultura a escenarios de cambio global. Aplicación de métodos participativos en la cuenca del río Júcar, Economia Agraria y Recursos Naturales* (in press).

Özerol, G., & Bressers, H. (2015). Scalar alignment and sustainable water governance: The case of irrigated agriculture in Turkey. *Environmental Science & Policy, 45*, 1–10.

Özerol, G., Bressers, H., & Coenen, F. (2012). Irrigated agriculture and environmental sustainability: An alignment perspective. *Environmental Science & Policy, 23*, 57–67.

Perry, C., & Steduto, P. (2017). *Does improved irrigation technology save water* (pp. 36–39). Cairo: FAO.

Petriccione, G., & dell'Aquila, C. (2011). Ortofrutta e catena del valore globale. *Agriregionieuropa, 7*(27), 10.

Playán, E., & Mateos, L. (2006). Modernization and optimization of irrigation systems to increase water productivity. *Agricultural Water Management, 80*(1–3), 100–116.

Qadir, M., Bahri, A., Sato, T., & Al-Karadsheh, E. (2007). Wastewater reuse production, treatment, and irrigation in Middle East and North Africa. *Irrigation & Drainage System, 24*, 37–51.

Ribes, A., Corre, L., Gibelin, A., & Dubuisson, B. (2016). Issues in estimating observed change at the local scale – a case study: The recent warming over France. *International Journal of Climatology, 36*(11), 3794–3806. https://doi.org/10.1002/joc.4593.

Richard-Ferroudji, A., Garin, P., Matignon, M., Maton, L., Montginoul, M., Rinaudo, J. D., et al. (2013). Engager des agriculteurs à répondre à l'injonction d'adapter la gestion de l'eau au changement climatique. Une démarche de prospective participative avec des agriculteurs usagers des nappes du Roussillon (France). In V. Ancey, B. Dedieu, M. Antona, et al. (Eds.), *Agir en situation d'incertitude en agriculture. Regards pluridisciplinaires au Nord et au Sud.* Peter Lang, Editions scientifiques internationales. 17, 419 p. ISBN 978-2-87574-037-3.

Roccu. (2013). *The political economy of the Egyptian revolution Mubarak, economic reforms and failed hegemony*. Basingstoke: Palgrave Pivot.

Sanchis-Ibor, C., García-Mollá, M., & Avellà-Reus, L. (2017a). Effects of drip irrigation promotion policies on water use and irrigation costs in Valencia, Spain. *Water Policy, 19*(1), 165–180.

Sanchis-Ibor, C., Boelens, R., & García-Mollá, M. (2017b). Collective irrigation reloaded. Re-collection and re-moralization of water management after privatization in Spain. *Geoforum, 87*, 38–47.

Seckler, D. (1996). *The new era of water resources management* (Research Report 1). Colombo: International Irrigation Management Institute (IIMI).

Sese-Minguez, S., Boesveld, H., Asins-Velis, S., van der Kooij, S., & Maroulis, J. (2017). Transformations accompanying a shift from surface to drip irrigation in the semi-arid Cànyoles watershed, Valencia, Spain. *Water Alternatives, 10*(1), 81–99.

Soto-García, M., Martínez-Álvarez, V., García-Bastida, P. A., Alcón, F., & Martín-Górriz, B. (2013). Effect of water scarcityand modernisation on the performance of irrigation districts in south-eastern Spain. *Agricultural Water Management, 124*, 11–19.

Tanouti, O. (2017). *La gestion intégrée des ressources en eau à l'épreuve du bassin versant – Le cas du bassin du Tensift au Maroc*. PhD thesis, University of Paris-Nanterre.

Venot, J. P., Kuper, M., & Zwarteveen, M. (2017). *Drip irrigation for agriculture: Untold stories of efficiency, innovation and development*. Abingdon: Earthscan from Routledge.

Ward, F., & Pulido-Velázquez, M. (2008). Water conservation in irrigation can increase water use. *PNAS, 105*, 18215–18220.

Willardson, L. S., Allen, R. G., & Fredericksen, H. D. (1994). *Elimination of irrigation efficiencies*. 13th technical conference USCID, Denver, CO.

World Bank. (2007). *Emerging public-private partnerships in irrigation development and management*. Washington, DC: The World Bank.

Index

F. Molle et al. (eds.), *Irrigation in the Mediterranean*, Global Issues in Water Policy 22, https://doi.org/10.1007/978-3-030-03698-0

Y

Printed by Printforce, the Netherlands